Andrzej Benedykt Koltuniewicz
Sustainable Process Engineering
De Gruyter Textbook

W0230057

Andrzej Benedykt Koltuniewicz

Sustainable Process Engineering

Prospects and Opportunities

DE GRUYTER

Author
Prof. Dr. Andrzej Benedykt Koltuniewicz
Warsaw University of Technology
Fac. of Chemical and Process Engineering
Warynskiego 1
00-645 WARSZAWA
Poland
A.Koltuniewicz@ichip.pw.edu.pl

ISBN 978-3-11-030875-5
e-ISBN 978-3-11-030876-1

Library of Congress Cataloging-in-Publication Data
A CIP catalog record for this book has been applied for at the Library of Congress.

Bibliographic information published by the Deutsche Nationalbibliothek
The Deutsche Nationalbibliothek lists this publication in the Deutsche Nationalbibliografie;
detailed bibliographic data are available in the Internet at http://dnb.dnb.de.

© 2014 Walter de Gruyter GmbH, Berlin/Boston
Cover image: Mathieu Aucher, Getty Images
Typesetting: PTP-Berlin, Protago-TEX-Production GmbH, Berlin
Printing and binding: CPI books GmbH, Leck
♾Printed on acid-free paper
Printed in Germany

www.degruyter.com

Acknowledgments

First of all, I would like to thank my colleagues from the Warsaw University of Technology for their help in writing this book. Special thanks to Dr. Katarzyna Dabkowska for writing part of 6.2 on the separation of enantiomers and Dr. Maciej Pilarek for writing part 6.1 about the relationship between process engineering and medicine, and Dr. Paweł Sobieszuk for helping to write part 4.1 about microreactors. I would like to thank Professor Ryszard Pohorecki for the fruitful discussions about the many impacts of process engineering and its role in sustainable development. Thanks again for your time and consultations.

Writing this book was supported by collected and updated materials to the lectures I have conducted over many years at the Faculties of Chemistry and Fundamental Problems of Technology at the Technical University of Wroclaw, and for several years at the Warsaw Technical University Faculty of Chemical and Process Engineering and the Department of Biology of the University of Warsaw.

I gained valuable insights in discussions with my colleagues: Professor Enrico Drioli and Professor Lidetta Giorno from University of Calabria, Professor Stan Kołaczkowski, Professor John Howell, Dr. Robert Field and Dr. Tom Arnot from the Chemical Engineering Department of Bath University, Professor Joao Crespo, University of Nova Lisboa, Professor Carme Guell from Tarragona University and Professor Siegfried Ripperger of Kaiserslautern University and Dresden University. They have all contributed to the formation of my scientific views and ideas, and thus to the creation of this book.

I really appreciate it, and I am very grateful and sincerely committed to you for this.

Andrzej B. Koltuniewicz

Contents

1 Inevitability of sustainable development

All living beings including humans are formed by their habitat in every respect, i.e. the physical, biological, social etc. Therefore human existence cannot be separated from the whole environment. The biosphere is a system, integrating all living beings with their mutual relations of lithosphere, hydrosphere and atmosphere. As a whole, the system is sometimes referred to as ecosphere. The great numbers of living creatures form the numerous food chains which are interconnected in a web of life. A necessary condition of existence of this network is solar radiation, which is absorbed by our planet in about 30 % of the energy that reaches earth's surface. Earth's mass remains almost constant, as do the total number of atoms on Earth, although the chemicals are subject to permanent change. The web of life must permanently adapt to the changes in amount and quality of seasonal and daily solar radiation. But the changes are the stimuli and even the driving force for all natural processes on earth. The network of life always finds a state of dynamic equilibrium.

Fig. 1.1. Earth (courtesy of NASA).

There is only one way to maintain life on planet Earth in the present form (Fig. 1.1). This is simply keeping this homeostasis, by the rational use of natural resources, such as water, air and the content of soil and the earth's crust. Otherwise, if the uncontrolled exploitation of resources and emissions of various pollutants in the near future were not taken into account, then nobody will be able to prevent an inevitable and uncontrollable chain of events. It would still be possible by using a variety of methods and achievements that are known in the fields of science and engineering. The phenomena of nature always act in accordance with the principles of physics. Process engineering is based on these basic foundations of physics; its main task was always the best industrial production. Process engineering has always focused on the issues of how to produce cheaper, faster, easier, whatever it may be. Now it must answer the question of how to produce more safely and for as long as possible, taking future generations into account.

All citizens of the world should think about future generations, which may not be doomed to extinction. Therefore, those who recognize imbalances in nature must sound the alarm before it is too late. To take appropriate effective remedies for the imbalance caused by human activity in nature, and which are the most visible, painful and spectacular, it is not essential to reduce the rate of development of civilization. Instead, we can use new knowledge which maintains a modern development.

Scientists from NASA managed to create a miniature world which is completely self-contained and requires only solar energy, just like Earth (Fig. 1.2). This is the model of our ecosphere, that is the totally enclosed ecosystem, but extremely simple, as it consists only of algae, bacteria, and shrimps. Algae are food for shrimps, and also produce oxygen through photosynthesis from the sun during the conversion of CO_2 and organic substances excreted by the bacteria. On the other hand, bacteria reciprocate these services, by converting all organic substances excreted by the shrimps into the ingredients consumed by algae. In this way, a cycle of matter is closed. It should be noted that the survival of all life forms maintains the environment, which consists of filtered seawater, and marble and coral. The only difference between the model and reality lies in the complexity of the system. The food chains on Earth are numerous, forming the extremely large systems. These systems are also definitely more sensitive to external factors, and therefore the conditions for maintaining them are much more complicated.

Fig. 1.2. Glass model of the ecosphere (courtesy of Ecosphere Associatges Inc.).

From the physical point of view, the biosphere is a semi-closed system. This means that it is a closed system for the flow of materials but open for the flow of solar energy which reaches the earth from outside. But the condition of steady-state is the same, i.e. balancing of material streams, according to the conservation laws of mass or energy.

To maintain the world in a stationary condition, the accumulation of any component of mass and energy must be zero. To attain zero accumulation, even for the smallest component, all mass streams in the web of life must circulate, thus must be renewable. Such language of physics justifies the possibility of the continuation of life of the three shrimps of the model as well as us all on Earth. The difference is that there are more people on Earth than shrimps in the ecosphere model. By 2030, it is estimated that the world population will stand at 8 billion [1, 2].

We can only hope that measures built on a solid foundation in many disciplines, and taken by the whole public at the same time, will take effect and possibly reverse unfavorable trends. Therefore, the dissemination of knowledge about the so-called sustainable development is one of those activities which lead to success. I have tried to show how important the role of process engineering in sustainable development is.

1.1 The real determinants of our ecosphere

Our environment is composed of air and water. Our ecosphere consists of three main systems, i.e.: (a) cycles of matter, water, minerals and organic matter, (b) energy flow, (c) the entire web of life, with its individual components. The limited space (see Tab. 1.1) and resources, in addition to the limited ability to maintain homeostasis mean that Earth is very sensitive to the uncontrolled management of raw materials and energy. Further, insufficient capacity to accumulate increasing amounts of pollutants in the ecosphere will certainly lead to irreversible degradation of the earth.

Table 1.1. The dimensions of Earth

Dimension	Value	Unit
Diameter at the equator	12,756.32	km
Diameter through the poles	12,715.43	km
Total surface area (100%)	510,072,000	km^2
Surface occupied by water (70.8%)	361,132,000	km^2
Surface area occupied by land (29.2%)	148,940,000	km^2
Total volume	1.08321×10^{12}	km^3
Total mass	5.9736×10^{24}	kg

The ecosphere spreads from the bottom of the sea up to 10 km into the atmosphere. The total amount of water on Earth is 1.4 billion km^3 (see Tab. 1.2), comprising three-quarters of the surface of Earth. It seems that this is a very large amount, but it only makes up 1.26 % of earth's volume. Furthermore, only a small portion of the volume, about 0.3 %, is available for human use. To picture this ratio more easily, imagine that

Table 1.2. Earth's water resources.

Total water content in earth's ecosphere	13 622 692 950	km^3	100%
Oceans	13 320 000 000	km^3	97,77802%
Glaciers and icebergs	290 000 000	km^3	2,12880%
Groundwater	8 400 000	km^3	0,06166%
Total amount of freshwater	4 000 000	km^3	0,02936%
Lakes	125 000	km^3	0,00092%
Inland seas	100 000	km^3	0,00073%
Moisture in the ground	66 700	km^3	0,00049%
Rivers	1 250	km^3	0,00001%

if all the water is evenly distributed over the earth the water level reaches more than 2.6 km but less than 8 m; the layer would be useful for people. So it is only 0.00378 % of the volume of the Earth (sic). What happens in this small part is therefore very important for our future.

As it is important to maintain the balance of the flow of material streams, and the narrow region which is associated with the contents of the individual components, the example of oxygen is convincing. The ecosphere maintains 20 % oxygen in the atmosphere, but if it were 25 % organic matter on the planet (e.g. forests) would spontaneously ignite. Otherwise, if it were only 15 % the fires could not be sustained. Similar theoretical divagations on the temperature of our globe have been conducted by famous scientists such as Joseph Fourier in 1824 and John Tyndale and Swante Arrhenius in 1858 and 1896, who discovered the greenhouse effect already in the 19th century. They discussed the greenhouse effect as follows: assuming that Earth is a perfect black body (black astral body), defined as having 100 % absorbance of the solar energy which reaches the surface of the earth, then the average temperature would be about 5 °C. It was found, however, that 70 % of the energy which reaches the earth is radiated back again. Thus it was easy to calculate, on the basis of the Boltzmann law, what the average temperature of the Earth should be: at most −19 °C. But in fact earth's mean temperature is 14 °C, which is 35 degrees more. Modern science has already diagnosed the greenhouse effect as a result of the presence of a layer of greenhouse gases in the atmosphere. Climate change, as ordinarily increasing the temperature will cause the melting of icebergs and glaciers, thereby increasing the level of the oceans, which in turn will result in more unpredictable phenomena such as desertification of arable land. Moreover, pollution also causes the formation of acid rain.

Ozone protects earth's surface from UV radiation, which causes skin cancer in humans and is harmful to the entire food chain in the oceans. Some gases have reduced the concentration of ozone, which may have unpredictable effects on the conservation of life on Earth, by increasing lethal UV radiation. However, the presence of ozone in the lower layers closer to Earth causes a number of diseases and destruction of plants.

It seems that water resources are immense because the surface occupied by the oceans is large, 70.8 %. Unfortunately however, less than 0.03 % of the total amount of water on Earth is available to people (see Tab. 1.2 for details) . The problem is that water is the environment and also the main raw material used by man with no restraint or reflection on its limited resources (see Tab. 1.2). Moreover, the consumption of water on Earth is increasing rapidly with the progress of civilization. As far as water consumption in 1680 was estimated at 86 km^3, by 1900 the amount had increased to 522 km^3, but by 1980 risen to 2,120 km^3, and in 2000 to 2,700 km^3 per year. The UN expects that by 2025, two thirds of the world's population will not have enough drinking water.

We could have nothing to fear, were it not for the actual environmental pollution, caused by mindless greed and selfishness hidden under the guise of economic principles and the laws of the market. After all, it has long been common knowledge that there are many well-known and relevant technologies to provide all people with an abundance of fresh water and air. As mentioned before, the sole aim should be to maintain a balanced environment.

1.2 Material problems of civilization

Approximately 30 % of the energy reaching the earth is absorbed in different ways. This makes it possible for different phenomena such as water circulation, air flow, and different forms of life, to take advantage of these phenomena. They are linked together to form food chains. The so-called web of life enables continuous exchange of the basic elements and conditions needed for life. Mankind is no exception, and his existence and activities can be carried out thanks to the stable conditions on Earth, as well as the entire web of life. If the conditions necessary for human life are changed, nature will probably cope even without us. We must be careful that human activities do not violate the conditions on Earth in a manner that endangers not only human beings but all living creatures. At present, the production of most goods and energy is widely based on oil and fossil fuels rather than renewable energy sources. In the future, when all resources are exhausted, the production of goods on Earth will have to rely solely on renewable natural products, i.e. those produced by solar energy and water. Water is involved in photosynthesis with carbon dioxide, converting solar energy into chemicals in the form of sugars, which circulate in many food chains. Thus, water as a carrier of solar energy must be completely renewable. However, we must be careful that water continues to be circulated as it is now in its natural cycles. This means that we should respect water and air resources in our own best interests. We are not able to control the individual creatures, food chains or even subsystems in our ecosphere, and this is why we care about biodiversity and nature as a whole. We must remember once and for all that the stage of world conquest is over. Instead, modern

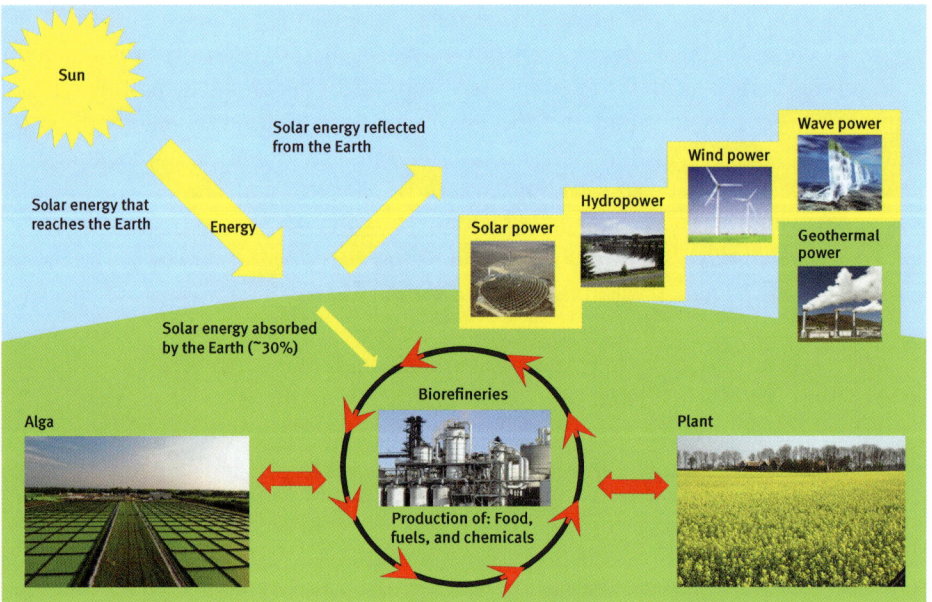

Fig. 1.3. The fate of solar energy on Earth in terms of sustainable development.

Homo sapiens should begin to think about the next phase of life in total harmony with nature (see Fig. 1.3).

In 2012, the human population exceeded 7 billion. Most of them (4.2 billion) living in Asia, followed by Africa (almost 1.1 billion), Europe (about 0.74 billion), Latin America (nearly 0.6 billion) and North America (nearly 0.35 billion). The average population density in the world in 2012 was 47 persons per square kilometer of land. Unfortunately, demographic trends do not go hand in hand with the development of civilization. Across the globe there are a lot of different threats to the environment. The most important are: climate change, including global warming due to the greenhouse effect, desertification and soil eutrophication and destruction of the ozone layer. In addition to acid rain, pollution of soil, water and air and the introduction to the food chains of many foreign substances, leading to the extinction of entire species, which can cause unpredictable and irreversible ecological changes. The total number of species on land is estimated at about 8.7 million, while the number of oceanic species is much lower, estimated at 2.2 million. Habitat destruction has played a key role in the extinction of species, especially those related to the destruction of forests and waters. Factors contributing to the loss of these habitats are overpopulation, global warming, deforestation and pollution. The loss of biodiversity is considered one of the most serious environmental problems today, recognized by the Food and Agriculture Organization of the United Nations [3]. According to some estimates, if current trends persist, as many as half of all plant species could face extinction in the very near future, i.e. within the next few years [4]. Most of these threats to the environment are the result of excessive and uncontrolled human expansion into, rather than coexistence with the environment. However, many problems and negative effects of human activities could be solved with the use of means available to modern civilization. However, this requires integration and cooperation in international forums. In addition, many issues require a multidisciplinary approac. These are issues such as biodiversity, climate change; hydrological changes which should be addressed in cooperation with a wide range of experts in various fields, such as climatologists, ecologists, agricultural specialists, etc.

Unfortunately, the global environmental problems are very diverse and also dependent on geographical location and level of technology, economy and finally on the level of civilization in the given country. Developing countries are likely to have problems with emissions of the most dangerous toxic substances such as lead, cadmium, mercury and many other specific toxic chemicals which are used in tanneries, dyehouses, textile factories, including waste from chemical manufacturing and landfill sites.

The developed countries have quite different problems relating to population density, urbanization, communication and industrialization, the intensive use of land such as monoculture. This leads to permanent changes in the environment such as eutrophication, soil erosion, contamination of water, and the extinction of some species. The changes in the soil, water and air are not so spectacular but more hidden, longterm and therefore even more dangerous because irreversible. The latent impact of heavy metals and organic substances used in agriculture and industry is seriously affecting the health of humans and animals. These hidden effects, such as carcinogenic, mutagenic, immunogenic, may become apparent only in following generations, when it is too late for prevention. Hidden effects are not so spectacular and are therefore difficult to identify without a systematic analysis of statistics on the presence and frequency of occurrence in given areas which are difficult to identify. The WHO has estimated that environmental exposure contributes to 19 % of cancer incidence worldwide. People have confidence in their expert's models, regulations, the limits of pollution. However, this weakens their vigilance and societal activity.

Air pollution is the introduction into the atmosphere of gaseous chemicals, particulates, or biological materials which cause discomfort, disease, or death to humans; damage to other living organisms and food crops, or to the natural environment. Stratospheric ozone depletion due to air pollution has long been recognized as a threat to human health as well as to earth's ecosystems. Water and soil pollution is mainly caused by heavy metals and organic contaminants from the oil industry. Globally, the most dangerous pollutants are from mining and the resultant tailings which emit lead, chromium, asbestos, arsenic, cadmium and mercury.

Soil pollution and eutrophication was already recognized as a pollution problem in European and North American lakes and reservoirs in the mid-20th century. Eutrophication is the ecosystem's response to the addition of artificial or natural nutritional substances, such as nitrates and phosphates from fertilizers or sewage, to an aquatic system. Eutrophication leads to insufficient levels of oxygen in the water. The World Resources Institute has identified up to 375 coastal zones in the world with a clear lack of oxygen, which causes the degradation of water. These zones are mainly in the coastal areas of Western Europe, the eastern and southern coasts of the US, and East Asia, especially in Japan.

In order to introduce a common fact or to define various harmful influences, the WHO has introduced a parameter named DALY (disability-adjusted life year), which is a measure of the impact of different factors, including environmental ones, on human health. This unit determines the average reduction of human life (in years) caused by disability (YLD – years of life with disability) and years of life lost (YLL – years of life lost). DALYs are calculated by adding these two components as follows:

$$DALY = YLL + YLD. \tag{1.1}$$

The ailments caused by various industries are summarized in Tab. 1.3 below for comparison, the set of the unit DALY creating the ranking of the most troublesome sectors.

Table 1.3. The most troublesome sectors, selected based on DALY.

Rank	Industry	Daly
1	Lead-acid recycling	4.80
2	Lead smelting	2.60
3	Mining and ore processing	2.52
4	Tannery operation	1.93
5	Industrial/municipal dump sites	1.23
6	Industrial estates	1.06
7	Artisanal gold mining	1.02
8	Product manufacturing	0.79
9	Chemical manufacturing	0.76
10	Dye industry	0.43

In order to reduce the influence of harmful factors on the environment, rules determining the upper limits of impurities have been introduced by law. The European Parliament set agreed upper limits of pollutants in the form of Directive 2001/81/EC. The aim of this directive is to limit emissions in order to improve the protection of the environment and human health against risks of adverse effects from acidification, soil eutrophication and ground-level ozone. In addition, the directive aims to limit critical levels and recognized health risks from air pollution by establishing national emission ceilings, taking the years 2010 and 2020 as benchmarks. Member states shall each year report their national emission inventories established in accordance with Article 7 to the Commission and the European Environment Agency. Emission projections shall include information to enable a quantitative understanding of the key socioeconomic assumptions. The protocols [5] with reporting requirements of the four main substances responsible for acidification, eutrophication and ground-level ozone pollution: sulphuric dioxide (SO_2); nitrogen oxides (NO_X); volatile organic compounds (VOC); and ammonia (NH_3). Since the substances concerned are transported in large quantities across national boundaries, individual countries could not, in general, meet the underpinning objectives of the NEC Directive to protect human health and the environment within their territory by national action alone.

The U.S. Environmental Protection Agency prepared in the same order a list of 126 harmful substances. The priority pollutants are a set of chemical pollutants which should be regulated, and for which the appropriate analytical test methods should be developed. The current list of 126 Priority Pollutants, shown below, can also be found in Appendix A to 40 CFR, Part 423. [6]

1.3 The environmental problems of the air

1.3.1 Greenhouse effect

As early as the 19th century, a famous scientist analyzed the greenhouse effect and noted the difference between the projected average temperature of the Earth ($-19\,°C$), and the actual average ($14\,°C$), as mentioned earlier in this book. Thus, the increase in average temperature in respect to the hypothetical has been named the "greenhouse effect". The average temperature of earth's surface is constantly increasing, which is known as global warming. The explanation of greenhouse effects is given by pure physics. According to Wien's law (see Section 2.4.7, Heat radiation), the parent solar energy which reaches Earth has a wavelength range within the UV range. These short waves pass easily through earth's atmosphere and are subsequently absorbed on the surface of the Earth. The heat radiation from earth's surface is characterized by much longer wavelengths than sunlight, and can therefore be easily absorbed by the layer of greenhouse gases surrounding the Earth. This layer re-emits electromagnetic radiation in all directions, and therefore also in the direction of Earth. Thus a kind of trap is formed for the solar energy reaching the earth in the form of UV radiation, which eventually results in the heating of the Earth, to a much greater extent than expected. Although water vapor in the clouds is mainly responsible for this phenomenon, human activities, primarily the burning of fossil fuels and clearing of forests, have intensified the natural greenhouse effect, causing global warming. The main cause of the greenhouse effect is contamination of the air and the emission of greenhouse gases into the atmosphere, e.g. carbon dioxide (CO_2), methane (CH_4), freons, originating mainly from burning fossil fuels (such as a coal, natural gas, oil, etc.). The exhaust gases emitted from our transport, i.e. emissions from cars, planes, boats, motorcycles, etc. are also the problem of anthropogenic activity (see Tab. 1.4). The global-warming potential (GWP) is a relative measure of the absorption of heat energy in the atmosphere

Table 1.4. The EPA list of 126 priority pollutants.

Acenaphthene	Methylene chloride	Bromoform
Acrolein	Methyl chloride	Aldrin
Acrylonitrile	Methyl bromide	Dieldrin
Benzene	Dichlorobromomethane	Chlordane
Benzidine	Chlorodibromomethane	4,4-DDT
Carbon tetrachloride	Hexachlorobutadiene	4,4-DDE
Chlorobenzene	Hexachlorocyclopentadiene	4,4-DDD
1,2,4-trichlorobenzene	Isophorone	Alpha-endosulfan
Hexachlorobenzene	Naphthalene	Beta-endosulfan
1,2-dichloroethane	Nitrobenzene	Endosulfan sulfate
1,1,1-trichloreothane	2-nitrophenol	Endrin
Hexachloroethane	4-nitrophenol	Endrin aldehyde
1,1-dichloroethane	2,4-dinitrophenol	Heptachlor
1,1,2-trichloroethane	4,6-dinitro-o-cresol	Heptachlor epoxide
1,1,2,2-tetrachloroethane	N-nitrosodimethylamine	Alpha-BHC
Chloroethane	N-nitrosodiphenylamine	Beta-BHC
Bis(2-chloroethyl) ether	N-nitrosodi-n-propylamine	Gamma-BHC
2-chloroethyl vinyl ethers	Pentachlorophenol	Delta-BHC
2-chloronaphthalene	Phenol	PCB−1242 (Arochlor 1242)
2,4,6-trichlorophenol	Bis(2-ethylhexyl) phthalate	PCB−1254 (Arochlor 1254)
Parachlorometa cresol	Butyl benzyl phthalate	PCB−1221 (Arochlor 1221)
Chloroform	Di-N-Butyl phthalate	PCB−1232 (Arochlor 1232)
2-chlorophenol	Di-n-octyl phthalate	PCB−1248 (Arochlor 1248)
1,2-dichlorobenzene	Diethyl phthalate	PCB−1260 (Arochlor 1260)
1,3-dichlorobenzene	Dimethyl phthalate	PCB−1016 (Arochlor 1016)
1,4-dichlorobenzene	benzo(a) anthracene	Toxaphene
3,3-dichlorobenzidine	Benzo(a)pyrene	Antimony
1,1-dichloroethylene	Benzo(b) fluoranthene	Arsenic
1,2-trans-dichloroethylene	Benzo(k) fluoranthene	Asbestos
2,4-dichlorophenol	Chrysene	Beryllium
1,2-dichloropropane	Acenaphthylene	Cadmium
1,2-dichloropropylene	Anthracene	Chromium
2,4-dimethylphenol	Benzo(ghi) perylene	Copper
2,4-dinitrotoluene	Fluorene	Cyanide, Total
2,6-dinitrotoluene	Phenanthrene	Lead
1,2-diphenylhydrazine	Dibenzo(h) anthracene	Mercury
Ethylbenzene	Indeno (1,2,3-cd) pyrene	Nickel
Fluoranthene	Pyrene	Selenium
4-chlorophenyl phenyl ether	Tetrachloroethylene	Silver
4-bromophenyl phenyl ether	Toluene	Thallium
Bis(2-chloroisopropyl) ether	Trichloroethylene	Zinc
Bis(2-chloroethoxy) methane	Vinyl chloride	2,3,7,8-TCDD

(see Tab. 1.5). It is the relative measure comparing the amount of heat trapped by a certain mass of the gas in question to the amount of heat trapped by a similar mass of carbon dioxide. GWP is calculated for a specific time interval, commonly over 20, 100, or 500 years [7]. The GWP depends on the following parameters: firstly on the given constituent of the atmosphere, secondly on the wavelength of the absorbed radiation and component lifetime in the atmosphere. Therefore, large values of GWP correspond to the components of the gas, which absorb a lot of energy in the infrared and have a long life. As can be seen from the table (Tab. 1.5), showing various greenhouse gases, all of them are more dangerous than CO_2. The total greenhouse emissions in 2011 included 6,702 million metric tons of CO_2 equivalent in the US. The biggest contributions were CO_2 (84 %), methane (9 %), nitrous oxide (5 %) and fluorinated gases (2 %). These emissions were mainly the result of electricity (33 %), transportation (28 %), industry (20 %), commercial residential (11 %) and agriculture (8 %) [8].

The most dangerous greenhouse effect is global warming of Earth. This will lead to the melting of glaciers in Greenland and Antarctica and cause sea levels to rise. As a result, many large cities in the world will be flooded. Those cities most vulnerable to flooding in Europe are London, Hamburg, Copenhagen, Stockholm, Padua, Venice, most of the area of Belgium, the Netherlands and Denmark. Long term changes in the environment change the global atmospheric circulation, leading to further climate change, causing an increase in precipitation over the oceans by about 10–15 %. Global warming will lead to changes in vegetation, changes in soil-forming processes and problems in the cultivation of plants that were previously specific and typical to their areas. Thus global warming will ultimately lead to desertification. It is believed that the only way to prevent these risks is to reduce greenhouse gas emissions through reducing deforestation and the extensive cultivation of various crops.

The ozone layer protects the earth from gamma radiation at an altitude of about 30 km from earth's stratosphere. Since the discovery of the "ozone hole" in 1985, its area has increased by 15 %. The reason for the formation of the "ozone hole" is the emission of chlorine, freon, aerosols and pollutants from industrial plants, which reduce the thickness of the layer of ozone (O_3) in the stratosphere. It is estimated that each year the ozone content is decreasing by approximately 0.2 % over the equator to about 0.4–0.8 %. The reduction of the ozone layer on the surface of Earth increases the amount of radiation (UV), which results in a higher incidence of many diseases such as skin cancer, eye disease, a weakened immune system. The ozone hole destroys plankton in marine waters and oceans, which produce most of the oxygen in our atmosphere. Prevention of growth of the ozone hole is possible via a reduction in emissions of greenhouse gases, mainly chlorofluorocarbon (CFC). Fluorinated gases are a very large group of gaseous compounds which are not found in nature at all (see Tab. 1.5). This group includes, among others hydrofluorocarbons, perfluorocarbons and sulfur hexafluoride [9].

1.3.2 Acid rain

Acid rain is a form of precipitation which has elevated levels of hydrogen ions (reducing pH). This can have a detrimental effect on plants, aquatic animals, people and infrastructure. Acid rain is caused by emissions of sulfur dioxides and nitrogen oxids, which react with water molecules in the atmosphere. Examples of human sources of such emissions are electricity generation, factories and motor vehicles. Electric power plants which use coal are the largest sources of gaseous emissions responsible for acid rain. Since 1970 the governments of several countries have made efforts to reduce the release of sulfur dioxide SO_x and nitrogen oxides NO_x into the atmosphere, with rather positive results. Sulfur and nitrogen oxides may also be produced naturally by lightning strikes and volcanic eruptions. Acid rain causes paint to peel, corrosion of steel structures, such as bridges, and erosion of stone statues. However, nitric acid in rainwater is an important source of fixed nitrogen for plant life. Acidic deposits have been found in glaciers and their age determined for thousands of years [10].

Another effect of acid rain is the acidification of soil and water, which is very harmful to vegetation and especially trees, contributing to leaf damage, excessive evaporation of water, and the disruption of photosynthesis. It kills a large number of trees and destroys undergrowth. Acid rain activates aluminum and cadmium, and causes accumulation of nitrate and sulfate, whereas the plant roots have a reduced ability to take in the most essential nutrients such as calcium, magnesium and potassium. Thus acid rain causes a shortage of these nutrients in plants. The reactions of sulfur and nitrogen oxides, which lead to the formation of corresponding acids, can be presented as follows [11]:

$$SO_2 + OH^- \rightarrow HOSO_2 \tag{1.2}$$

$$HOSO_2 + O_2 \rightarrow HO_2 + SO_3 \tag{1.3}$$

$$SO_3 + H_2O \rightarrow H_2SO_4 \tag{1.4}$$

$$NO_2 + H_2O \rightarrow HNO_3. \tag{1.5}$$

Both lower pH and higher aluminum concentration in surface waters occur as a result of acid rain and can cause damage to fish and other aquatic animals. At a pH below 5, most adult fish are killed. Generally, the biodiversity in lakes and rivers which are acidified is continuously declining [12]. Thus, the chemistry and biology in soil can also be seriously damaged by acid rain, since only a few bacteria are able to tolerate low pH. The enzymes of these bacteria are also damaged.

The hydronium ions of acid rain also mobilize toxins such as aluminum and cause run-off of essential nutrients and minerals such as magnesium [13]. Forests in high mountains are particularly vulnerable to the effects of acid rain, especially from the clouds and fog with which they are usually surrounded. The use of lime and fertilizers to replace lost nutrients can minimize the impact of acid rain on crops [14, 15].

Acid rain does not have a direct impact on the health of humans and animals because the acids are too diluted. However, sulfur and nitrogen oxides can have a negative impact and can contribute to heart and lung diseases and health problems such as asthma and bronchitis.

Table 1.5. The properties of greenhouse gases

Name	Formula	Lifetime (years)	Radiative Efficiency ($W\,m^{-2}\,ppb^{-1}$)	GWP100-
Carbon dioxide	CO_2	12	1.4×10^{-5}	1
Methane	CH_4	12	3.7×10^{-4}	25
Nitrous oxide	N_2O	114	3.03×10^{-3}	298
CFC-11	CCl_3F	45	0.25	4,750
CFC-12	CCl_2F_2	100	0.32	10,900
CFC-13	$CClF_3$	640	0.25	14,400
CFC-113	CCl_2FCClF_2	85	0.3	6,130
CFC-114	$CClF_2CClF_2$	300	0.31	10,000
CFC-115	$CClF_2CF_3$	1700	0.18	7,370
Halon-1301	$CBrF_3$	65	0.32	7,140
Halon-1211	$CBrClF_2$	16	0.3	1,890
Halon-2402	$CBrF_2CBrF_2$	20	0.33	1,640
Carbon tetrachloride	CCl_4	26	0.13	1,400
Methyl bromide	CH_3Br	0.7	0.01	5
Methyl chloroform	CH_3CCl_3	5	0.06	146
HCFC-22	$CHClF_2$	12	0.2	1,810
HCFC-123	$CHCl_2CF_3$	1.3	0.14	77
HCFC-124	$CHClFCF_3$	5.8	0.22	609
HCFC-141b	CH_3CCl_2F	9.3	0.14	725
HCFC-142b	CH_3CClF_2	17.9	0.2	2,310
HCFC-225ca	$CHCl_2CF_2CF_3$	1.9	0.2	122
HCFC-225cb	$CHClFCF_2CClF_2$	5.8	0.32	595
HFC-23	CHF_3	270	0.19	14,800
HFC-32	CH_2F_2	4.9	0.11	675
HFC-125	CHF_2CF_3	29	0.23	3,500
HFC-134a	CH_2FCF_3	14	0.16	1,430
HFC-143a	CH_3CF_3	52	0.13	4,470
HFC-152a	CH_3CHF_2	1.4	0.09	124
HFC-227ea	CF_3CHFCF_3	34.2	0.26	3,220
HFC-236fa	$CF_3CH_2CF_3$	240	0.28	9,810
HFC-245fa	$CHF_2CH_2CF_3$	7.6	0.28	1030
HFC-365mfc	$CH_3CF_2CH_2CF_3$	8.6	0.21	794
HFC-43-10mee	$CF_3CHFCHFCF_2CF_3$	15.9	0.4	1,640
Sulphur hexafluoride	SF_6	3,200	0.52	22,800
Nitrogen trifluoride	NF_3	740	0.21	17,200
PFC-14	CF4	50,000	0.10	7,390
PFC-116	C_2F_6	10,000	0.26	12,200
PFC-218		2,600	0.26	8,830

Name	Formula	Lifetime (years)	Radiative Efficiency (W m^{-2} ppb^{-1})	GWP100-
PFC-318		3,200	0.32	10,300
PFC-3-1-10		2,600	0.33	8,860
PFC-4-1-12		4,100	0.41	9,160
PFC-5-1-14		3,200	0.49	9,300
PFC-9-1-18		> 1,000d	0.56	> 7,500
CF3SF5		800	0.57	17,700
HFE-125		136	0.44	14,900
HFE-134		26	0.45	6,320
HFE-143a		4.3	0.27	756
HCFE-235da2		2.6	0.38	350
HFE-245cb2		5.1	0.32	708
HFE-245fa2		4.9	0.31	659
HFE-254cb2		2.6	0.28	359
HFE-347mcc3		5.2	0.34	575
HFE-347pcf2		7.1	0.25	580
HFE-356pcc3		0.33	0.93	110
HFE-449sl (HFE-7100)		3.8	0.31	297
HFE-569sf2 (HFE-7200)		0.77	0.3	59
HFE-43-10pccc124 (H-Galden 1040x)		6.3	1.37	1,870
HFE-236ca12 (HG-10)		12.1	0.66	2,800
HFE-338pcc13 (HG-01)		6.2	0.87	1,500
PFPMIE		800	0.65	10,300
Dimethylether		0.015	0.02	1
Methylene chloride		0.38	0.03	8.7
Methyl chloride		1.0	0.01	13

1.3.3 Main air pollutants

Certain substances which come from air pollution are also responsible for the acidification and eutrophication of soils [16]. They are: carbon dioxide, nitrogen oxides (NO_x), sulfur oxides (SO_x), volatile organic compounds (VOCs) and ammonia (NH_3). These five substances are also considered to be the main pollutants. The Clean Air Act established in the US outlines the requirements for national air quality standards for six common air pollutants. These are fine particles of the solid contaminants PM25 and PM10, which express their average size in microns, and also carbon monoxide, sulfur oxides, nitrogen oxides and lead. These chemicals threaten the environment and the health of humans, animals and plants. The individual sectors emit different levels of these impurities, which can be observed by the respective charts below. The National Ambient Air Quality Standards (NAAQS) established the regulations on theses emissions.

Carbon monoxide [17] emission is caused mainly by vehicles, i.e. more than 40 Mt, whereas mostly (38 Mt) they are supplied with gasoline (see Fig. 1.4). Carbon monoxide causes harmful health effects by reducing the oxygen supply to the organs, myocardial ischemia as a result of decreased oxygen supply to the heart (often accompanied by chest pain, angina pectoris) and tissues leading to death. Even very short exposure is harmful because it weakens the ability to respond to increased oxygen demand.

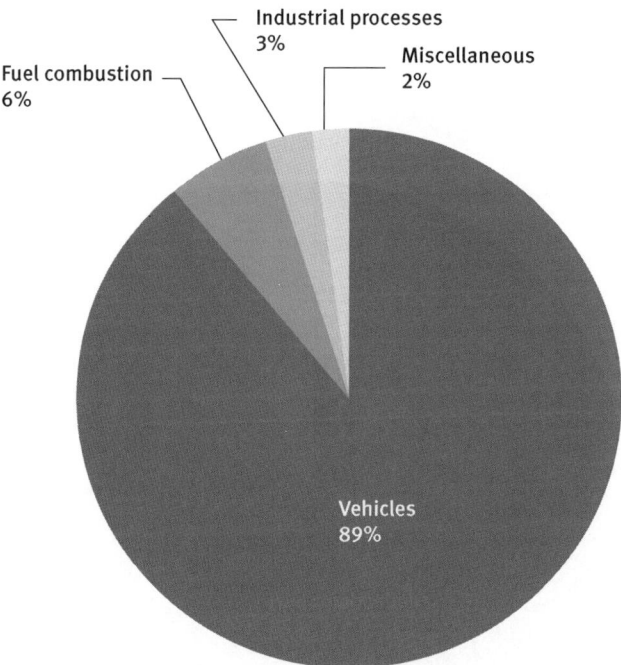

Fig. 1.4. The CO emissions of various sectors of the economy.

NO_x emissions [18] are dangerous and represent a significant risk to public health (Fig. 1.5). NO_x emissions are a group of highly reactive gases, and generally create more NO_x. The ambient air quality standards therefore use NO_2 as an indicator for the larger group of nitrogen oxides. NO_2 is formed from emissions from cars, trucks and buses, power plants and off-road equipment. The main source of the emissions is transportation vehicles, i.e. more than 10 Mt, fuel combustion around 5 Mt, and industrial processes only 1 Mt (Fig. 1.6).

The main emission of sulfur oxides comes from the burning of fuels [19]. In the US this is well documented, reaching more than 9 Mt/y, but most of these emissions are caused by power generation (7.6 Mt/y) (Fig. 1.7). The rest is produced by industry (around 0.9 Mt/y) and vehicles (0.7 Mt/y). Emissions which lead to high concentrations of SO_2 generally also lead to the formation of other SO_x. SO_2 is the component of

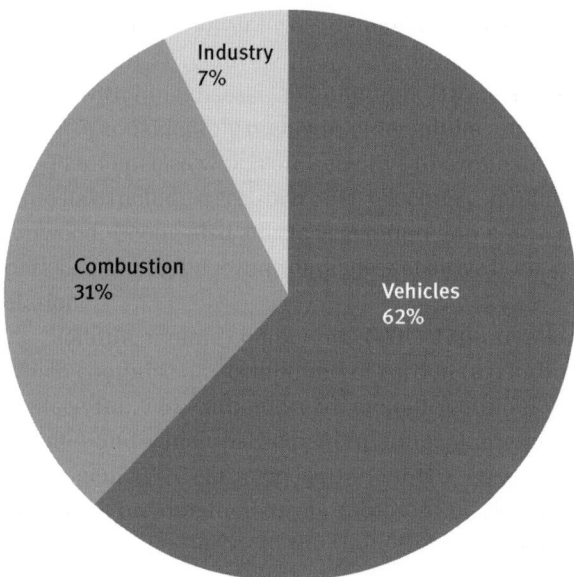

Fig. 1.5. The main emitters of NOx.

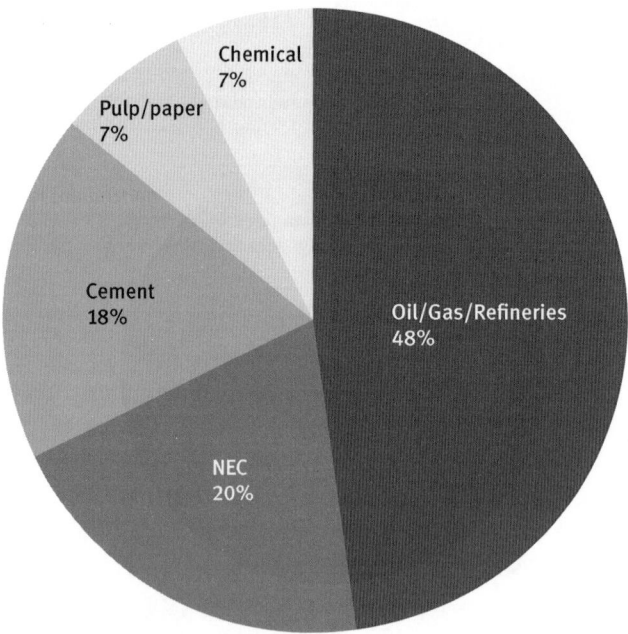

Fig. 1.6. Emissions of NOx by various industrial sectors.

greatest concern and is therefore used as the indicator for the larger group of gaseous sulfur oxides. Other gaseous sulfur oxides (e.g. SO_3) are found in the atmosphere at concentrations much lower than SO_2. This may have the important co-benefit of reducing the formation of fine sulfate particles, which pose significant public health threats. SO_x can react with other compounds in the atmosphere to form small particles. These particles penetrate deeply into sensitive parts of the lungs and can cause or worsen respiratory disease, such as emphysema and bronchitis, and can aggravate existing heart disease, leading to increased hospital admissions and even premature death. One sulfur oxide exerts rather low effects on human health. Short-term exposure to SO_2 leads to adverse respiratory affects including bronchoconstriction and increased asthma symptoms. These effects are particularly important for asthmatics.

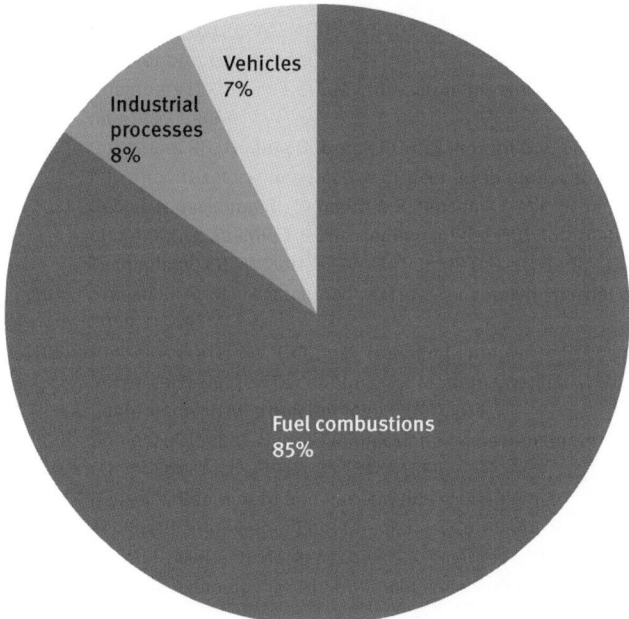

Fig. 1.7. Sulfur dioxide emissions by various sectors of the economy.

Ozone [20] occurs in two regions of Earth's atmosphere, i.e. on the lower and upper layers of the troposphere, which is known as the ozonosphere (O_3). Ozone acts as a protective layer about 30 km above the earth. Ozone in the lower layer is the main component of smog, which occurs frequently in the summer.

Both ozone in the troposphere and ground-level ozone are formed by means of chemical reactions between nitrogen oxides (NO_x) and volatile organic compounds (VOC). Emissions from industrial facilities and electric utilities, motor vehicle exhaust, gasoline vapors, and chemical solvents are some of the major sources of NO_x and VOC.

Ozone in the lower layers definitely has harmful effects on humans, animals and plants, even at relatively low concentrations. The negative health effects occur in adults with lung disease and children. Ozone is associated with increased mortality, which is considerably higher in older adults and in the summer.

Plant species which are sensitive to ozone and potentially at increased risk from exposure include trees such as black cherry, quaking aspen, ponderosa pine and cottonwood. Ozone also affects vegetation and entire ecosystems, including forests and parks. In particular, ozone damages sensitive flora, especially during their intense vegetation.

Volatile organic compounds (VOC) [21] are a large group of organic compounds which comprise any compound of carbon (excluding carbon monoxide, carbon dioxide (see Tab. 1.5), carbonic acid, metallic carbides or carbonates, and ammonium carbonate) and which participate in atmospheric photochemical reactions. Global anthropogenic VOC emissions are about 110 Mt/year. According to the EPA, 78 % of VOC emissions are due to natural processes and only 22% are caused by human activity. Anthropogenic sources in the United States emit about 21% of the total global emissions of VOCs, followed by Russia, China, India and Japan. Globally, the major sources of VOC emissions are fuel burning fires and wood burning savannah which account for over 35 % of total global emissions. Manufacturing was also found to be a significant source of VOC emissions. According to NEI data, the national total estimated VOC emissions from anthropogenic sources (see Fig. 1.8), excluding wildfires and prescribed burns, decreased by 35 percent between 1990 and 2005 (i.e. from 23 to 15 Mt/y).

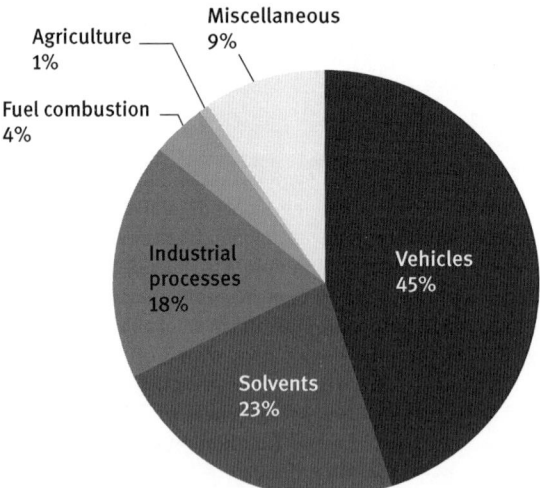

Fig. 1.8. Anthropogenic VOC emissions by various sectors of the economy.

Ammonia (NH_3) is a decomposition product of protein, urea, amides, and uric acid. This gas shed by air can disrupt the nitrogen cycle in nature. The total global emission of ammonia is estimated at about 62 Mt/y, of which about 42% come from animal production (see Fig. 1.9). Emissions are estimated at 8 million tons in Europe, of which about 72% are caused by livestock. Ammonia is a gas toxic to both animals and humans, and also has a negative impact on the environment, not only in the immediate vicinity of livestock buildings, but also in areas of forest and grassland.

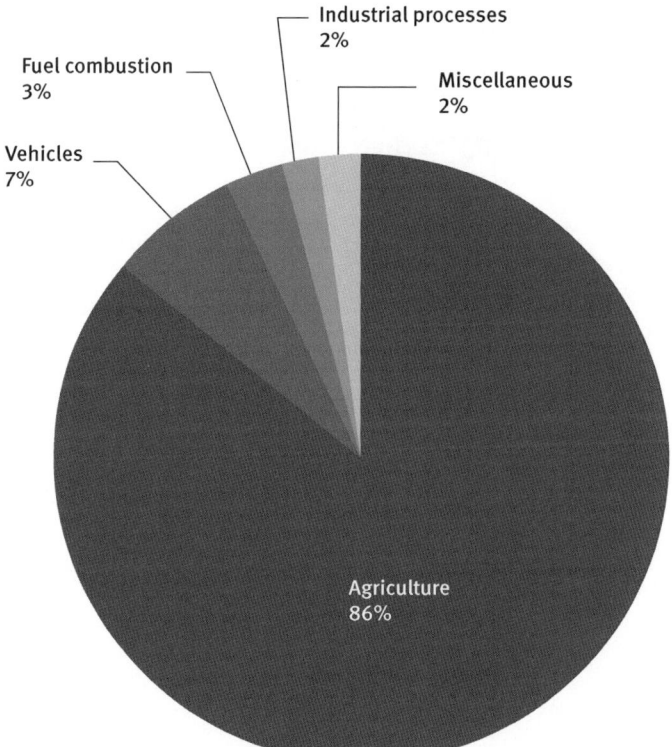

Fig. 1.9. Ammonia emissions by various sectors of the economy.

Particulate matter, also known as particle pollution or PM, is a complex mixture of extremely small particles and liquid droplets (see Fig. 1.10). Particle pollution is made up of a number of components, including acids (such as nitrates and sulfates), organic chemicals, metals, and soil or dust particles. Particle pollution includes inhalable coarse particles, with diameters larger than 2.5 micrometers and smaller than 10 micrometers, and fine particles with diameters less than 2.5 micrometers. The health effects of particles on humans are: premature death in people with heart or lung disease, nonfatal heart attacks, irregular heartbeat, aggravated asthma, decreased lung function, and increased respiratory symptoms, such as irritation of the airways, coughing or difficulty breathing. The ecological effects of particle pollution are acidification of lakes and streams, changes to the nutrient balance in coastal waters and large river basins, depletion of the nutrients in soil, damage to sensitive forests and farm crops, and reducing ecosystem biodiversity. More details on the effects of particulate pollution and acid rain are shown in the charts below.

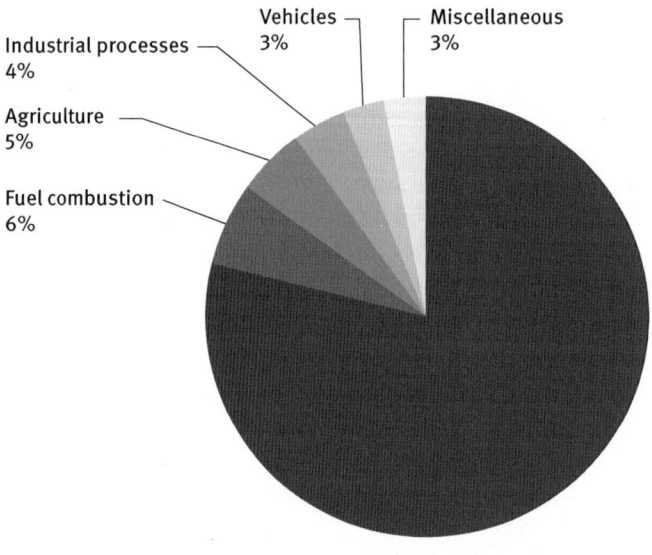

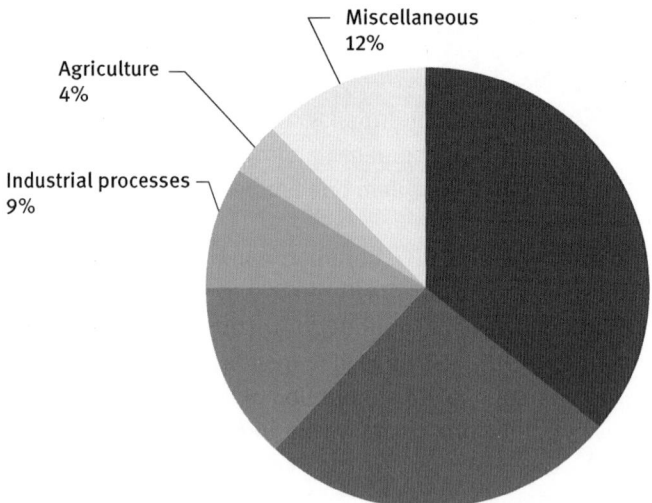

Fig. 1.10. Particulate matter emissions by various sectors of the economy: (a) PM10, (b) PM2.5.

The main sources of air pollution in Europe are unevenly distributed among the states, but we feel their effects equally (see Figs. 1.11 and 1.12).

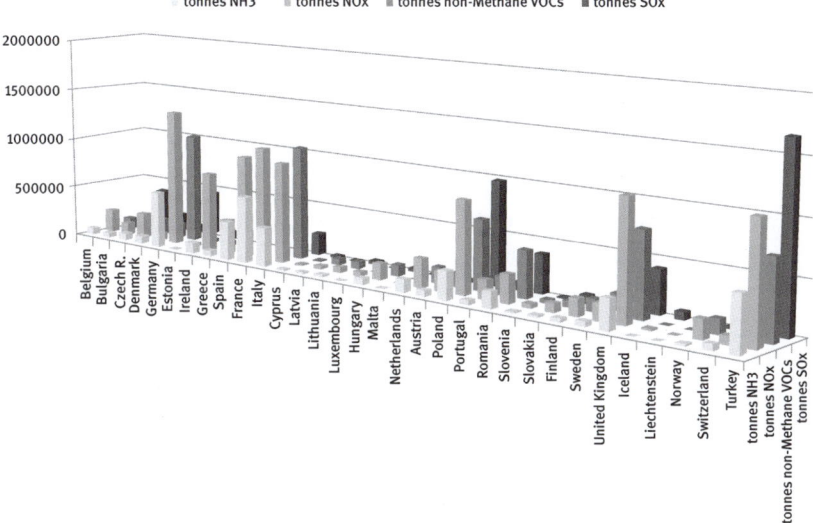

Fig. 1.11. European national air contaminants.

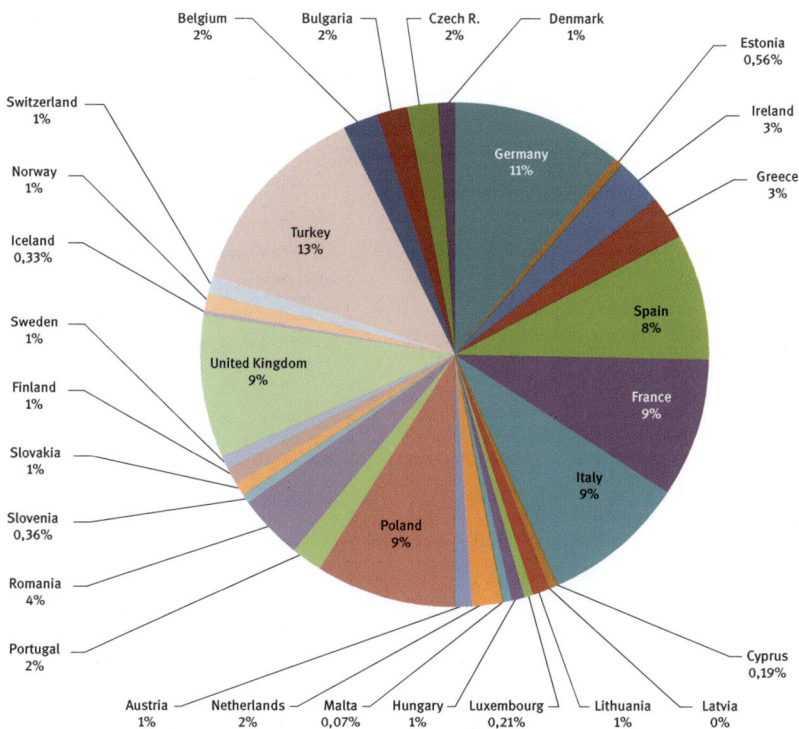

Fig. 1.12. European national share in percentiles of total air contamination.

1.4 The environmental problems of water

Eighty percent (80 %) of all pollution in seas and oceans comes from human activities on land, caused mainly by agriculture, urbanization, and various technologies. Therefore it can be concluded that human activity is the main cause of pollution. By 2010, 80 percent of people live less than 60 km from the coast, and 75 % of the world's biggest cities lay above the sea. The most intensive anthropogenic pollution may result from sectors of the economy such as industry, agriculture, urban development and transport. Every day, more than 40 km^2 of the oceans and seas are covered with a layer of oil spills, which comes from 0.1–0.2 % of oil production, which in 2012 amounted to 30 billion barrels. Hormonal drugs in common use as contraceptives cause feminization of marine animals which may seem harmless, but leads to their extinction. The most common anthropogenic pollution of surface water includes pesticides, surfactants, petroleum hydrocarbons, phenols, chlorinated derivatives of biphenyl and heavy metals such as lead (Pb), copper (Cu), chromium (Cr), cadmium (Cd), mercury (Hg) and zinc (Zn), and heated water (thermal pollution) which are particularly dangerous for surface water with low flow or stagnant water.

All chemicals used in agriculture must be used in large amounts to ensure their effectiveness at the right time of vegetation, regardless of variable weather conditions. The excess of all these chemicals then flows down rivers and streams to the oceans. They can certainly cause irreversible and unpredictable changes in ecosystems. This is because they accumulate in animals, and the impact of these substances on living organisms is hidden, being mutagenic, carcinogenic and damaging to immune systems. Heavy metals and organic substances emitted from various anthropogenic sources, once introduced into the food chain, will never leave. Seafood provides the largest percentage of the world's protein consumed by humans. In addition, most of the world's major fisheries are operated and maintained above their maximum sustainable yield capacity.

It is ironic that water pollution, which is the result of "apparent savings", turns out to be eventually very costly. Death and disease caused by polluted waters costs the U.S. economy alone up to $ 12.8 billion per year. It has been noted that only $ 7.2 billon were costs for the treatment of liver disease due to contamination of the sea.

Most anthropogenic pollution is toxic to aquatic organisms. Impurities are very persistent in the aquatic environment and inhibit chemical and biochemical processes.

1.4.1 The main effects of water pollution

The attempt to define irritants caused by the different sectors was undertaken on the basis of drinking water contaminants listed by the EPA (see Tab. 1.6). In order to fully exploit the data in this table the following analysis was prepared. Those compounds whose activity manifests itself even in the smallest amounts (zero or at most of order of magnitude of ppb), and therefore the most dangerous to life and health, were selected from this list. Afterwards the set of the most serious diseases caused by substances used was established. In this way, the set of the most dangerous xenobiotics which are released into the aquatic environment as well as the different emitting sectors has been prepared. Cancer is most often caused by 30 different substances from the list, mainly from agriculture (especially biocides, 7 items from the table), together with 7 petrochemical substances, 4 chemical products and one from treatment plants (5). 21 substances from the EPA list cause liver disease such as 7 items from agriculture, 4 different chemicals and 5 petrochemicals. Kidney disease (11 substances) is caused by substances used in the manufacturing sector, such as agriculture (7), chemicals (4) and petrochemicals (5). Diseases of the reproductive system (12) are mainly caused by landfill (8), and substances used in agriculture (5). Finally, a blood disorder is caused mainly by substances derived from the electronics industry and refineries. All of these human activities contribute to environmental pollution in various forms and amounts of impurities. However, earth's subsystems, such as air, water and soil are open and therefore different xenobiotics and pollutants are reciprocally transported between them. The contaminants spread even more rapidly in air and water because they are liquids.

Water pollution (e.g. lakes, rivers, oceans, aquifers and groundwater) also spreads when various xenobiotics are discharged into water without proper treatment. The presence of impurities causes undesirable changes in nature because substances disrupt existing systems. Water pollution affects plants and organisms living in these bodies of water. In almost all cases the effect is not only detrimental to individual species and populations, but also to the whole ecosystem.

Interactions between groundwater and surface waters are very complex. Therefore, groundwater pollution is sometimes referred to as surface water pollution. Contaminants enter the water system as a result of leaching from soil and waste dumps or the permanent drainage of sewage and incidental spillage. Groundwater contamination depends on soil properties, geological and hydrogeological conditions and the nature of the contamination.

Table 1.6. The most acute contaminants in drinking waters (based on the EPA list)

Contaminant	Location	Health symptoms
1,1,2-Trichloroethane 50	Factories	Immune system
1,1,2-Trichloroethane 50	Factories	Kidneys
1,1,2-Trichloroethane 50	Factories	Liver
1,1-Dichloroethylene 7 ppb	Chemical	Liver
1,2,4-Trichlorobenzene 70 ppb	Textile	Glands
1,2-Dibromo-3-chloropropane (DBCP)	Agriculture	Cancer
1,2-Dibromo-3-chloropropane (DBCP)	Agriculture	Reproductive organs
1,2-Dichloroethane	Chemical	Cancer

Contaminant	Location	Health symptoms
1,2-Dichloroethane	Chemical	Cancer
1,2-Dichloropropane	Paint	Cancer
2,4,5-TP (Silvex) 50 ppb	Agriculture	Cancer
Acrylamide	Wastewater	Cancer
Acrylamide	Wastewater	Cancer
Acrylamide	Wastewater	Nervous system
Alachlor	Agriculture	Cancer
Alachlor	Agriculture	Eyes
Alachlor	Agriculture	Kidneys
Antimony 6ppb	Petrochemicals	Blood
Antimony 6ppb	Electronics	Blood
Arsenic	Electronics	Blood
Arsenic	Glass	Cancer
Arsenic	Electronics	Skin
Atrazine 3ppb	Agriculture	Blood
Atrazine 3ppb	Agriculture	Reproductive organs
Benzene	Petrochemical	Blood
Benzene	Petrochemical	Cancer
Benzo(a)pyrene PAH	Petrochemical	Cancer
Benzo(a)pyrene PAH	Petrochemical	Reproductive system
Berylium 4ppb	Aerospace	Intestinal lesions
Bromate	Water treatment	Cancer
Cadmium 5 ppb	Paint	Kidneys
Cadmium 5 ppb	Bateries	Reproductive organs
Carbon tetrachloride	Chemical	Cancer
Carbon tetrachloride	Chemical	Liver
Chlordane	agriculture	Cancer
Chlordane	Agriculture	Liver
Chlorobenzene	Chemical	Kidneys
Chlorobenzene	Chemical	Liver
Di(2-ethylhexyl) phthalate	Rubber	Cancer
Di(2-ethylhexyl) phthalate	Chemical	Cancer
Di(2-ethylhexyl) phthalate	Rubber	Liver
Di(2-ethylhexyl) phthalate	Chemical	Liver
Di(2-ethylhexyl) phthalate	Rubber	Reproductive organs
Di(2-ethylhexyl) phthalate	Chemical	Reproductive organs
Dichloromethane	Pharmaceuticals	Cancer
Dichloromethane	Pharmaceuticals	Liver
Dinoseb	Agriculture	Reproductive organs
Dioxin (2,3,7,8-TCDD	Combustion,	Cancer
Dioxin (2,3,7,8-TCDD	Chemicals	Reproduction organs
Diquat 20 ppb	Agriculture	Cataracts
Endrin 2 ppb	Agriculture	Liver
Epichlorohydrin	Wastewater	Cancer
Epichlorohydrin	Chemical	Stomach
Ethylbenzene	Petrochemical	Kidneys
Ethylbenzene	Petrochemical	Liver

Contaminant	Location	Health symptoms
Ethylene dibromide	Petroleum	Cancer
Ethylene dibromide	Petrochemical	Cancer
Ethylene dibromide	Petrochemical	Liver
Ethylene dibromide	Petrochemical	Reproductive organs
Ethylene dibromide	Petrochemical	Stomach
Haloacetic acids (HAA5)	Water treatment	Cancer
Heptachlor 0.4 ppb	Agriculture	Cancer
Heptachlor 0.4 ppb	Agriculture	Liver
Heptachlor epoxide 0.2 ppb	Agriculture	Cancer
Heptachlor epoxide 0.2 ppb	Agriculture	Liver
Hexachlorobenzene	Petroleum	Cancer
Hexachlorobenzene	Petroleum	Kidneys
Hexachlorobenzene	Petroleum	Liver
Lead	Plumbing	Blood
Lindane 0.2 ppb	Agriculture	Kidneys
Lindane 0.2 ppb	Agriculture	Liver
Mercury 2 ppb	Agriculture,	Kidneys
Mercury 2 ppb	Refineries	Liver
Methoxychlor 40 ppb	Agriculture	Reproductive organs
Pentachlorophenol	Biocide	Cancer
Pentachlorophenol	Biocide	Kidneys
Pentachlorophenol	Biocide	Liver
Polychlorinated biphenyls PCB	Landfills	Cancer
Polychlorinated biphenyls PCB	Landfills	Immune system
Polychlorinated biphenyls PCB	Landfills	Reproductive organs
Selenium 50	Mining	Fingernails
Selenium 50	Mining	Hair
Simazine 4 ppb	Agriculture	Reproductive organs
Tetrachloroethylene	Factories	Cancer
Tetrachloroethylene	Dry cleaning	Liver
Thallium 0.5 ppb	Electronics	Fingernails
Thallium 0.5 ppb	Glass drugs	Fingernails
Toluene	Petrochemical	Kidneys
Toluene	Petrochemical	Liver
Toxaphene	Agriculture	Kidneys
Toxaphene	Agriculture	Liver
Toxaphene	Agriculture	Thyroid
Trichloroethylene	Degreasing	Cancer
Trichloroethylene	Degreasing	Liver
Vinyl chloride	Pipes	Cancer
Xylenes	Petrochemical	Nervous system

1.4.2 Thermal pollution of water

Thermal pollution is the rise or fall in temperature of natural waters due to the influence of man. Thermal pollution, in contrast to chemical impurities, changes the physical properties of water. A common cause of thermal pollution is the use of water as a coolant by power plants and industrial manufacturers. Elevated water temperature reduces the concentration of oxygen, which supports the thermophilic organisms. Killing higher creatures changes the whole food chain and reduces the biodiversity of species. Urban temperatures may also increase in surface water or can also be caused by the release of very cold water from reservoirs and tanks to warmer rivers.

1.4.3 Organic contaminants in water

More and more organic matter reaches the oceans each year as a result of oil spills. This is a threat to the microfauna and microflora which produce oxygen vital to most living beings. Alone in the 2011, 30 billion barrels of oil were taken over the world, 159 liters per barrel. Thus, at least 8,200,000 barrels of oil were accidentally spilled into the sea, which is the lower limit of the estimated amount of about 0.1–0.25 % of annual production. This is equivalent of 16 disasters the size of an average oil tanker sinking (such as the "Prestige" in 2002).

Organic contaminants of water contain a number of organic compounds, such as byproducts of disinfection of drinking water; waste processing (which may contain substances which require oxygen), fats and greases, insecticides and herbicides, a huge range of organohalides and other chemical compounds. Petroleum hydrocarbons, including fuels (gasoline, diesel, jet fuel and heating oil) and lubricants (motor oil), and fuel combustion byproducts, from the run-off of rain. Volatile organic compounds (VOCs), such as industrial solvents, chlorinated solvents (dense nonaqueous phase liquids, or DNAPLs, which fall to the bottom of tanks), polychlorinated biphenyls (PCBs), THMs, perchlorates, and various compounds used for personal hygiene and cosmetics. Polyaromatic hydrocarbons (PAHs) are often associated with oil spills from tankers and pipelines, and are introduced into the marine food chains for many years [22]. They are the following: acenaphthene, acenaphthylene, anthracene, benzanthracene, benzopyrene, benzopyrene, benzofluoranthene, benzoperylene, benzofluoranthene, benzofluoranthene, chrysene, coronene, dibenzanthracene, fluoranthene, fluorine, indeno (1, 2, 3 ...) pyrene, phenanthrene, pyrene and naphthalene. The toxicity of PAHs depends in each case on the molecular structure (i.e. the number of rings). Benzopyrene is very famous because it stands out from the very dangerous chemicals found in cigarettes and has carcinogenic, mutagenic and teratogenic effects. The PAHs such as benzanthracene and chrysene, benzofluoranthene, benzofluoranthene, benzofluoranthene, benzopyrene, benzoperylene, coronene, dibenzoanthracene ($C_{20}H_{14}$), indeno (1, 2, 3-cd) pyrene ($C_{22}H_{12}$) and ovalene have similar properties [23–26].

99 % of the forms of life on Earth are found in the oceans. However, less than 10 % have been investigated so far. Less than half of one percent of marine habitats is protected, but 50 % of the earth is beyond any jurisdiction. New examples confirming the negative human impact on marine waters are continuously revealed. Populations of commercially attractive fish have fallen by 90 % over the past century. Nearly 60 % of the world's remaining reefs are in decline, and will disappear over the next three years. Eutrophication of waters greatly increases phytoplankton in a water body as a response to increased levels of nutrients. Negative environmental effects include hypoxia, the depletion of oxygen in the water, which reduces in the numbers of specific fish and other animal populations. Further species (such as Nomura's jellyfish in Japanese waters) may experience an increase in population which negatively affects other species.

Table 1.7. The US national multipollutant emissions comparison: percent contribution by source sector in 2008

Pollutant	Agriculture	Industry	Transport	Combustion	Dust	Solvent	Mix
Ammonia	86,3	2,1	7,2	2,6	0	0	1,9
Carbon monoxide	0	2,7	88,4	6,4	0	0	2,4
Nitrogen oxides	0	7	62	30,3	0	0	0
Sulfur oxides	0	8,1	6,7	84,9	0	0	0,2
VOCs	0,7	18,4	45,5	3,8	0	23,1	8,6
PM10	4,8	3,7	3,2	6,4	79,2	0	2,7
PM25	4,4	9,5	12,6	26	35,7	0,1	11,7
Lead	0	25,6	57,8	15,1	0	0,5	1,1

1.4.4 Inorganic contaminants in water

Inorganic water pollutants include acids caused by industrial discharge such as sulfur dioxide from power plants, ammonia from food processing waste, chemical waste as industrial byproducts, fertilizers containing nutrients such as nitrates and phosphates which are in storm water run-off from agriculture and urban sites, as well as commercial and residential use, heavy metals from motor vehicles, mine drainage, sediments in run-off from construction sites (see Tab. 1.7 and Fig. 1.13).

One inorganic contaminant is arsenic. Arsenic occurs naturally in rocks and soils, water and air, as well as in plants and animals. It can be released into the environment by volcanic activity, erosion of rocks and forest fires, or by human activity. About 90 % of industrial arsenic in the US is currently used as a wood preservative, but it is also used in paints, dyes, medicines, and semiconductors. High levels of arsenic can also result from certain fertilizers and animal feeding operations. Industrial practices

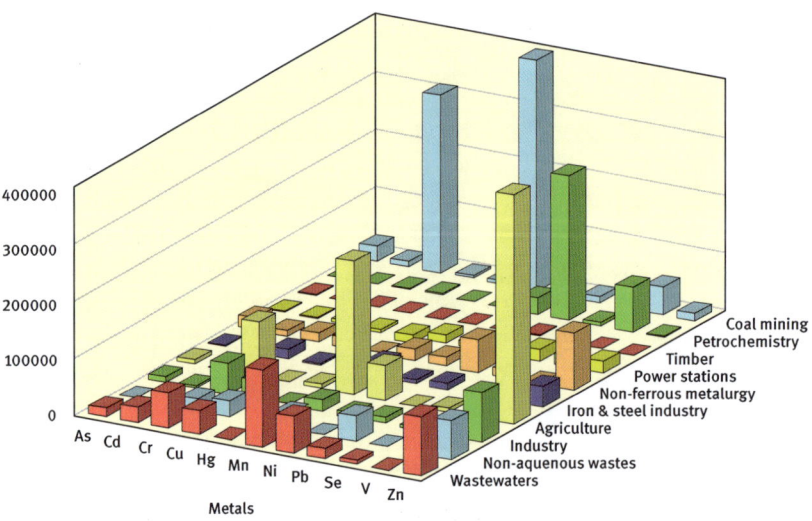

Fig. 1.13. Water contamination with heavy metals by different production. sectors

such as copper smelting, mining and coal burning also contribute to increasing in the amount of arsenic in our environment. Higher concentrations of arsenic are usually found in ground water than in surface water such as lakes and rivers, which are the main sources of drinking water. Large scale extraction of groundwater for municipal water supply wells and private drinking water can cause the release of arsenic from rock formations. European countries have more systems with arsenic concentrations above EPA standards of 10 parts per billion (ppb) than in the United States.

Another metallic contaminant is chromium. Approximately 136,000 tons of hexavalent chromium was produced in 1985. The hexavalent chromium compounds are chromium trioxide and various salts of chromate and dichromate and chromic acid. The total chromium concentration containing both chromium 6 and 3 permitted by the EPA is 0.1 mg per liter (mg/l = ppm) for tap water. This standard was established in 1991 and is based on the multiple adverse effects on organisms such as mutagenic, immunogenic, reproductive, and moreover obvious toxicity effect and also skin reactions. Inhaled hexavalent chromium is recognized as a human carcinogen. Problematic exposure is known to occur among workers who handle chromate-containing products and those who weld or grind or braze stainless steel. Hexavalent chromium compounds are used as chromate pigments in dyes, paints, inks, textile dyes, leather tannings and wood preservatives. Hexavalent chromium is used in making stainless steel, anticorrosive agents in paint and conversion coatings as a decorative or protective coating. Chromic acid is used for electroplating metal parts. The use of hexavalent chromium in electronic equipment is largely prohibited in the European Union by the Restriction of Hazardous Substances Directive.

Lead emissions in 2008 in the US (see Fig. 1.14) were mainly caused by the transport of 586 t, almost all of which was attributed to planes (579 t). At that time, only half of this amount (259 t) was issued by the industry, most of which were nonferrous metals (101), stainless steel (78), chemicals (12), cement (8), battery production, metal products and ammunition. Lead can have adverse effects on the blood, as well as the nervous, immune, renal and cardiovascular systems. Early childhood and prenatal exposure is associated with deceleration of cognitive development, learning difficulties, and other effects. Exposure to lead can occur from breathing contaminated air or from the workplaces where lead or its materials are processed, as well as from lead-based paints. Exposure to high amounts of lead can cause gastrointestinal symptoms, severe damage to the brain and kidneys, and may cause reproductive effects. Large doses of some lead compounds have also caused cancer in lab animals.

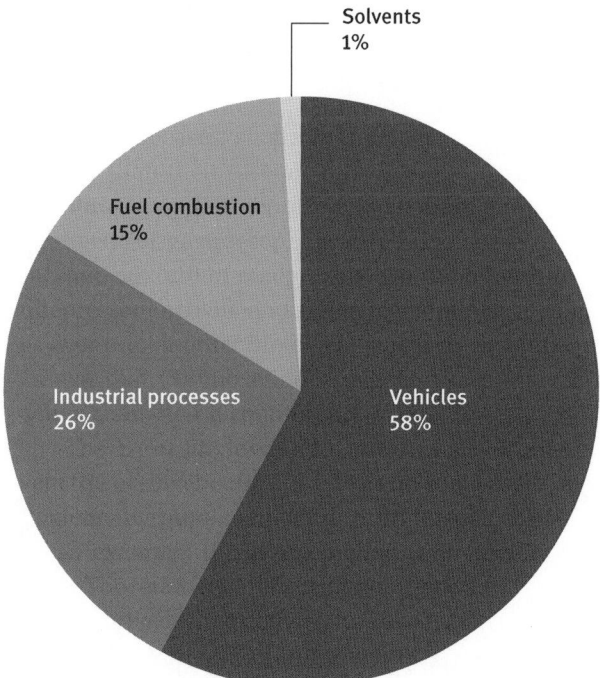

Fig. 1.14. Lead (Pb) emissions by various sectors of the economy.

1.5 The environmental problems of soil

Soil is mainly contaminated by waste in post-industrial areas, abandoned human and animal habitats and landfills. Landfills may contain chemical waste, waste from households, hospitals, and industrial processes. The biggest landfills are situated in

Africa, Eastern Europe and Northern Asia. As stated, the main impurities include lead, chromium, cadmium, arsenic, cyanide, dioxins, mercury, sulfur dioxide, volatile organic compounds and other particulates. The US Bureau of Labor Statistics classifies the following chemicals: basic chemicals, including pigments, dyes, and petrochemicals, synthetic materials, such as plastic products, paints, cleaning agents and other chemicals, including film, paint and explosives. It should be emphasized that landfill is full of valuable materials which could easily be re-used. This applies to materials such as metals and plastics. Aluminum resources are easier to obtain than in their ores i.e. bauxites. Plastic after grinding and melting can be used as construction materials, railway sleepers, beams, slabs. This method is called recycling and mining landfill is very attractive from an economic point of view, and also helps to protect the environment. Note, however, that many of the waste are very harmful as electronic waste, because it contains many hazardous materials. Therefore, the use of these recovered materials is very desirable and is a component of sustainable development.

The civilization of modern man is also characterized by the fact that we buy a lot of unnecessary things under the influence of irrational incentives caused by advertising or fashion. This results in the emergence of continuous and endless stream of new products which soon become waste too. It is estimated that in America alone in the decade 1997–2007 about 500 million computers were replaced. Electronic waste is composed of a mixture of different materials such as metals, chemicals and plastics which means they require special solutions. The waste must be dismantled and the components extracted before recycling or disposal. Many of these methods, even in e-waste disposal sites, are unsafe and may release hazardous elements to the environment.

400 metric tons of mercury each year is released into the air as a result of gold-processing in artisanal gold mining, according to EPA estimates. Hexavalent chromium is found at a many polluted tannery sites and is a carcinogen which causes lung cancer and potentially increases the risk of stomach cancer when ingested. The toxic properties of lead are well documented but it is still used for various important purposes in the world economy, as it is corrosion-resistant and malleable. More than 500 locations in the worldare contaminated by lead, and it was estimated that approximately 16 million people are at risk because of the smelting of ores, mining and processing industry, as well as recycling and the production of lead-acid batteries. Lead pollution comes from e-waste and chemical pesticides. It is used in many industrial processes all over the world for the production of alloys, paints, ammunition and as a lubricant [27].

Petrochemicals refer to a wide range of chemicals which are used in adhesives, carpets, cosmetics, paints, rubber, textiles, fertilizers and plastics. Petrochemical processing is particularly special, because fossil fuels are used as the building blocks of the chemicals. Therefore, petrochemical processing is often performed in oil-producing regions. Therefore, a high proportion of global pollution is due to the processing of fossil fuels, coal, gas and oil for many products.

Dyes are used primarily in the production of consumer products, including paints, textiles, printing inks, paper, and plastics. Natural dyes extracted from vegetables, fruit and flowers have been used since 3,500 BC to color fabrics and other materials [28]. These dyes were replaced by chemical dyes which bind better to fabric, providing and retaining richer color throughout washing and exposure [29].

Biocides are substances for preserving a variety of goods during storage. Biocides prevent the growth of unwanted plants, fungi and animals. They are used in agriculture and various industries [30, 31]. During the production of biocides a high volume of impurities is generated [32]. Pesticides are widely used throughout the world, but the main risk areas are located in Eastern Europe, Central and South America and South Asia. China is currently the largest manufacturer and exporter of pesticides in the world. Nearly eight million people are at risk due to pollution resulting from the production of pesticides, contaminated sites and agricultural practices.

The gradual increase and enrichment of ecosystems by nutrients such as nitrates (N) and/or phosphates (P) is termed eutrophication. Eutrophication, or more precisely hypertrophication, is the ecosystem's response to the addition of artificial or natural substances, such as nitrates and phosphates, through fertilizers or sewage, to an aquatic system. Increased nitrates in soil are frequently undesirable for plants. Many terrestrial plant species, such as the majority of orchid species in Europe are endangered as a result of soil eutrophication [33]. Inputs of atmospheric N are also a source of essential nutrients. This fertilizer effect results in increased plant growth and an increased demand for other plant nutrients. The negative impacts of eutrophication are adverse changes in ecosystems such as excessive growth of primitive forms of life, depletion of oxygen in water and as a result, the reduction of specific populations of fish and other animals [33]. Monoculture is the agricultural practice of producing or growing a single crop or plant species over a wide area and for a large number of consecutive years. It is widely used in modern industrial agriculture and its implementation has allowed for large harvests requiring minimal labor. The practice of monocultures can lead to the quicker spread of diseases, where a uniform crop is susceptible to a pathogen. Polyculture, which is the mixing of different crops, has natural variation and the likelihood that one or more of the crops will be resistant to any particular pathogen. Studies have shown planting a mixture of crop strains in the same field can be effective at combating disease [34].

1.6 Measures to maintain the quality of the environment

1.6.1 Environmental conventions

The relationship between air pollution and human activity were not well understood until the end of the 19th century; only the "Great Smog" in London caused a political response. London has long been known for its fog, but when the "Great Smog" fell on the city in December 1952 its effects were surprisingly severe. In the first weeks about 4,000 people died and a further 8,000 died in the weeks following the Smog, causing great concern to the public. The fog was so severe that it halted public transport and public life. Under the influence of this dramatic event, within four years, the British Parliament passed the Clean Air Act in 1956. This was the first time in the world that human activity was associated with disorders of the environment. The Act introduced a number of measures to reduce air pollution, in particular through the introduction of "smoke control areas" where only smokeless fuel was permitted, to reduce the amount of pollutant gases and sulfur dioxide from domestic ovens.

A dozen years later, the Clean Air Act was signed by President Richard Nixon on December 31, 1970 with a view to supporting the development of American economy and industry while improving human health and the environment. In the seventies, the US Congress passed three important acts: the 1970 Clean Air Act, the Clean Water Act of 1972 and the Resource Conservation and Recovery Act, RCRA, which were the primary sets of federal laws which regulated water quality issues in the United States.

In 1979, a new Convention on Long-Range Transboundary Air Pollution, often abbreviated as air pollution or the CLRTAP was adopted, which sought to protect the human environment by gradually reducing and preventing air pollution, including long-range transboundary air pollution. In Europe, the implementation of the Convention was entrusted to the European Monitoring and Evaluation Programme EMEP (short for Co-operative Programme for Monitoring and Evaluation of the Long-range Transmission of Air Pollutants in Europe), led by the United Nations Economic Commission for Europe (UNECE). The Convention was opened for signature on 11.13.1979, but came into force on 03.16.1983. The Convention now has 51 Parties which have agreed to comply. The EMEP results are published on the EMEP website [35]. Further steps in this direction are given in Annex 16, Vol. 2 to the International Convention of Civil Aviation (Chicago, 1944), signed in Montreal in 1981 for aircraft engine emissions.

The Vienna Convention for the Protection of the Ozone Layer is a multilateral environmental agreement was adopted in 1985, and entered into force in 1988. In terms of universality, it is one of the most successful treaties of all time, because it has been ratified by 196 countries (all members of the United Nations). The Vienna Convention coordinates international efforts to protect the ozone layer from destruction. However, the law does not include the reduction of CFC emissions, which are the main chemicals responsible for ozone depletion. Still, they are defined in the Montreal Protocol [36] on Substances that Deplete the Ozone Layer (a protocol to the Vienna Convention for the Protection of the Ozone Layer). This is an international treaty designed to protect the ozone layer by gradually reducing the production of numerous substances considered to be responsible for ozone depletion. The Agreement was opened for signature on 16 September 1987 and entered into force on 1 January 1989 at the first meeting in Helsinki in May 1989. Since that time seven amendments have been passed: in 1990 (London), 1991 (Nairobi), 1992 (Copenhagen), 1993 (Bangkok), 1995 (Vienna), 1997 (Montreal) and 1999 (Beijing). It is believed that if this international agreement is fully implemented, the ozone layer will be recovered by 2050 [37]. The Montreal Protocol is considered the most successful international agreement. The two ozone Treaties have been ratified by 197 states and the European Union making them the most widely ratified treaties in United Nations history [38].

The problem of global climate change as a result ofincreasing global warming due to anthropogenic activities was addressed by the UN Framework Convention on Climate Change (UNFCCC or FCCC). International treaty adopted at the UN Conference on Environment and Development (UNCED), informally known as the Earth Summit, held in Rio de Janeiro from 3 to 14 June 1992 [39]. The treaty is the "stabilization of greenhouse gas concentrations in the atmosphere at a level that would prevent dangerous anthropogenic interference with the climate system".

The Clean Water Act (CWA) is the primary federal law in the United States governing water pollution. Adopted in 1972, the law has set targets to eliminate large quantities of toxic substances from surface water to meet the standards necessary for human sports and recreation by 1983, eliminating additional water pollution by 1985 [40]. Major changes were implemented in the Clean Water Act of 1977 and the law on the quality of water in 1987. In addition, the international community has accepted conventions on global water pollution in the act for marine waters in London in 1972 [41]. The convention on the protection and use of transboundary waters and international lakes is known as the Water Convention. The aim of this convention is to improve national endeavors and measures for the protection and management of transboundary surface waters and groundwater at the international level. Parties are obliged to cooperate and establish joint bodies. The Convention contains provisions on: the monitoring, research, development, consultations, warning and alarm systems, mutual assistance and access, as well as the exchange of information.

The next step in global environmental policy was the Treaty of Basel to control the international trade in hazardous wastes, which was signed in March 1989 by 34 countries and the EC. All signatories agreed in principle to ban the entire trade in hazardous wastes and established the relevant notification procedures.

The Aarhus Convention was signed on 25 June 1998 in the Danish city of Aarhus (The UNECE Convention on Access to Information, Public Participation in Decision-making and Access to Justice in Environmental Matters). The Aarhus Convention grants the public rights regarding access to information, public participation and access to justice in governmental decision-making processes on matters concerning the local, national and transboundary environment. It focuses on interactions between public opinion and public authorities. It entered into force on 30 October 2001 when the EU began applying Aarhus-type principles in its legislation, notably the Water Framework Directive (Directive 2000/60/EC). Liechtenstein, Monaco, and Switzerland have signed the convention but have not ratified it. As of May 2013, it had been ratified by 45 states and the European Union and Central Asia. The real integration and interaction between nations and companies towards Sustainable Development was commenced at the U.N. Conference on Environment and Development UNCED "Earth Summit", which was held in Rio de Janeiro, Brazil, 1992 (Agenda 21) [42].

1.6.2 Carbon neutral policy

The "carbon neutral" or "zero net carbon emissions" policy refers to achieving zero net carbon emissions by balancing a measured amount of sequestration, or by buying enough carbon credits to make up the difference. The concept of carbon neutrality can be extended to other greenhouse gases (GHG) in terms of equivalence with respect to carbon dioxide, which is emitted in the largest quantities. It should be noted that other gases are responsible for the greenhouse effect to an even greater extent than carbon dioxide. They are defined by the Kyoto Protocol, namely: methane (CH_4), nitrous oxide (N_2O), hydrofluorocarbons (HFCs), perfluorocarbons (PFCs) and sulfur hexafluoride (SF_6). The carbon neutral policy is implemented in two ways:

Firstly by balancing the carbon dioxide emissions into the atmosphere from burning fossil fuels, by the use of renewable energy so that carbon emissions are offset by the use of renewable energy sources (mainly plants which absorb carbon dioxide through photosynthesis), also called the carbon economy.

Secondly, by carbon credits and emissions trading, in which more developed economies pay to offsett the emissions of some underdeveloped economies. thus contributing to the removal of 100 % carbon dioxide emitted into the atmosphere, for example through the implementation of "carbon projects". This should lead to the prevention of the financing of future greenhouse gas emissions.

The first method is more reasonable because it causes changes in technology, interest in renewable energy and as a result really contributes to reducing harmful emissions. The second method is very problematic and may even lead to stabilization of the current state of differences and contrasts in the level of economic development between rich and poor countries, causing poor countries to become increasingly poor and lacking in the resources to change a large number of obsolete technologies, and thus unable to reduce emissions in the future. In this sense, the second method is not very rational because it does not lead to the desired effect. It should be noted that this policy has many opponents.

In 2000, the first Climate Justice Summit took place in Hague, the Netherlands, parallel to the Sixth Conference of the Parties (COP 6) to the United Nations Framework Convention on Climate Change (UNFCCC). The Summit's mission stated: "We affirm that climate change is a rights issue. It affects our livelihoods, our health, our children and our natural resources. We will build alliances across states and borders to oppose climate change-inducing patterns and advocate for and practice sustainable development" [43].

Signatories of the organization Climate Justice [43], which brings together 40 nongovernmental organizations, discussed the priorities for action on climate change by recommending the withdrawal of subsidies to fossil fuels and their exploration and production, support and use. Instead, the report calls for structural changes in production and consumption, as well as support for initiatives on sustainable energy. These organizations advocate that instead of paying the money, which relieves some countries of responsibility for the future, it should be considered necessary to introduce a number of civilization changes, i.e. in economics, politics, law, and lifestyle of modern man. Only then would it be possible to change methods of production. The history of international conventions aimed at protection of the environment can be traced according to the following Tab. 1.8.

Table 1.8. Chronology of international environmental conventions

Migratory Bird Treaty Act	1918
International Convention for the Regulation of Whaling (ICRW), Washington	1946
European Agreement Concerning the International Carriage of Dangerous Goods by Road (ADR)	1957
Vienna Convention on Civil Liability for Nuclear Damage	1963
Treaty Banning Nuclear Weapon Tests in the Atmosphere, in Outer Space and Under Water	1963
Convention on Fishing and Conservation of Living Resources of the High Seas	1966
International Convention for the Conservation of Atlantic Tunas (ICCAT)	1966
International Convention for the Prevention of Pollution of the Sea by Oil (1954, 1962, 1969)	1969
Ramsar Convention Convention on Wetlands as Waterflow Habitat	1971
Anti-Ballistic Missile Treaty (ABMT)	1972
Convention on the International Trade in Endangered Species of Wild Flora and Fauna (CITES)	1973
Convention on the Prevention of Marine Pollution by Dumping Wastes and Other Matter	1975
Barcelona Convention for Protection against Pollution in the Mediterranean Sea	1976
Convention for the Conservation of Antarctic Seals	1976
Conventions within the UNEP Regional Seas Programme	1976
Convention for the Conservation of Antarctic Seals	1978
Convention on the Prohibition of Military or Any Other Hostile Use of Environmental Modification Techniques	1978
Kuwait Regional Convention for Co-operation on the Protection of the Marine Environment from Pollution	1978
Convention on the Conservation of Migratory Species of Wild Animals (CMS)	1979
Volatile Organic Compounds Protocol	1979
Convention for the Conservation of Antarctic Marine Living Resources (CCAMLR)	1980
Convention for the Protection and Development of the Marine Environment of the African Region	1981
Convention for the Protection of the Marine Environment of the South-east Pacific	1981
Convention for the Conservation of Antarctic Marine Living Resources[1]	1982
Convention on the Conservation of European Wildlife and Natural Habitats	1982
Regional Convention for the Conservation of the Red Sea and the Gulf of Aden Environment	1982

Bonn Agreement (European Community on oil spills)	1983
Convention for the Protection of the Marine Environment of the Caribbean Region.	1983
Convention on Certain Conventional Weapons	1983
Convention on Long-Range Transboundary Air Pollution	1983
FAO International Undertaking on Plant Genetic Resources	1983
International Tropical Timber Agreement3 (expired)	1983
The 1984 Geneva Protocol on the Long-range Transmission of Air Pollutants in Europe (EMEP)	1984
Convention of the Protection Environment of the Eastern African Region	1985
Sulphur Emissions Reduction Protocols 1985 and 1994	1985
FAO International Code of Conduct on the distribution and use of Pesticides	1985
Convention for the Protection of the Natural Resources and Environment of the South Pacific	1986
Convention on Assistance in the Case of a Nuclear Accident or Radiological Emergency	1986
Convention on Early Notification of a Nuclear Accident (Notification Convention)	1986
Convention on Wetlands of International Importance Waterfowl Habitat	1987
Vienna Convention for the Protection of the Ozone Layer including the Montreal Protocol	1987
China Australia Migratory Bird Agreement	1988
Convention on the Control of Transboundary Movements of Hazardous Wastes and their Disposal	1989
Convention on Civil Liability for Damage Dangerous Goods by Road, Rail, and Vessels (CRTD)	1989
Montreal Protocol on Substances That Deplete the Ozone Layer.	1989
Convention on the Movements and Management of Hazardous Wastes within Africa.	1991
Nitrogen Oxide Protocol	1991
Espoo Convention Convention on Environmental Impact Assessment in a Transboundary Context	1991
Convention on Protection of the Marine Environment of the North-east Atlantic(OSPAR Convention)	1992
Convention on Biological Diversity (CBD)	1992
Convention on the Protection and Use of and International Lakes (ECE Water Convention)	1992
Convention on the Transboundary Effects of Industrial Accidents	1992
Convention on the Protection of the Black Sea against Pollution.	1992
Convention on the Protection of the Marine Environment of the Baltic Sea Area	1992
Framework Convention on Climate Change (UNFCCC)	1992
United Nations Framework Convention on Climate Change	1992
Geneva Protocol (Prohibition of the Use in War of Asphyxiating, Poisonous or other Gases and of Bacteriological Methods of Warfare)	1993
Convention on Nuclear Safety	1994
Convention to Combat Desertification (CCD)	1994
International Tropical Timber Agreement, (ITTA)	1994
North American Agreement on Environmental Cooperation	1994
United Nations Convention on the Law of the Sea	1994
Alpine Convention together with its nine protocols	1995

Waigani Convention Movement and Management of Hazardous Wastes within the South Pacific	1995
Comprehensive Nuclear Test Ban Treaty (CTBT)	1996
Comprehensive Test Ban Treaty	1996
UNECE Convention on Access to Information, Public Participation in Decision-making and Access to Justice in Environmental Matters, Aarhus Convention	1998
Protocol on Environmental Protection to the Antarctic Treaty	1998
Heavy Metals Protocol	1998
Directive on the legal protection of biotechnological inventions	1998
Protocol on Environmental Protection to the Antarctic Treaty	1998
Convention on the Procedure for Hazardous Chemicals and Pesticides in International Trade	1998
Multi-effect Protocol (Gothenburg protocol) Abate Acidification, Eutrophication and Ground-level Ozone	1999
Cartagena Protocol on Biosafety GMO	2000
European Agreement Concerning the International Carriage of Dangerous Goods by Inland Waterways	2000
Stockholm Convention Stockholm Convention on Persistent Organic Pollutants	2001
ASEAN Agreement on Transboundary Haze Pollution	2002
POP Air Pollution Protocol	2003
Declaration of Regional Cooperation for the Sustainable Development of the Seas of East Asia	2003
Kyoto Protocol - greenhouse gas emission reductions	2005
Energy Community (Energy Community South East Europe Treaty, ECSEE)	2006
Framework Convention for the Protection of the Marine Environment of the Caspian Sea	2006
International Treaty on Plant Genetic Resources for Food and Agriculture	2006
Western Regional Climate Action Initiative	2007
Agreed Measures for the Conservation of Antarctic Fauna and Flora	2011
Asia-Pacific Partnership on Clean Development and Climate	2011
Biological Weapons Convention (Prohibition of Bacteriological and Toxin Weapons and their Destruction)	2013
Carpathian Convention Framework on the Protection and Sustainable Development of the Carpathians	2013
Chemical Weapons Convention	2013
Convention on Cluster Munitions	2013
International Convention for the Prevention of Pollution from Ships	2013

1.6.3 Green chemistry concept

The idea of green chemistry is to develop new products, reaction media, conditions and/or utility of materials. Green chemistry was introduced by the EPA in 2002 [44]. More specifically, green chemistry is the design of chemical products which reduce or eliminate the use or generation of hazardous substances by offering environmentally friendly alternatives instead. Sustainable chemistry technologies can be categorized into the following three focus areas: the use of alternative synthetic pathways; the use of alternative reaction conditions; and the design of safer chemicals which are less toxic than current alternatives or inherently safer with regard to accident potential. There are twelve principles of green chemistry, namely:

1. Prevent waste, by designing chemical syntheses to avoid waste to treat or clean up.
2. Design safer chemicals and products to be fully effective, with no toxicity.
3. Design less hazardous chemical syntheses to use and generate no substances toxic to humans and the environment.
4. Use renewable rather than depleting feedstock. Renewable feedstock is usually made from agricultural products or is waste, whereas depleting feedstock is made from fossil raw materials.
5. Use catalysts rather than stoichiometric reagents, which are used excessively and work only once, to minimize waste.
6. Avoid chemical derivatives by using blocking or protecting groups or any temporary modifications if possible. Derivatives use additional reagents and generate waste.
7. Maximize atom economy. Design syntheses so that the final product contains the maximum proportion of the starting materials. There should be a minimal amount of atoms wasted.
8. Use safer solvents and reaction conditions.
9. Increase energy efficiency at ambient temperature and pressure whenever possible.
10. Design degradable chemicals and products which will break down to harmless substances after use to avoid accumulation in the environment.
11. Analyze through real-time monitoring and control during syntheses to minimize or eliminate the formation of byproducts to prevent pollution.
12. Minimize the potential for accidents by designing safer chemicals to minimize the potential for chemical accidents, explosions, fires, and releases to the environment.

The green chemistry concept is addressing the pollution prevention problem at the molecular level by focusing on chemicals. In this sense, green chemistry is complementary to clean technologies, which is based on processes rather than the materials.

1.6.4 Clean technologies

The Commission of the European Communities considers clean technologies ("any technical measures taken at various industries to reduce or even eliminate at source the production of any nuisance, pollution, or waste, and to help save raw materials, natural resources and energy") a main objective of proactive strategies. Clean technologies were recently adjoined (ADEME 1998) to sustainable development policy. This was changed subsequent to the European Community's adherence to the concept of sustainable development. The important social impacts of clean technologies are also economical, employment and safety issues. Thus, the European Commission provided the definition: "Clean technologies are new industrial processes or modifications of existing ones intended to reduce the impact of production activities on the environment, including reducing the use of energy and raw materials." To support this definition, the main attributes of clean technologies were precisely formulated:

1. Conservation of raw materials
2. Optimization of production processes
3. Rational use of raw materials
4. Rational use of energy
5. Rational use of water
6. Disposal or recycling of unavoidable waste
7. Accident prevention
8. Risk management to prevent major pollution
9. Restoring sites

Clean technologies offer a realistic perspective of sustainable development by means of improving the environment, efficiency and saving resources, energy and production costs by focusing on process engineering. Clean technologies are mainly based on separation techniques for the removal, recovery, reuse, or recycling of various substances and material streams. The European Commission has defined the concept of clean technologies as "Any technical measures taken at various industries to reduce or even eliminate at source the production of any nuisance, pollution, or waste and to help saving raw materials, natural resources and energy" [45]. The main attributes of clean technologies are the following:

1. Reduction of waste at source
2. Conservation and rational use of raw materials
3. Conservation and rational use of energy
4. Optimization of production processes.

All of these attributes meet modern membrane processes, as discussed extensively in the book Membranes for Clean Technologies [131]. The first two demands are simultaneously reducing waste and the efficient use of raw materials. Reduction of waste at source and conservation and rational use of raw materials may be achieved by separation processes applied to RECOVER, REUSE, RECYCLE waste from streams, mining waters, tailings, leachates etc. of different material such as: water, unreacted starting materials and substrates, catalysts, solvents, surfactants, detergents, adsorbents, cooling agents, metal ions, heavy metals, organic compounds, dyes, finishing agents, process liquids, such as bleaching tannery solutions, acids and salts from pickling and galvanic baths [46, 132]. The role of conservation and rational use of energy (attribute 3, clean technology) is a) new energy sources such as fuel cells (ion-selective membrane and catalyst), b) new fuels (bio-fuels, bio-diesel-membranes and membrane reactors for trans esterification of fatty acids with alcohols), c) energy saving (new solutions and devices for energy recovery systems.

The optimization of production processes may be achieved by the integration of different technologies (materials management) [47]. This is especially essential in biorefineries, where several products are manufactured simultaneously. The optimization process by the application of membranes has found new methods such as: (a) the substitution of the conventional separation processes using low-energy alternative membrane processes, (b) the integration process ("hybrid processes"), and even (c) the integration of various technologies (materials management) so that waste from one technology were appropriate resources for other technologies. Hybrid processes are a way of allowing for the complete elimination of waste from industrial processes. This can be done by the uptake of a variety of useful components from wastewater, integrating different processes and technologies.

Conservation and rational use of energy may be achieved by new energy sources such as fuel cells (catalytic membranes and ion selective membranes), processing new

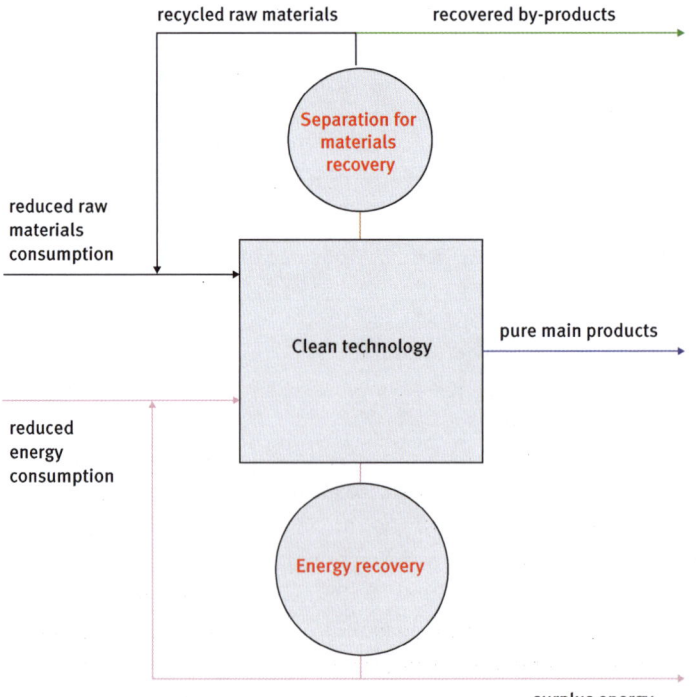

Fig. 1.15. Main attributes of clean technologies.

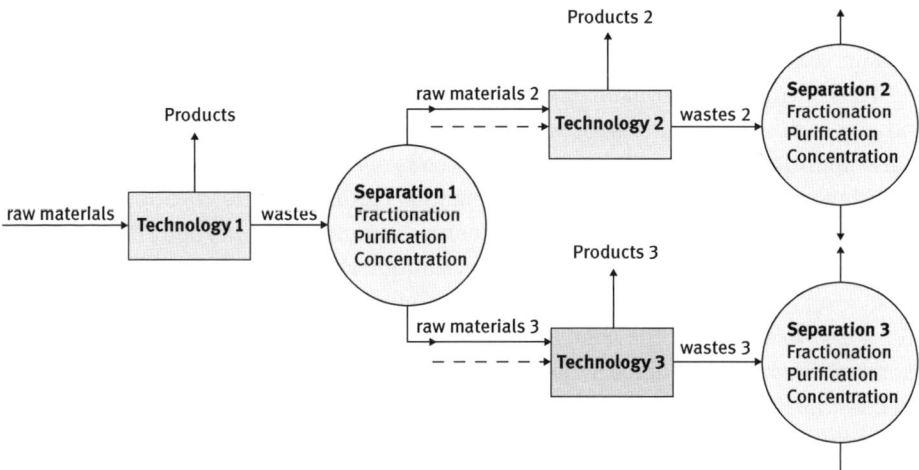

Fig. 1.16. The clean technologies achieved by their integration.

fuels such as bio-fuels, bio-diesels (membranes and membrane reactors for transesterification of fatty acids with alcohols) and energy saving such as new solutions for work, pressure, and energy recovery

Clean technologies may be arranged as a single web, where producers and customers of waste are in the same network. Some adjustment may be needed (separation processes see Fig. 1.4)

1.6.5 Sustainable development

"A sustainable development symbolizes a mode of technology that would satisfy the needs of today's generation without compromising the chances of the later generations' ability to satisfy their own needs." This was the concluding statement of the the Brundland report [48]. The majority of countries admitted the concept of sustainable development since the debate at the World Conservation Strategy in 1980, the report ("Our Common Future") of the World Commission on Environment and Development in 1987, and Agenda 21 in 1992. A strategy for sustainable development was agreed upon in Goteborg on June 2001 but this attitude has its origins in the past.

Sustainable development stems from the earlier concepts based on the similarities between biological and industrial systems [49] such as symbiosis, metabolism and industrial ecosystems. Further examples are (1) Ecologically Conscious Management [50], (2) Life Cycle Analysis, (3) Efficient Design for Longevity, Sustainable Product Development [51], Green chemistry and cleaner technologies. The concept of green chemistry was introduced by the EPA in 2002 [44]. Green chemistry is an attempt to prevent pollution at the molecular level by focusing on chemicals. Green chemistry allows for the development of new environmentally friendly products, catalysts and reaction conditions.

The so-called Industrial Ecology focuses largely on the physical flow of substances and the physical transformation processes, as introduced by Lester, who made the observation that industry is similar to biological systems. Ayres also developed the concept of industrial metabolism, taking into account the material flow. The industrial metabolism was defined as the whole integrated collection of physical processes which convert raw materials and energy, plus labor, into finished products and waste in a more or less steady-state condition. Tibbs, Graedel and Ayres pointed out a longer-term vision of production within a living system: "To manage the earth's resources in such a way as to approach and maintain a global carrying capacity for our species which is both desirable and sustainable over time, given continued evolution of technology and quality of life." Proactive pollution prevention approaches aim to achieve sustainable production capabilities, where environmental and economic systems are balanced. Waste minimization is one proactive approach. The main objective of proactive strategies is the development of clean technologies.

The principles of sustainable development must be based on real assumptions and physical (conservation) and biological laws; the biogeochemical cycles; the ecological interdependencies of species; and the anthropogenic influence on the ecosphere.

The principles of sustainable development were formulated as follows:

1. To eliminate our contribution to systematic increases in concentrations of substances from the earth's crust by substituting certain minerals that are scarce in nature with others that are more abundant, using all mined materials efficiently, and systematically reducing dependence on fossil fuels.

2. To eliminate our contribution to systematic increases in concentrations of substances produced by society by systematically substituting certain persistent and unnatural compounds with ones that are normally abundant or break down more easily in nature ("green chemistry"), and using all substances produced by society efficiently.

3. To eliminate our contribution to the systematic physical degradation of nature through over-harvesting, introductions and other forms of modification. This means drawing resources only from well-managed eco-systems, systematically pursuing the most productive and efficient use both of those resources and land, and exercising caution in all kinds of modification of nature.

4. To contribute as much as we can to the meeting of human needs in our society and worldwide, over and above all the substitution and dematerialization measures taken in meeting the first three objectives. This means using all of our resources efficiently, fairly and responsibly so that the needs of all people on whom we have an impact, and the future needs of people who are not yet born, stand the best chance of being met.

During the 2000 Seville Conference, industry representatives proposed the Integrated Pollution Prevention and Control (IPPC) Directive with the requirements for Best Available Techniques (BAT). "Best" means most effective in achieving a high general level of protection of the environment as a whole. As stated in BAT reference documents, the directive should be descriptive rather than prescriptive. Paragraph 11 in Article 2 of the Integrated Pollution Prevention and Control Directive defines "Best Available Technique" as "the most effective and advanced stage in the development of activities and their methods of operation which indicate the practical suitability of particular techniques for providing in principle the basis for emission limit values designed to prevent and, where that is not practicable, generally to reduce emissions and the impact on the environment as a whole." The purpose of the Directive is to achieve integrated prevention and control of pollution arising from industrial activities, leading to a high level of protection of the environment as a whole. The application of the best available techniques enables improvement of environmental performance. Specifically:

1. Acidification resulting from emissions into the air
2. Soil and water eutrophication resulting from emissions to air or water
3. Diminution of oxygen content in water
4. Global warming
5. Depletion of the ozone layer
6. Emission of particles into the air, especially microparticles and metals
7. Formation of photochemical ozone
8. Discharge of persistent, bio-accumulative and toxic substances into water or into the soil
9. Generation of waste, in particular hazardous waste
10. Vibrations, noise and odors
11. Over-exploitation of raw material and water resources.

1.6.6 How to achieve sustainable development

All phenomena on earth (and elsewhere too) are subject to the laws of conservation of mass and energy. Therefore, the need for sustainable development is not a viewpoint or the subject of a never-ending discussion. It is appropriate to discuss the ways of introducing sustainable development rather than whether it should be done. However, this is a very difficult challenge for representatives of a democratic society with NGOs against the profit of the ad hoc, justifying them by means of almighty arguments referring to the law of the free market. The lobby promoting the higher benefits by means of older technologies was always stronger than naive idealism presenting "wishful thinking". Therefore it is not enough just to know how sustainable development should be implemented, although it is seems to be the most essential. Although, this is so much to do in different areas of life, even assuming that all people are honest and they do not think only about personal interest, but about society as a whole.

The primary task for politicians and economists is promoting a society which would prefer the prospect of stable development of ordinary life for future generations, than ad hoc profits. Politicians have to try to find ways for the consensus of the common regulations to apply worldwide. This may change the conditions of a free market in a certain sense. Also necessary are new ideas and regulations on the subject of new crops. Politicians should anticipate and counteract the effects of monoculture in the case of over-development of nonfood agricultural production, associated with eutrophication and desertification in the end of hunger. These are the basic problems of the new agriculture, which will be subject to economic pressures associated with the conversion of food production for large monocultures of profitable cultivation of raw materials for green industry and biorefineries in the future.

This may be very dangerous for environmental reasons and will contribute to the extinction of many species. Eutrophication is mainly caused by sewage but may also occur as a result of overuse of fertilizers. It is possible everywhere, since nutrients enter the groundwater and then rivers, lakes and finally accumulate in the oceans. The eutrophication process is strongly accelerated by human activities such as the discharge of industrial and municipal waste, biocides, xenobiotics and toxins as a result of agricultural intensification. Xenobiotics are chemical compounds found in the body, which it either does not produce or cannot normally taken from food i.e. are not accessible. Any unavoidable new xenobiotics should be completely and rapidly degradable.

The ideal image of sustainable development will never be realized if we do not know how to define the right direction. The right direction and the rate of pursuit of sustainable development is being grossly distorted by the so-called realists who wish to have only short-term profits, and speak of economic principles and steadfast market laws. We all behave in such way (to some extent) by consenting to the proposed compromises, but we will all bear the consequences, because the laws of nature which are subordinate to the physical laws are even more unbending.

Consent may apply only for the limited period of time required for the preparation of the new strategy and new technologies. This time is always necessary, but must not be drawn out to infinity. Therefore, the most important roles are those of the politicians convinced of the expediency of sustainable development and the lawyers preparing the relevant provisions, and only afterwards is it time for the economists and engineers who know how to do this.

In the future industry should produce exclusively such materials which circulate in appropriate food chains. New biological processes require the use of algae and photo-bioreactors of new generation. Therefore the first step towards sustainable production would be reliance on renewable energy, raw materials and products. Raw materials for production should be biological materials, such as enzymes, microbials; plant tissues produced from photosynthesis and also waste. Bioproducts will be biochemicals, biofuels and biopolymers. Increased efficiency of such processes can be achieved through the use of integrated circuits and hybrids.

Energetics and industry also need to be completely adapted to the metabolism of the entire ecosphere, which should be confined to energy from renewable resources and must be able to circulate in nature for all mass components. Such a condition is a direct result of obvious conservation laws. Clean energy can be produced during the activities of algae which can split water into oxygen and hydrogen. Hydrogen is the cleanest fuel on Earth because during its combustion only water arises, and during its manufacture pure oxygen is produced. However, biohydrogen production is still in its infancy because of its low efficiency and low economy algae culture. Parallel to the development of algae other technologies for hydrogen production based on organic carbon from waste should be strengthened.

Currently, the application for production of power and different materials such as polymers and chemicals from renewable resources instead of crude oil and fossil fuels has become a fashionable biorefineries concept. It seems that is not enough, because all products should be produced in this way, and also, and above all, all the food for humans and animals. The interdisciplinary approach is necessary for biofuels production. Biofuels such as bioethanol and biodiesel should be produced from wastes and therefore they need high-performance, low-cost cellulase enzymes and/or cellulolytic organisms; the separation of lignin from cellulose; the optimization of simultaneous fermentation of hexoses and pentoses; and the purification of ethanol and recovery of other valuable byproducts. Biodiesel production needs sustainable production of oilseed crops and the development of technologies to allow use of waste oils.

Meanwhile, the big drawback of biorefineries is that they derive their raw materials from large monoculture crops, which use the same large areas to grow plants of one species with similar soil requirements over a long period. Monocultures have a demonstrated negative impact, causing a rapid depletion of nutrients and exhaustion of soil fertility. In nature, plants and animals feed each other to develop chemicals and minerals. For example, legumes enrich the soil with nitrogen and manure provides many other nutrients. Eliminating these natural cycles of the different ecosystems requires fertilizers and biocides and other xenobiotics which are used to increase yields. Furthermore, monocultures are particularly susceptible to diseases which can spread much faster on a single area of the crop compared to the biodiversity of the ecosystem. Moreover the uniform crop plants in one area require the use of large amounts of different chemicals which leads quickly to depleted and deficient soil. Organisms have not been adapted to xenobiotics during evolution. Xenobiotics are mostly human-made, such as drugs, various biocides and preservatives.

So far, the only gradual and limited changes in the direction of sustainable development may be fruitful and predictable. These are diversification of industrial production, energy, and more efficient use of raw materials. There are several reasons for the reluctant introduction of the desired changes although they are absolutely necessary. The economic and political changes are far beyond the range of influence of the average citizen in a short time. Therefore, many experts from different parties must synchronize their interdisciplinary efforts to follow the agreed direction. However, the technological changes are to some extent within the capabilities of engineers, scientific minds, those working in measurable areas. They can analyze the current situation and pave some directions for the development of their areas.

Process engineering provides new opportunities for sustainable development through the invention, investigation and development of effective processes in different sectors. This will help to improve environmental conditions and simultaneously reduce production costs. This is an essential message of this book. However, a humanistic issue runs continuously through the entire book too: that we can only develop in harmony with the natural rules of the world, and this knowledge must encourage mankind to aspire to sustainable development, regardless of what is actually happening now.

2 Past and present of process engineering

2.1 The origins and domains of process engineering

2.1.1 Early history of process engineering

The origin of unit processes can be attributed to certain magical practices from time immemorial, such as alchemy, which survived for several centuries until the 21st century. Some inventions have unknown origin, but their use is obvious, because they come from use for clothes, such as washing, dyeing, tanning hides, or for foods, drugs and perfumes, from plants and from animal sources. Most unit processes have also been used in daily practice in homes for centuries, for food preservation, preparation of stores for the winter, cleaning, washing, coloring, preparing fabrics and hides. The extraction process was used to obtain colors and precious metals from ore dust as evidenced by the oldest archaeological finds. The leaching process was mainly used to obtain potash (potassium carbonate) from the ashes of terrestrial and marine plants. Potash was used as a bleaching agent for fabrics, and from about 500 AD for the manufacture of soap and glass. Ancient culture delivers a variety of written material in the form of philosophical treatises. Distillation has been known since Greek times and was described by Alexander Aphrodisias at the end of the 2nd century AD. Archimedes of Syracuse (287–212 BC), was known as a Greek philosopher and mathematician. He was a versatile investigator, interested in many areas of science, such as mechanics, optics, astronomy etc. It must be admitted that much of the scientific work was for military applications. Archimedes formulated the law of hydrostatics defining buoyancy in the old form as follows: "The body immersed in a liquid or gas, seemingly losing the weight so the weight of liquid or gas that is ousted by the body". It is often used in the calculation of process engineering in the processes of sedimentation, hydraulic classification, fluidization, bubblers, spray drying, hydraulic transport etc. The distinguishing of real precious metals from their fakes was probably the first valuable use of Archimedes law:

$$F_w = P - W = (\rho - \rho_c) \cdot V \cdot g. \tag{2.1}$$

Leonardo da Vinci was a master of all areas of engineering and his achievements in the field of process engineering include: designing mills and energy-fueled engines and water pumps to extract water from great depths. The multi-talented Isaac Newton (1642–1727) was the pioneer of modern scientific theories. He believed that the object of study can only be the phenomenon, and the goal is to find the relationship between the phenomenon and the law. He formulated the first scientific basis for process engineering in the 17th century, defining the law of cooling and the basics of differential calculus:

$$\frac{dQ}{dt} = -k \cdot \Delta T. \tag{2.2}$$

In addition to his well-known theory of universal gravitation, Isaac Newton gave rise to the modern science of physics by addressing various issues which are very helpful in the work of engineers now, such as similarity theory, mathematical analysis and rheology.

Relationships between engineering and science have always been very strong. When a problem arises a scientist asks why, and tries to find the most general solution. Meanwhile, the engineer wants to know how to solve this problem practically and how to implement the solution. In other words, scientists try to explain existing phenomena, whereas engineers use the available resources, not just science, to create solutions to new problems.

Leonhard Euler (1707–1783), is known as a creator of mathematical analysis, proposed equations describing the conservation of mass and momentum for inviscid fluids. Pierre Simon de Laplace (1749–1827), known for his transformation, contributed to the solution of the transport equation for certain boundary conditions, which was of great importance for process engineering (transport equation of momentum, heat and mass transport). Antoine-Laurent de Lavoisier (1743–1794), a famous French scientist of chemistry and biology, named oxygen and hydrogen, introduced the metric system and clearly formulated the principle of mass conservation. John Dalton was an English physicist (and also chemist), and published Atomic Weights (1805), which is the basis for mass balances in chemical engineering. At the same time he dealt with flows of real fluids, as did Gotthilf Heinrich Ludwig Hagen (1797–1884), a German physicist and engineer, and Jeanean Louis Marie Poiseuille (1797–1869), a French physician and physiologist, also trained in physics and mathematics. He was interested in the flow of human blood in the narrow tubes of blood vessels. They described the law on the pressure drop in laminar flow. Pressure drop during flow was also an interest of Henry Darcy (1803–1858), a French scientist, known for his invention of the Pitot tube and the first considerations of the laminar boundary layer. He was influential regarding turbulence in fluid flow. Osborne Reynolds (1842–1912) studied the same difficult issue. He was an Irish physicist, member of the Royal Academy of Sciences and professor who popularized the theory of similarity and dynamics of fluid flow in pipes. In 1883, Reynolds formulated the basic relations for the transition from laminar flow to turbulent. Osborne Reynolds was a prominent innovator in the understanding of fluid dynamics. Separately, his studies of heat transfer between solids and fluids brought improvements in boiler and condenser design. Reynolds analogy is popularly known to relate turbulent momentum and heat transfer. Turbulence and the flow regime are very important for all processes, i.e. mass transport, heat transport and momentum transport.

In this period researchers were involved in issues beyond single fields of classic science such as mathematics, physics chemistry, biology, and dealt with simply solving real problems, as has been found in their work. Surprisingly, there are significant parallels between engineering and medicine. Both disciplines rely on solving problems using knowledge, experience and intuition. In addition, they are characterized by the same pragmatism: to find solutions before the phenomenon is fully explained. With the development of each narrow discipline, when the accumulated experience, new ways of solving problems and the new research methods grow, general knowledge also expands and can be applied to other fields, creating multidisciplinary areas. So it becomes just in the classical disciplines such as mathematics, physics, biology, chemistry, which combine in different multidisciplinary sciences such as biotechnology, nanotechnology. Process engineering, which was historically regarded as a branch of chemical engineering, is now in such a position. Process engineering is currently focused on the design, operation, control, optimization and mathematical modeling in a broad area of industrial processes including chemical, physical and biological processes, as previously chemical engineering only focused on chemical processes.

Multidisciplinary research was already performed by the majority of famous scientists in the 19th century. Adolf Eugen Fick (1829–1901) was a German physicist and mathematician before eventually realizing his talent in medicine. In 1855 he formulated the law of gas diffusion across membranes, and he also tested dialysis of solutions through artificial membranes formed from collodion (1865). The work of Van't Hoff in these areas has helped to create the discipline of physical chemistry, which is the foundation of process engineering. He described the phenomenon of osmosis, which had previously been discovered by Abbe Nollet in 1747 using natural membranes made from animal bladders. Jacobus Henricus van't Hoff Jr. (1852–1911), was a Dutch researcher on physics and organic chemistry and the first winner of the Nobel Prize in Chemistry. He made discoveries in chemical kinetics, chemical equilibrium, osmotic pressure, and stereochemistry. Van't Hoff's discoveries in chemical kinetics are also the foundation of modern process engineering. In 1864 Peter Waage and Cato Gulberg (brothers-in-law) formulated "The law of mass action", which states that the reaction rate is proportional to the amounts of reactive substances. This law was verified by Jacobus Henricus van't Hoff. Walther Hermann Nernst (1864–1941), was a German physical chemist and physicist who made a great contribution to the foundation of mass transport kinetics as he formulated the concept of the diffusion boundary layer (in 1904 and still valid today) when studying mass transport phenomena on electrodes. Moreover, he is known for his laws, such as "Nernst distribution right", important for extraction equilibrium, the third law of thermodynamics, for which he won the 1920 Nobel Prize in Chemistry, and theories on chemical affinity (Nernst–Ettingshausen effect), important for electronics. Paul Richard Heinrich Blasius (1883–1970), a German physicist, dealt with fluid dynamics. Michael Faraday (1791–1867), was an English scientist who contributed to the fields of electromagnetism and electrochemistry. His main discoveries include that of electromagnetic induction, diamagnetism and electrolysis. Faraday conducted the first rough experiments on the diffusion of gases, a phenomenon that was first pointed out by John Dalton, and the physical importance of which was more fully brought to light by Thomas Graham and Joseph Loschmidt. Faraday was the first to report what later came to be called metallic nanoparticles. In 1847 he discovered that the optical properties of gold colloids differed from those of the corresponding bulk metal. This was probably the first reported observation of the effects of quantum size, and might be considered to be the birth of nanoscience.

Since this time a lot of handbooks and publications regarding process engineering as a scientific discipline have been produced, with the real and important prospect of participating in technological development in many industrial sectors, not only in chemistry, as its former name (chemical engineering) suggested. Contemporary process engineering has gained a basis which enables it to contribute towards various areas such as clean technologies, environmental protection (as part of environmental engineering) and the production of pharmaceuticals; in medicine, diagnostics, and the production of artificial organs; in agriculture, forestry and farming in the dosing of chemicals. However such a state would not be possible without the greatest development during the industrial revolution.

2.1.2 Industrial era of process engineering

The industrial revolution, which began in the eighteenth century in England and Scotland, was associated with the transition from an economy based on agriculture to an economy based mainly on production on a large industrial scale. The new era known as the age of steam and electricity started with several inventions of the steam piston engine by James Watt in 1763 and application of an electric current. Sadi Carnot (1824) described the thermodynamics of combustion reactions in steam engines in his work "On the Motive Power of Fire". Rudolf Clausius (1850) began to apply the principles developed by Carnot to chemical systems. Further, the rapid development of technology, to which we owe such achievements as the invention of the internal combustion engine by Jean Joseph Etienne Lenoir (1807), and an external combustion engine by Robert Stirling in 1816 (British Patent No. 4081), an electric motor by Michael Faraday in 1823 and finally electric light by Thomas Edison in 1879.

The earliest industrial applications of process engineering concerned production such as sugar, paper, fertilizers, biocides, oil and coal, fuels, plastics, and also explosive materials. Further, textiles, mining, metallurgy, ceramics and chemicals which were related to new inventions in various fields had an impact on the development of process engineering. Nicolas LeBlanc (1780), developed a process for making soda ash from common salt, used to make glass, soap, paper and more. Norbert Rillieux (1843) invented the first successful multiple vacuum process for producing sugar. In the same year, 1879, Bouchardat created a form of synthetic rubber, producing a polymer of isoprene. These inventions led to the rapid development of various technologies and the increase in the consumption of fossil fuels such as coal, oil, gas and metal ores.

Now, process engineering is the basis for manufacturing a variety of commodities which are found in everyday life, specifically inorganic and organic industrial chemicals, ceramics, fuels, petrochemicals, food, paints and other coatings, such as inks, sealants and adhesives, plastics and elastomers, oleochemicals, detergents and detergent products (soap, shampoo, cleaning fluids), fragrances and flavors, additives, dietary supplements and pharmaceuticals etc. Process engineering largely involves the optimization, design, control and maintenance of processes based on physical, chemical or biological phenomena for large-scale manufacture. The future of products includes high performance materials needed for aerospace, automotive, biomedical, electronic, environmental, space and military applications. Examples include ultra-strong fibers, fabrics, and dye-sensitized composites for new fuels and solar cells. Additionally, process engineering is often intertwined with biology and biomedical engineering. The new products of process engineering are artificial organs such as kidneys, lungs, pancreas, liver and skin. Other products are bio-compatible materials for implants and prosthetics, gels for medical applications, pharmaceuticals, and films with special dielectric, optical or spectroscopic properties for opto-electronic devices. Devices for controlled release are used in medicine and also to deliver biocides and other chemicals such as pheromones in agriculture, forestry and cattle breeding. Controlled release prevents overdosing of chemicals to soil, water and air. The line between chemical engineering and other disciplines has become progressively thinner. Sustainable development requires innovations in chemistry, physics, biology and mathematics.

Intensive industrialization has led to the transformation of chemical engineering to process engineering, which relates to more industrial sectors than just chemistry. This has created the concept of unit processes to satisfy the needs of new sectors, based on integrated knowledge from different fields. For example, fractional distillation was used in crude oil and the production of rectified spirit in the fermentation industry. It was the same with extraction and multi-absorption. The term "unit processes" was introduced in order to use them in a variety of technologies to provide a repeatable process. The number of unit processes is small and may be optimized and improved with respect to the specific technology. Often the same process was carried out using very different industrial equipment, such as tray columns and packed columns. Next, the same equipment was used for quite different unit processes. For example, the columns which are used for the multi-stage distillation process are also used for absorption and extraction. The experience gained in these various technologies has enriched the knowledge of process engineering and allowed improvement of these processes, and thus technology. Moreover, it was possible to focus on the essential elements of the process dry as the use of multi-stage cascades and the kinetics intensification of the processes.

At the end of the 19th century very strong interest can be seen in education in the field of process engineering. In 1882 the course "Chemical Technology" was offered at University College London. However, in 1885 Henry Edward Armstrong offered a course in "chemical engineering" at Central College (later Imperial College). Ivan Levinstein defined chemical engineering as the conversion of laboratory processes into industrial in 1886. In 1888, Departments of Chemical Engineering were established in Glasgow and West of Scotland Technical College. In the same year, Lewis M.

Norton started a new course in Chemical Engineering at the Massachusetts Institute of Technology. In 1890, Professor Mrs. Ellen Swallow Richards introduced the word ecology at the Massachusetts Institute of Technology (MIT). The American Institute of Chemical Engineers (AIChE), a professional organization for chemical engineers, was established in early 1908. 20 years later the Institution of Chemical Engineers (IChemE) was founded in England in 1922. A Ph.D. in chemical engineering was first awarded in the US at Yale University in 1983, to Josiah Willard Gibbs. Willard Gibbs proposed a graphical analysis of multi-phase chemical systems and developed process engineering using the thermodynamics of Clausius. E. Sorel developed the mathematical basis for interpretation of distillation of binary solutions in 1889. Ludwig Prandtl (1875–1973), a physicist, revolutionized fluid mechanics (1894) with the concept of a boundary layer. This boundary layer theory made it possible to resolve mass and heat transfer kinetics in a number of unit processes such as distillation, absorption, extraction, adsorption and the majority of membrane processes as the concentration polarization problem.

The Swedish inventor Alfred Nobel established the prizes in 1895. The prizes in physics, chemistry, physiology or medicine, literature, and peace were first awarded in 1901. Tab. 2.1 presents selected inventions, important for the development of process engineering, although awards of process engineering, as such, have not been granted so far.

The kinetics of chemical and biochemical reactions is a very important issue in process and bioprocess engineering. Svante August Arrhenius (1859–1927), one of the founders of process engineering, was a Swedish chemist and physicist, and developed the theory of electrolytic dissociation. He investigated the properties of toxins and antitoxins, led study on the temperature of the sun and planets and studied the phenomenon of the aurora borealis. In 1903, he received the Nobel Prize in Chemistry for developing the theory of electrolytic dissociation. In 1907 he created the theory of panspermia on the origin of life on Earth. The rate constant of chemical reaction is calculated using Arrhenius' formula:

$$k = A \cdot e^{-E_a/RT}, \tag{2.3}$$

where A – a frequency factor dependent on collisions, E_a – activation energy, T – temperature.

The Arrhenius equation is a simple but remarkably accurate formula for the temperature dependence of the reaction rate constant. The equation was first proposed by the Dutch chemist J. H. van't Hoff in 1884 five years earlier, but in 1889 Arrhenius provided a physical justification and interpretation for it. Currently, it is best seen as an empirical relationship.

Irving Langmuir (1881–1957) was an American scientist of physics, chemistry and electricity. In the 1910s and early 20s, as the first in the world, he led a systematic study of adsorption monolayers. The Langmuir adsorption equation or Hill-Langmuir equation relates to the coverage or adsorption of molecules on a solid surface at a fixed

Table 2.1. Nobel Prize awards concerning process engineering

Jacobus Henricus Van't Hoff	(Chemistry) Discovery of the laws of chemical dynamics and osmotic pressure in solutions.	1901
Antoine Henri Becquerel, Maria Curie Sklodowska, and Pierre Curie	(Physics) Leaching of uranium ores.	1903
George De Hevesy	(Chemistry) For his work on the use of isotopes as tracers in the study of chemical processes.	1943
Willard Frank Libby	(Chemistry) Method of using "carbon 14" for age determination in archaeology, geology, geophysics and other branches of science.	1960
Lars Onsager	(Chemistry) Reciprocal relations (bearing his name), which are fundamental for the thermodynamics of irreversible processes.	1968
Dudley R. Herschbach, Yuan T. Lee, and John C. Polanyi	(Chemistry) Dynamics of elementary chemical processes	1986
Richard R. Ernst	(Physics) Development of the methodology of high resolution nuclear magnetic resonance (NMR) spectroscopy.	1991
Peter Agree and Roderick MacKinnon	(Chemistry) Discovery of water channels in cell membranes.	2003

temperature. The equation was developed by Irving Langmuir in 1916. The equation is stated thus:

$$q = q_{max} \frac{b \cdot C_s}{1 + b \cdot C_s}. \tag{2.4}$$

Herbert Max Finlay Freundlich (1880–1941, Minneapolis) was a German-American chemist who proposed a different adsorption isotherm to Langmuir. The Freundlich isotherm has no asymptotes and sorption capacity increases the concentration of the solution as follows:

$$q = q_{max} \cdot C^{1/n}. \tag{2.5}$$

Lanmuir also suggested another very important idea for process engineering, which is defined as a cascade reactor with perfect mixing. The number of virtual variation of the reactor corresponds to the residence time distribution (RTD), not only in chemical reactors, but the flow in all the tanks. The concept was first proposed by MacMullin and Weber in 1935, but was not used extensively until P.V. Danckwerts analyzed a number of important RTDs in 1953.

Frederick George Donnan (1870–1956) was an Irish physicist and chemist known for his work on membrane equilibrium (Donnan equilibrium), describing ionic transport in cells. In 1911 Donnan formed "distribution laws". Ernst Kraft Wilhelm Nusselt (1882–1957) was a German engineer, who received his doctorate from the Munich Technical University in 1907. Nusselt devised the basic heat-transfer relationship which later bore his name. During heat transfer at a boundary within a fluid, the Nusselt number is the ratio of convective to conductive heat transfer across the interface between both phases.

$$Nu = \frac{\text{convective heat transfer coefficient}}{\text{conductive heat transfer coefficient}} = \frac{\alpha \cdot l}{\lambda}. \tag{2.6}$$

Arthur Dehon Little (1863–1935), was an American engineer in the field of chemical engineering at the Massachusetts Institute of Technology (MIT). He played a large role in the development of chemical engineering by proposing the concept of "unit operations" in order to explain the processes in industrial technologies in 1916. Warren L. McCabe and Warren L. Thiele developed a method determining the number of theoretical plates of the rectification column for a binary mixture on the basis of equilibrium between vapor-liquid, and the molar volume of the distillate and reflux streams in 1925. In 1935, Ralph Higbie proposed the penetration theory of mass transfer on the assumption that contact between the gas and the liquid is constant. In the case of columns, this assumption was accurate, but the theory has also been used for all other mass transfer processes. Peter Victor Danckwerts (1916–1984) used the Higbie theory as a basis; however he resigned from that assumption of a constant contact time. Instead, he introduced statistical distribution of the elements of the residence times at the interface. This distribution depends on the rate of surface renewal at the interfacial area. Both theories are discussed in the next chapter of this book (Sections 2.5.7.3 and 2.5.7.4). Danckwerts made a large contribution to chemical engineering by developing an original method for describing the kinetics of processes on surfaces between phases called the surface renewal theory. Surfaces between phases where mass transfers occur are coated with microscopic elements of various ages. Each element can be detached from the surface and replaced with a new one. In each element the nonstationary mass transport is held. The average flux, from the entire surface, is integral mean value from age distribution. Age functions are calculated on the assumption that the surface renewal process is stochastic; it is independent of the age and location of items. The impressive growth and development of process engineering (previously known as chemical engineering) came in the 20th century. There were many fundamental works in the field. Some of them are shown in the table below (Tab. 2.2).

Table 2.2. The most important books published in the field of chemical engineering

Newton, Isaac	Philosophiae naturalis principia mathematica.	1687	London
Nollet, JA	Lecons de physique experimentale	1748	Paris
Davis, George E.	A Handbook of Chemical Engineering.	1904	Davis Bros
Robinson, Clark	Elements of Fractional Distillation	1922	McGraw-Hill

Lewis WK, McAdams WK, and Walker WL	The principles of chemical engineering	1923	McGraw-Hill
Lewis WK, Radash and Gilliand, ERA	Industrial Stoichiometry	1923	McGraw-Hill
McCabe, W and Badger, WL	Elements of Chemical Engineering	1931	McGraw-Hill
McAdams	Heat Transmission	1933	McGraw-Hill
Perry, J	Chemical Engineers' Handbook	1934	McGraw-Hill
Vilbrandt, F	Chemical Engineering Plant Design	1934	McGraw-Hill
Groggins, PH	Unit Processes in Organic Chemistry	1935	McGraw-Hill
Sherwood, T	Absorption and Extraction	1937	McGraw-Hill
Sherwood, T and Reed, C	Applied Mathematics in Chemical Engineering	1939	McGraw-Hill
Hougen, OA and Watson, K	Chemical Process Principles	1943	Wiley
Shreve, N	Chemical Process Industries	1945	McGraw-Hill
Othmer, DF	Encyclopedia of Chemical Technology	1947	Wiley
Smith, JM	Introduction to Chemical Engineering Dynamic	1949	McGraw-Hill
Coulson, JM and Richardson, JF.	Chemical engineering	1954	McGraw-Hill
Treybal, RE	Mass Transfer Operations	1955	McGraw-Hill
McCabe, W and Smith, J	Unit Operations of Chemical Engineering	1956	McGraw-Hill
Bird, BR, Stewart, WE, Lightfoot, EN	Transport Phenomena	1960	Wiley
Zenz, F	Fluidization and Fluid-Particle Systems	1960	Pemm Co.
Himmelblau, DM	Basic Principles and Calculations in Chemical Engineering	1962	Prentice Hall
Levenspiel, O	Chemical Reaction Engineering	1962	Wiley
Coughanowr, DR and Koppel, L	Process Systems Analysis and Control	1965	McGraw-Hill
Wen, CY and Fan, LT	Models for Flow Systems and Chemical Reactors	1975	Dekker
Seider, WD and Myers, AL	Introduction to Chemical Engineering and Computer Calculations	1976	Prentice Hall.
Sandler, SI	Chemical, Biochemical, and Engineering Thermodynamics	1977	Wiley

Fedler, R and Rousseau R	Elementary Principles of Chemical Processes	1978	Wiley
Fogler, HS	Elements of Chemical Reactions	1986	Prentice Hall
Yang, RT	Gas Separation by Adsorption Processes	1986	Imp. College Press
Shuler, M	Bioprocess Engineering: Basic concepts	1992	Prentice Hall
Schmidt, LD	The Engineering of Chemical Reactions	1998	Oxford Univ. Press
Seader, JD	Separation Process Principles	1998	Wiley

In 1988 Peter Agre identified aquaporins which build channels to allow the flow of water through the cell walls. It turned out that the mechanism of water transport in aquaporins is a brand new one-dimensional single file diffusion. In this transport method, one molecule after another wanders through the channels of the cell wall without colliding, which is much more effective than the usual intermolecular diffusion. In 2003 he and Roderick MacKinnon received the Nobel Prize in Chemistry for discoveries concerning channels in cell membranes, in which water and ions flow. The new diffusion mechanism, which was discovered in aquaporins, could revolutionize the performance of the membrane processes. For instance, much greater membrane permeability in these processes may ease future water deficiency crisis.

Further developments of civilization such as urbanization, population growth and rising expectations of a healthy and comfortable life in the future will induce the need for change in future processes. This in turn will result in even greater demand for energy, but also increase the fear of environmental pollution. So far, the phenomenon of hunger has been overcome in the world, but there is a steadily growing deficit of clean water and concern about the fate of future generations. This means that "sustainable development" has growing global support.

2.2 Principles, system and methodology

2.2.1 Unit processes concept

The individual processes used by chemical engineers are called unit processes. Unit processes are the components of each technology. Each unit process is subject to the same three primary physical laws which underly the principles of chemical engineering. They are 1) Conservation of mass, 2) Conservation of momentum and 3) Conservation of energy. The rate of the process is a key factor, for the most economical technology is based on principles of thermodynamics, reaction kinetics, fluid mechanics and mass transport phenomena.

Unit processes are grouped together in various configurations for the purpose of chemical synthesis and/or chemical separation, and also in other sectors of the industry. Hybrid processes are a combination of different unit processes forming a hybrid system usually cheaper than the original processes. The entire production chain must be planned and controlled for cost and efficiency. Each physical process means an orderly string of consecutive changed states occurring sequentially and progressively. The idea of "unit processes" is used in chemical engineering to determine the reproducible elements of the technologies. Each technology can be represented in terms of the system as a network formed by the flow of material, energy and information together in blocks which correspond to unit processes (see Fig. 2.1)

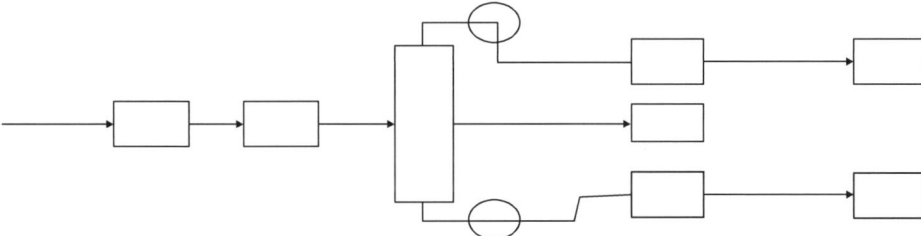

Fig. 2.1. Technology formed by unit processes.

It is more rational to focus on unit processes rather than the same technologies, since the number of possible technologies is unimaginably large, almost infinite, compared to the small number of unit processes (see Fig. 2.2). For this reason, chemical engineering became an autonomous field of knowledge, independent of chemical technology. It should be noted that this "emancipation" contributed to the development of process engineering to a very large extent, because the small number of unit processes which formed these technologies permitted focusing on the optimization and relevant device development. In addition, unit processes can refer to fundamental knowledge, enriching it with new experiences.

Moreover, they can create hybrids which are more efficient than their component processes. It just so happens, that chemical engineering derives mainly from physics rather than chemistry. Therefore the name chemical engineering has more historical meaning and should now rather be replaced by the broader modern concept of process engineering. Process engineering develops design tools to create technology by taking into account the natural laws of physics. Process engineering provides a new opportunity for sustainable development based on multidisciplinary achievements in nanotechnology, biology, chemistry and physics. Process engineering can offer a realistic prospect for sustainable development by improving production and the environment, by separation process improvements and savings in resources, energy and

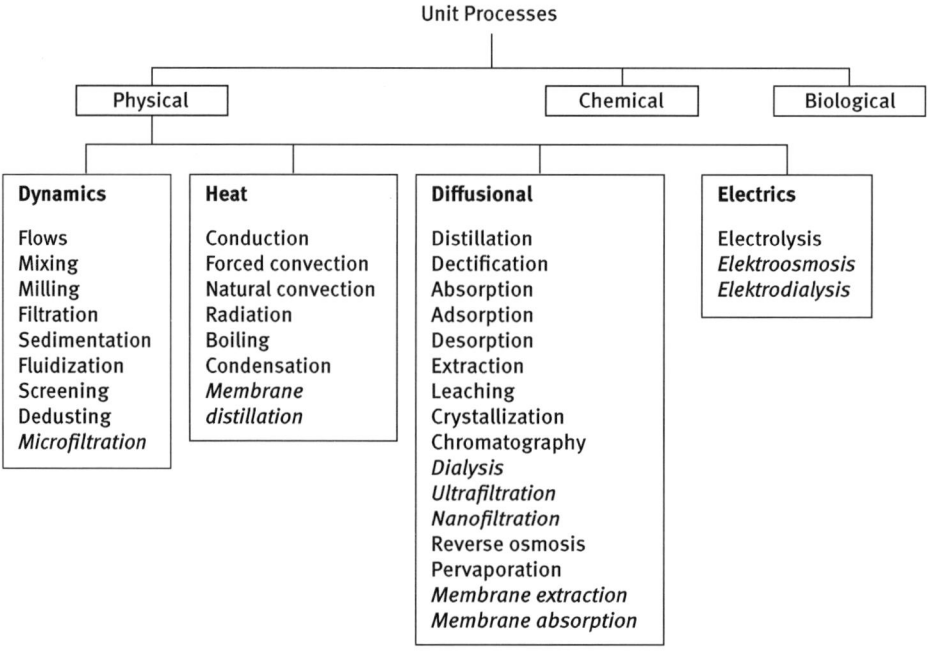

Fig. 2.2. Unit processes. Membrane processes are marked in italics.

costs. Process engineers ensure that the processes are operated safely, sustainably and economically.

Unit processes constitute the fundamental principles of process engineering. All processes in nature tend towards states of equilibrium but do not exist in equilibrium. Therefore the "distance" from equilibrium is the driving force of the process. Thermodynamic equilibrium – the term used in thermodynamics – refers to the state in which the macroscopic parameters, such as pressure, volume and all of the functions are constant over time. In a thermodynamic equilibrium there are no chemical reactions (chemical equilibrium), no macroscopic flow of particles, there are no unbalanced forces (mechanical equilibrium), and there is no flow of energy (thermal equilibrium). The "phase rule", which was defined by Gibbs, applies to the state of thermodynamic equilibrium:

$$s = \alpha - f + 2, \tag{2.7}$$

where: s – number of degrees of freedom, the number of intensive variables that can be changed without a qualitative change in the system (without changing the number of phases in equilibrium); α – the number of separate components, i.e., those which can be determined by means of chemical dependencies; f – the number of phases, and thus a homogeneous material chemically and physically (for example solution, gas phase, crystals of a specific composition).

2.2.2 Conservation laws

As previously mentioned each unit operation follows the same three primary physical laws which underlie chemical engineering design:
1. conservation of mass;
2. conservation of momentum;
3. conservation of energy.

The rate (kinetics) of flow and production of mass and energy is a key factor for the most economical operation and is based on principles of thermodynamics, reaction kinetics, fluid mechanics and transport phenomena. Unit processes are stationary if the incoming and outflowing streams are equal (see Fig. 2.3), and the process is independent of time. In nonstationary processes, there is accumulation. This also applies to all systems and subsystems.

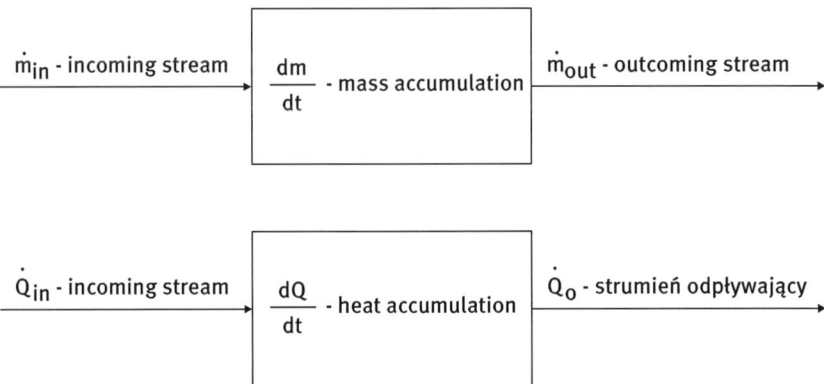

Fig. 2.3. Graphical representation of the conservation law of mass and heat.

Unit processes are stationary if the process is independent of time. Then, regardless of the system, which may be one device or a large technological system composed of multiple unit processes, the inflowing streams are balanced by the outflowing streams, and accumulation within any system is zero:

$$\frac{dm}{dt} = 0 \tag{2.8}$$

$$\frac{dQ}{dt} = 0. \tag{2.9}$$

If the flow system has a constant flow rate (v), and constant volume (V), the residence time of the fluid elements can be calculated from the formula

$$\tau = \frac{V}{\dot{v}}. \tag{2.10}$$

The residence time of the fluid elements in the process unit must be determined on the basis of the kinetics of the process

$$\tau \geq t_{process} . \tag{2.11}$$

At this point, the introduction of the concept of residence times should be explained. Remember that at constant volume and constant flow rate with incompressible fluid, the mean residence time is constant in the apparatus. This value is only the mean from distribution ("the first momentum of distribution"). In real systems, the various fluid elements, i.e. the viscous and compressible fluids, as opposed to ideal systems, which are nonviscous and noncompressible, the different residence time in a given volume remains. Therefore there is a time distribution around the mean value. This means that if some elements leave the apparatus in shorter time (than average), the others must stay longer. This phenomenon of residence time distribution around the expected value is known as longitudinal mixing.

2.2.3 Analogies between transport of momentum, heat and mass

Isaac Newton was the first scientist who introduced the notion of viscosity to fluid flow mechanics in his famous phenomenological equation, known as Newton's law:

$$\tau = -\mu \cdot \frac{du_x}{dy} . \tag{2.12}$$

The formal similarity of this equation to Fourier's and Fick's law is not a simple coincidence. In this case, a reduction in the momentum of fluid elements (P) in the direction (y), perpendicular to the direction of motion (x) under the influence of viscosity (μ). Shear stress (τ) can after all be considered "stream of momentum" according to the following formula:

$$\tau = \frac{d(mu)_x}{A \cdot dt} . \tag{2.13}$$

Where: A_y is the surface area perpendicular to the flow direction (x), time (t). The formal similarity of this equation to Fourier's equation is obvious from the analogy of streams. Jean Baptiste Joseph Fourier (1768–1830) formulated the heat transport equation by conduction, where heat flux (q) is defined analogously to the momentum flux:

$$q_x = \frac{d(Q)_x}{A_y \cdot dt}, \tag{2.14}$$

where Q_x is Here and Q_x is the thermal energy flowing through the conduction in the x-direction for Ay surface at time t. Fourier was a French mathematician (known for the Fourier transform and also the Fourier series). This law states that the time rate of heat transfer through a material is proportional to the negative gradient through which heat is flowing:

$$q_x = -\lambda \frac{dT}{dx}.$$

(2.15)

Newton's law of viscous fluid flow is the analogue to Fourier's law.

Two further analogies which are important for process engineering are Darcy's law and Fick's law. Darcy's law describes volumetric transport under the influence of the pressure gradient:

$$J_{Vx} = -\frac{K}{\mu} \cdot \frac{dP}{dx} \quad \left[\frac{m^3}{m^2 \cdot s} \right],$$

(2.16)

where, as previously (i.e., momentum flux and heat flux), the volumetric flux can be expressed by the Darcy formula

$$J_{Vx} = \frac{d(V)_x}{A_y \cdot dt}.$$

(2.17)

And yet a very important analogy for determining mass transport under the influence of the concentration gradient (i.e. potential field concentration) was formulated by Fick. Mass flow is expressed (in moles) as follows:

$$J_{mx} = \frac{d(m)_x}{A_y \cdot dt}.$$

(2.18)

Molecular mass transport under the influence of a concentration gradient describes Fick's law:

$$J_{mx} = -\frac{dC}{dx} \quad \left[\frac{mol}{m^2 \cdot s} \right].$$

(2.19)

All these analogies on the transport of momentum, heat and mass volume can be presented in the form of a transport equation, which can be written in vector form as follows:

$$\vec{J} = -\vec{\nabla \Phi}.$$

(2.20)

Please note that both the streams and gradients are vectors (i.e. they have specific directions). Minuses in all these equations indicate that the flows are in opposite directions in relation to the respective gradients, which are known as driving forces. In other words, the fluxes always flow in the direction of the larger to the smaller of respective potential (temperature, concentration, pressure etc), which are scalars on their own. These analogies apply to mass flow in a potential field concentration, which is proportional to the negative gradient of concentration. In all equations we are dealing with a general rule which says that the transport (momentum, heat, volume and mass) in a potential field is always in the opposite direction to the gradient of the field at the selected point. This important observation is called the general analogy of heat transport of mass and energy and is the foundation of modern process engineering.

When fluid is in motion the flow parameters are a function of time and position. The approach developed by Leonhard Euler introduces density, velocity and pressure, which are functions of the position of the time:

$$\rho = \rho(x, y, z, t), \tag{2.21}$$

$$u = u(x, y, z, t), \tag{2.22}$$

$$P = P(x, y, z, t). \tag{2.23}$$

Another approach was proposed by Joseph Louis Lagrange (1736–1813), in which fluid is divided into infinitesimal volume elements, which may be called "liquid elements". The position of each particle of fluid is clearly defined by the coordinates x, y and z, which are dependent on the time (t) and the position of the particle at the initial moment (t_0):

$$x = x(t, x_0, y_0, z_0, t_0), \tag{2.24}$$

$$y = y(t, x_0, y_0, z_0, t_0), \tag{2.25}$$

$$z = z(t, x_0, y_0, z_0, t_0). \tag{2.26}$$

In this way, we can focus on what is happening at a specific point in space, not on what is happening with a certain element of fluid.

2.2.4 Onsager theorem and analogies between different processes

In real processes, energy is released and dispersed in a continuous manner if the driving force is fixed and the amount of entropy produced is determined using the dissipation function

$$\Phi = T \cdot \frac{dS}{dt}. \tag{2.27}$$

For all irreversible processes the dissipation function can be expressed as the sum of the products of all streams coupled to each other and the respective driving forces:

$$\Phi = \sum J_i \cdot X_i. \tag{2.28}$$

It can be assumed that close to equilibrium streams are in a linear relationship between the respective driving forces. Phenomenological equations represent streams in the form of the products of the driving forces which are the appropriate gradients (of the field potentials such as temperature concentration, pressure etc). The kinetic coefficients characterizing the property of the medium in which the transport takes place are as follows:

$$J_i = L_i \cdot X_i. \tag{2.29}$$

The fourth law of thermodynamics states that the phenomenological matrix of kinetic coefficients is symmetrical:

$$L_{ij} = L_{ji}. \tag{2.30}$$

It is known that the temperature difference leads to the movement of heat from the hotter to the colder body:

$$J_T = L_T \cdot X_T. \tag{2.31}$$

As the pressure differential leads to the fluid flow from the place of higher pressure to the place of lower pressure:

$$J_P = L_P \cdot X_P. \tag{2.32}$$

It was found experimentally that there is a so-called coupled transport; if the differences are both pressure and temperature then the pressure gradient (X_P) also has an effect on the heat transfer:

$$J_T = L_{TT} \cdot X_T + L_{TP} \cdot X_P. \tag{2.33}$$

The temperature gradient (X_T) then also contributes to mass transport:

$$J_P = L_{PP} \cdot X_P + L_{PT} \cdot X_T. \tag{2.34}$$

If we now consider that the heat flow per unit pressure driving force (L_{TP}) is exactly equal to the volumetric flow rate per unit of temperature (L_{PT}), this is just a special case of symmetry of kinetic coefficients, an example of the fourth principle of thermodynamics, for which Onsager received the Nobel Prize in 1968:

$$L_{TP} = L_{PT}. \tag{2.35}$$

Similar commutability relationships exist between other pairs of kinetic coefficients describing coupled transport under the influence of the driving forces such as concentration gradient, electrical potential gradient etc. Onsager formulated the principle of alternation of kinetic coefficients for systems coupled with the increasing number of driving forces on the basis of statistical mechanics.

2.2.5 Phenomenological transport equations in unit processes

As mentioned in the previous subsection on the history of process engineering, the laws relating to the movement of momentum (Newton's law), heat (Fourier's law), volume (Hagen-Poiseuille law) were all discovered independently of one another and at different times. The attention was focused on the streams and "driving forces" causing the appropriate transport. Later it was noticed that although the same processes are physically different, their mathematical description is analogous. Although the same equation applies to a variety of transport streams flowing under the influence of different driving forces, the essence of the transport is the same. In this way an analogy was found between the transport of momentum, heat and mass transfer, which was the foundation of modern process engineering. This analogy permitted focusing on solving the equations of one field by the use of solutions from other areas (see Tab. 2.3).

Table 2.3. Analogies between heat, mass, and momentum transfer

Field	Potential	Flux	Stationary process	Nonstationary process
Temperature	$T(x, y, z, t)$	Heat	$q = -\lambda \dfrac{dT}{dx} \left[\dfrac{J}{m^2 s} \right]$	$D_h \dfrac{\partial^2 T}{\partial x^2} = \dfrac{\partial T}{\partial t}$
Concentration	$C(x, y, z, t)$	Mass	$J_m = -D \dfrac{dC}{dx} \left[\dfrac{mol}{m^2 s} \right]$	$D \dfrac{\partial^2 C}{\partial x^2} = \dfrac{\partial C}{\partial t}$
Pressure	$P(x, y, z, t)$	Volume	$J_v = -K \dfrac{dP}{dx} \left[\dfrac{m^3}{m^2 s} \right]$	$K \dfrac{\partial^2 P}{\partial x^2} = \dfrac{\partial P}{\partial t}$
Velocity	$u(x, y, z, t)$	Momentum	$R = -\mu \dfrac{du}{dx} \left[\dfrac{N \cdot s}{m^2 s} \right]$	$\mu \dfrac{\partial^2 C}{\partial x^2} = \dfrac{\partial C}{\partial t}$

The potentials of corresponding fields are scalars (pressure, concentration, temperature and velocity), but the gradients are vector quantities and have their direction and sense. Hence the opposite sign follows between the streams and the gradients of corresponding fields. It's obvious that the fluxes always flow in the direction of decreasing field potentials, such as temperature, concentration and pressure. In nonstationary conditions the parameters describing the potentials of appropriate fields, namely pressure, concentration, speed or temperature, are functions of time and place. Since the transport equations are so important for process engineering, we can try to derive them on the basis of one of the (three) main axioms of physics, which is called the "conservation law". This law says that for any area in three-dimensional space, the sum of all flow streams plus accumulation in this closed area must be always equal to zero.

Derivation of the transport equation will allow us to determine the function of the potential of space-time field, just as the temperature $T(x, y, z,$ and $t)$ during heat conduction. It can be assumed that heat transfer occurs because of the temperature gradient, the so-called control area, i.e. any small space with dimensions differential dx, dy and dz. The flux of heat flow in the x direction according to Fourier's formula:

$$q = -\lambda \frac{dT}{dx} \left[\frac{J}{m^2 s} \right].$$ (2.36)

Directed toward x, the heat flow rate is then

$$Q_x = q_x \cdot dy \cdot dz = -\lambda \frac{dT}{dx} \cdot dy \cdot dz.$$ (2.37)

However, the heat differential changes in the x, y and z-direction are given by

$$dQ_x = dq_x \cdot dy \cdot dz \cdot dt = -\lambda \frac{\partial^2 T}{\partial x^2} dx \cdot dy \cdot dz \cdot dt.$$ (2.38)

$$dQ_y = dq_y \cdot dx \cdot dz \cdot dt = -\lambda \frac{\partial^2 T}{\partial y^2} dx \cdot dy \cdot dz \cdot dt$$ (2.39)

$$dQ_z = dq_z \cdot dx \cdot dy \cdot dt = -\lambda \frac{\partial^2 T}{\partial z^2} dx \cdot dy \cdot dz \cdot dt. \tag{2.40}$$

Heat accumulation in the control volume element dx dy dz differential which has a density and specific heat can be calculated from the definition of the specific heat:

$$dQ_a = c_P \cdot \rho \cdot \frac{\partial T}{\partial t} dt \cdot dx \cdot dy \cdot dz. \tag{2.41}$$

Assuming heat flow through the control element can write the balance of flows in all directions, with the possibility of heat accumulation inside the element (see Fig. 2.4)

$$dQ_x + dQ_y + dQ_z + dQ_a = 0. \tag{2.42}$$

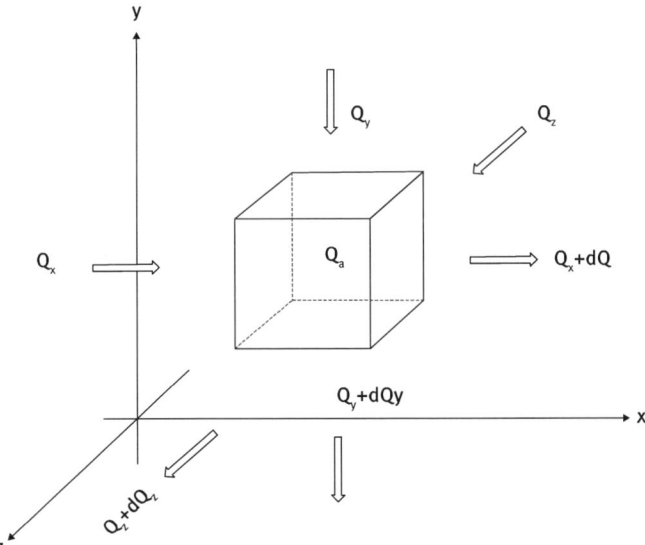

Fig. 2.4. Graphical representation of heat balance.

Based on heat balance in the three-dimensional space, including the heat accumulation inside the control element, we obtain the following equation:

$$-\lambda \left(\frac{\partial^2 T}{\partial x^2} + \frac{\partial^2 T}{\partial y^2} + \frac{\partial^2 T}{\partial z^2} \right) = c_P \cdot \rho \cdot \frac{\partial T}{\partial t}. \tag{2.43}$$

This equation simplifies considerably, after reducing the space to one dimension:

$$-D_h \cdot \frac{\partial^2 T}{\partial x^2} = \frac{\partial T}{\partial t}, \tag{2.44}$$

where D_h is the so-called "heat diffusivity":

$$D_h = \frac{\lambda}{c_P \cdot \rho}. \tag{2.45}$$

Finally, Fourier's second law was obtained as an example of how to derive a phenomenological equation, of heat transport in this case. Other phenomenological equations can be derived in a similar way. Their solutions represent functions, space-potentials, corresponding fields. These solutions may be obtained by assuming appropriate boundary conditions. These are partial differential equations, second order with respect to the spatial variables (x, y, z), and the first order with respect to the time variable, and therefore we need to define three limiting conditions, i.e. the two boundary conditions (on site) and one initial condition (on time). The solution of the transport equation with different boundary conditions was proposed by Carslaw and Jaeger [52]. Horatio Scott Carslaw (1870–1954), was a Scottish mathematician and John Conrad Jaeger (1907–1979) was an Australian professor of mathematics and engineering.

2.2.6 Solution of transport equations by the Laplace transform method

The partial differential equations of the second order have no direct solutions in the field of real numbers. Laplace transformation eliminates the time variable (t), and the equation is reduced to an ordinary differential equation which can be solved analytically. Laplace transform for an arbitrary function is defined as the integral:

$$\pounds \{f(t)\} = \int_0^\infty f(t) \cdot e^{-p \cdot t} dt. \tag{2.46}$$

After Laplace transformation, one-dimensional transport equation, moves for ordinary differential second order equation into the area of complex numbers:

$$-D_h \cdot \overline{\frac{\partial^2 T}{\partial x^2}} = f(p). \tag{2.47}$$

Methods for solving differential equations of any order, such as the method of substitution, are well known, however their solutions are in the area of complex numbers. On receiving the solution to ordinary equations in the field of complex numbers, we cannot directly use this solution to calculate the temperature field in time and space. Therefore, we must perform a "reverse procedure" in order to find the right solution in the field of real numbers, for example based on the tables of Laplace transformations, which usually occur in all technical manuals. Reverse procedure, that is, the designation of a function in the field of real numbers from the transform:

$$\pounds^{-1} \{T(t, p)\} \overleftrightarrow{\text{reverse action}} T(t, x). \tag{2.48}$$

In this way, by applying the Laplace transform, we can get a solution to the partial second-order equation in the field of real numbers. Another technique for solving transport equations is with numerical methods.

Prior to solution of the transport equation, the appropriate normalization of the variable T is used, in order to limit its field of numbers $0 \leq \theta \leq 1$, this facilitates the use of tables in order to determine the value of transforms. For this purpose, the following substitution is carried out:

$$\theta\,(x, t) = \frac{T\,(x, t) - T_p}{T_k - T_p}.$$ (2.49)

As a result a normalization equation is obtained for the other variable, but linearly related to the temperature:

$$D_h \cdot \frac{\partial^2 \theta}{\partial x^2} = \frac{\partial \theta}{\partial t}.$$ (2.50)

Laplace transform of the function with respect to any variable (in this case it is time, t), defined as follows:

$$\overline{\theta\,(x, t)} = \int_0^\infty \theta \cdot \exp\,(-p \cdot t) dt = \bar{\theta}\,(x, p).$$ (2.51)

Continued transformation will be determined by an upper mark. According to the algorithm shown above, in the first step we transform both sides of the partial differential equation:

$$D_h \cdot \overline{\frac{\partial^2 \theta}{\partial x^2}} = \overline{\frac{\partial \theta}{\partial t}}.$$ (2.52)

Transform of the first derivative of the time variable (t) is calculated, "by parts" according to the known formula of mathematical analysis:

$$\int u' \cdot v \cdot dt = u \cdot v - \int u \cdot v' \cdot dt.$$ (2.53)

Thus, Laplace transform of the first derivative of the temperature normalized after time is eventually calculated by the following formula:

$$\overline{\frac{\partial \theta}{\partial t}} = \int_0^\infty \frac{\partial \theta}{\partial t} \cdot \exp\,(-p \cdot t) \cdot dt = [\theta \cdot \exp\,(-p \cdot t)]_0^\infty - p \cdot \int_0^\infty \theta \cdot \exp\,(-p \cdot t) \cdot dt,$$ (2.54)

from which we obtain:

$$\overline{\frac{\partial \theta}{\partial t}} = -\theta_{t=0} + p \cdot \bar{\theta}.$$ (2.55)

As a result of the transformation we have received ordinary differential equation instead of the partial equation:

$$D_h \cdot \frac{d^2 \bar{\theta}}{dx^2} - p \cdot \bar{\theta} = -\theta_{t-0}.$$ (2.56)

This is a heterogeneous differential equation of the second-order, which can be easily solved by the so-called substitution method. According to this algorithm, first we seek solutions of ordinary homogeneous equation:

$$D_h \cdot \frac{d^2 \bar{\theta}}{dx^2} - p \cdot \bar{\theta} = 0.$$ (2.57)

The solution of the homogeneous equation is always only an exponential form:

$$\bar{\theta} = \exp\left(r \cdot x\right). \tag{2.58}$$

Here the letter "r" is the designated root of the characteristic equation. To find the characteristic equation we have to calculate the second derivative of the general solution:

$$\frac{d^2\bar{\theta}}{dx^2} = \frac{d}{dx}\left(\frac{d\bar{\theta}}{dx}\right) = \frac{d}{dx}\left(\frac{d}{dx}\left(\exp\left(r \cdot x\right)\right)\right) = \frac{d}{dx}\left(r \cdot \left(\exp\left(r \cdot x\right)\right)\right) = r^2 \cdot \exp\left(r \cdot x\right). \tag{2.59}$$

After substituting these expressions into the homogeneous equation we get

$$D_h \cdot r^2 \cdot \exp\left(r \cdot x\right) - p \cdot \exp\left(r \cdot x\right) = 0. \tag{2.60}$$

After dividing by the "exponent" (always different from zero), we obtain the characteristic equation:

$$D_h \cdot r^2 - p = 0, \tag{2.61}$$

from which we can determine two roots "r":

$$r_1 = +\sqrt{\frac{p}{D_h}} \quad \text{and} \quad r_2 = -\sqrt{\frac{p}{D_h}}. \tag{2.62}$$

Thus, the general solution of the homogeneous equation is

$$\bar{\theta} = B_1 \cdot e^{x \cdot \sqrt{\frac{p}{D_h}}} + B_2 \cdot e^{-x \cdot \sqrt{\frac{p}{D_h}}}. \tag{2.63}$$

We can obtain the particular solution by adding the term previously rejected from the heterogeneous equation to the general solution:

$$\bar{\theta} = B_1 \cdot e^{x \cdot \sqrt{\frac{p}{D_h}}} + B_2 \cdot e^{-x \cdot \sqrt{\frac{p}{D_h}}} + \theta_{t=0}. \tag{2.64}$$

B_1 and B_2 are constants determined after determining the boundary conditions suitable physical meaning of a differential equation. Two conditions of space must be specified (boundary conditions), and one on time (usually its initial condition). Unambiguous formulation of the boundary conditions determines the form of the function, which is the solution of the differential equation. In order to further solve the differential equation a specific example must be used. Now we can determine the boundary conditions for the following case: the temperature of the rod of the length (L) is equal (T_p), up to a time $t = 0$. The rod was then subjected to an elevated temperature (T_k) at one end. The other end is perfectly thermally insulated. The conditions are formulated for the actual and normalized temperature in the the following Tab. 2.4.

Table 2.4. Boundary conditions for solving heat transport conditions.

Boundary conditions	Time	Space	Actual temperature	Normalized temperature	Laplace transformation of the normalized temperature
Initial condition	$t = 0$	$0 < x < L$	$T(x, t) = T_p$	$\theta(x, t) = 0$	$\bar{\theta}(x, t) = 0$
Boundary condition	$t > 0$	$x = 0$	$T(x, t) = T_k$	$\theta(x, t) = 1$	$\bar{\theta}(x, t) = \dfrac{1}{p}$
Boundary condition	$t > 0$	$x = L$	$\dfrac{\partial T}{\partial x} = 0$	$\dfrac{\partial \theta}{\partial x} = 0$	$\dfrac{\partial \bar{\theta}}{\partial x} = 0$

The relationship between temperature and the normalized Laplace transform will always be the same as that for any constant if the boundary conditions are determined by the value of a constant temperature, rather than functions. If $\theta = $ const, then transform as a constant can be calculated according to the rule

$$\bar{\theta} = \int_0^\infty \theta \cdot \exp(-p \cdot t)dt = \theta \cdot \int_0^\infty \exp(-p \cdot t)dt = \theta \cdot \left[-\frac{\exp(-p \cdot t)}{p} \right]_0^\infty$$

$$= \theta \cdot \left[-\frac{\exp(-p \cdot \infty)}{p} + \frac{\exp(-p \cdot 0)}{p} \right] = \theta \cdot \left[-0 + \frac{1}{p} \right] = \frac{\theta}{p}. \tag{2.65}$$

Constants B_1 and B_2 in solving transport equations are determined after taking the boundary conditions into account. Taking the initial condition (the first boundary condition) into account leads to the conclusion that

$$\theta_{t=0} = 0. \tag{2.66}$$

On the basis of the second boundary condition we get the following equation:

$$B_1 + B_2 = \frac{1}{p}. \tag{2.67}$$

The use of the third boundary condition requires the designation of the derivative on normalized temperature:

$$B_1 \cdot \sqrt{\frac{p}{D_h}} \cdot e^{x \cdot \sqrt{\frac{p}{D_h}}} - \sqrt{\frac{p}{D_h}} \cdot B_2 \cdot e^{-x \cdot \sqrt{\frac{p}{D_h}}} = 0. \tag{2.68}$$

The solution of these two equations allows us to determine the constants B_1 and B_2:

$$B_1 = \frac{1}{p} \cdot \frac{e^{-L \cdot \sqrt{\frac{p}{D_h}}}}{e^{L \cdot \sqrt{\frac{p}{D_h}}} + e^{-L \cdot \sqrt{\frac{p}{D_h}}}} \quad \text{and} \quad B_2 = \frac{1}{p} \cdot \frac{e^{L \cdot \sqrt{\frac{p}{D_h}}}}{e^{L \cdot \sqrt{\frac{p}{D_h}}} + e^{-L \cdot \sqrt{\frac{p}{D_h}}}}. \tag{2.69}$$

The inclusion of these constants in solving particular equations (ordinary differential second-order in the field transform) gives the following solution in complex numbers:

$$\bar{\theta}(x, p) = \frac{e^{(L-x)\cdot\sqrt{\frac{p}{D_h}}} + e^{-(L-x)\cdot\sqrt{\frac{p}{D_h}}}}{e^{L\cdot\sqrt{\frac{p}{D_h}}} + e^{-L\cdot\sqrt{\frac{p}{D_h}}}}. \tag{2.70}$$

The solution of the transport equation in the field of complex numbers allows you to find the so-called original. In the course of the reverse procedure, this is a result of a specific spatiotemporal distribution of the actual temperature:

$$\theta(x, t) = \sum_{n=0}^{n=\infty} (-1)^n \cdot \left[\text{erfc}\left(\frac{2\cdot L\cdot n + x}{2\cdot\sqrt{D_h\cdot t}}\right) + \text{erfc}\left(\frac{2\cdot(L-x)\cdot(n+1)}{2\cdot\sqrt{D_h\cdot t}}\right)\right], \tag{2.71}$$

where the function erfc(s) is called the error function, which is defined by the following formula:

$$\text{erfc}(y) = \frac{2}{\sqrt{\pi}} \int_y^\infty e^{-\xi^2}\, d\xi. \tag{2.72}$$

This completes the solution of the transport equation, using the Laplace transform method.

2.2.7 Solving the transport equation for semi-permeable surfaces (membranes)

Semi-permeable membranes are also called selectively permeable membranes. Semi-permeable membranes are such surfaces which enable certain particles, molecules, ions and electrons to pass through, while other appropriate particles are retained on them. Membranes occur naturally, as cell walls, but they are also artificially manufactured for use in a variety of important separation processes.

Such processes are discussed later in this book in Chapter 4. However, regardless of the type of transport mechanism and driving force, in such processes, the substances which are rejected by the membrane are always collected on the membrane surface. This causes additional resistance for mass transport and interferes with the transport of the second component which passes through the membrane. Thus we have two types of components in membrane processes. Those which pass through the membrane (A, usually solvents), and the second (say B) which are rejected by the membrane, creating the concentration polarization layer. From a physical point of view the mass transport in membrane processes, this is quite a different type than those in unit processes (such as absorption, distillation, crystallization etc.). It is not concerned at all with mass transport in the membrane, but coupling of the mass transport of the two components on the membrane surface, where one of them will be rejected. In this case, the transport equation can be written as follows:

$$D\frac{\partial^2 C}{\partial x^2} - J\frac{\partial C}{\partial x} = \frac{\partial C}{\partial t}, \tag{2.73}$$

where C is the concentration of the rejected component (A), D is its diffusivity, and J is convective flux of the component (B), which passes through the membrane. Even at first glance, the form of this equation differs from the usual transport equation which was solved earlier. This difference is an additional convection element J, which corresponds to the velocity of interface movement through the surroundings. This equation cannot be solved as before. It can, however, be reduced to the same form by appropriate substitution to eliminate the convection term. In order to eliminate the convective term in the transport equation, the following substitution can be performed:

$$C(x, t) = \xi(x, t) \cdot e^{A \cdot x + B \cdot t}. \tag{2.74}$$

Now find the appropriate values of the constants A and B, according to the following procedure. The first derivative must be calculated, as follows:

$$\frac{\partial C}{\partial x} = \left[\frac{\partial \xi}{\partial x} + \xi \cdot A \right] \cdot e^{A \cdot x + B \cdot t}. \tag{2.75}$$

Next compute the second partial derivative of x:

$$\frac{\partial^2 C}{\partial x^2} = \left[\frac{\partial^2 \xi}{\partial x^2} + 2 \frac{\partial \xi}{\partial x} \cdot A + \xi \cdot A^2 \right] \cdot e^{A \cdot x + B \cdot t}. \tag{2.76}$$

Finally we calculate the first derivative with respect to time:

$$\frac{\partial C}{\partial t} = \left[\frac{\partial \xi}{\partial t} + \xi \cdot B \right] \cdot e^{A \cdot x + B \cdot t}. \tag{2.77}$$

After substituting all these derivatives for the transport equation and rearranging, we can write the transport equation of a new variable:

$$D \frac{\partial^2 \xi}{\partial x^2} + [2AD - J] \frac{\partial \xi}{\partial x} + \left[A^2 D - JA - B \right] \xi = \frac{\partial \xi}{\partial t}. \tag{2.78}$$

Expressions in square brackets must be zero in order to obtain the transport equation without convection member. Therefore, we can write and solve a system of two equations with two unknowns, A and B:

$$\begin{cases} 2AD - J = 0 \\ A^2 D - JA - B = 0. \end{cases} \tag{2.79}$$

The solution to this system of equations is as follows:

$$A = \frac{J}{2D}, \tag{2.80}$$

$$B = \frac{J^2}{4D}. \tag{2.81}$$

Thus, if the equation of convective transport member will make the substitution

$$C(x, t) = \xi(x, t) \cdot e^{\frac{J}{2D} \cdot x + \frac{J^2}{4D} \cdot t}. \tag{2.82}$$

We can get the form of the equation of transport without the convection constituent, which can be solved using previously described methods:

$$D\frac{\partial^2 \xi}{\partial x^2} = \frac{\partial \xi}{\partial t}. \tag{2.83}$$

The solutions to the transport equation with different boundary conditions were proposed by Carslaw and Jaeger [52].

The most general case of mass transport for semi-permeable membranes is shown in Fig. 2.5.

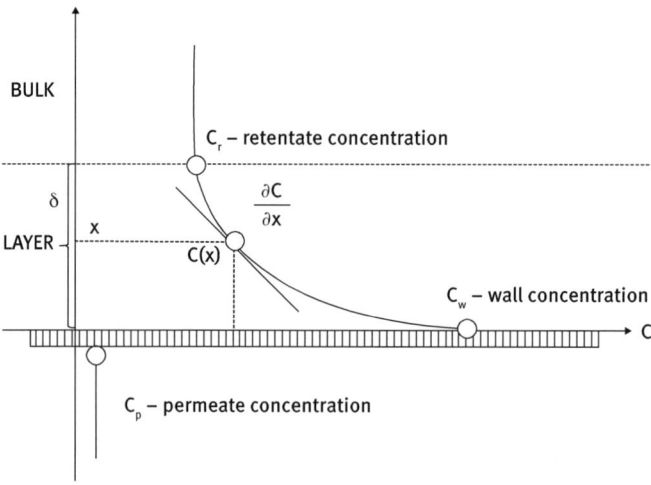

Fig. 2.5. Concentration-polarization and mass transfer at the semi-permeable surface.

2.2.8 Solution of transport equations by numerical methods

Numerical methods of solving differential equations involve replacing infinite small differentials by finite differences with defined sizes. The accuracy of such an approximation is greater, the smaller the increments of the variables used in the definition of derivatives are. A spectacular example of solving differential equations is Schmidt's graphical method [59], where transport equation can be expressed by a new numerical form. The first step is to normalize the variable temperature to such a form in order to have a value in the range of 0–1 as follows:

$$\theta\,(x, t) = \frac{T\,(x, t) - T_p}{T_k - T_p}. \tag{2.84}$$

To obtain another transport equation,

$$D_h \cdot \frac{\partial^2 \theta}{\partial x^2} = \frac{\partial \theta}{\partial t}. \tag{2.85}$$

Now we can replace derivatives by finite increments by using the definition of a derivative, which can be written thus:

$$\frac{\partial \theta}{\partial t} = \lim_{\Delta t \to 0} \frac{\theta (x, t + \Delta t) - \theta (x, t)}{\Delta t} \cong \frac{\theta (x, t + \Delta t) - \theta (x, t)}{\Delta t}.$$ (2.86)

The second derivative of the temperature of the location can also be replaced by appropriate increments based on its definition

$$\frac{\partial^2 \theta}{\partial x^2} = \frac{\partial}{\partial x} \left(\frac{\partial \theta}{\partial x} \right) = \lim_{\Delta x \to 0} \frac{\dfrac{\theta (x + \Delta x, t) - \theta (x, t)}{\Delta x} - \dfrac{\theta (x, t) - \theta (x - \Delta x, t)}{\Delta x}}{\Delta x}.$$ (2.87)

It is clear that the number two is written differently in numerator and denominator in order to mark double derivative. After the substitution of differentials with differences in the transport differential equation, the differential equation of the following form can now be introduced:

$$D_h \cdot \frac{\theta (x + \Delta x, t) - 2 \cdot \theta (x, t) + \theta (x - \Delta x, t)}{\Delta x^2} = \frac{\theta (x, t + \Delta t) - \theta (x, t)}{\Delta t}.$$ (2.88)

After simplifying and grouping the increments of independent variables, the following equation is obtained:

$$\frac{D_h \cdot \Delta t}{2 \cdot \Delta x^2} \cdot \left[\frac{\theta (x + \Delta x, t) + \theta (x - \Delta x, t)}{2} - \cdot \theta (x, t) \right] = \theta (x, t + \Delta t) - \theta (x, t).$$ (2.89)

The ratio of increments (without restricting their size) can always be selected so that the expression before the square brackets is equal to 1:

$$\frac{D_h \cdot \Delta t}{2 \cdot \Delta x^2} = 1.$$ (2.90)

Thus the transport equation becomes simpler, and can be used to determine the temperature profile in any given space and time:

$$\frac{\theta (x + \Delta x, t) + \theta (x - \Delta x, t)}{2} = \theta (x, t + \Delta t).$$ (2.91)

From this formula it follows that the temperature at a given instant in time (t) and place (x) can be determined based on temperatures at the neighboring places, at previous instants in time (see Fig. 2.6).

Figure 2.7 shows a graphical construction for determining the temperature profile at any time and place. The numbers correspond to the following moments: 1 corresponds to the time $t_1 = 1\Delta t$, 2 corresponds to the time $t_2 = 2\Delta t$, etc. Thus, by combining the appropriate temperature profiles we can get the numbers at the right moment. Please note that the accuracy of the method itself is not limited as it is dependent on our choice of the numerical values for Δt and Δx.

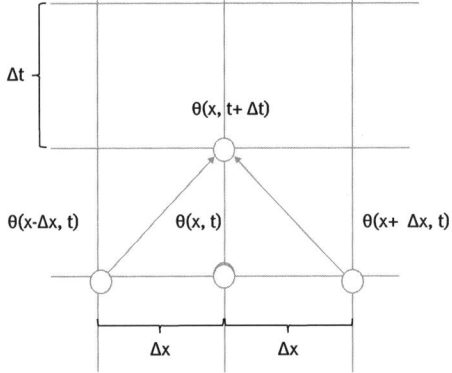

Fig. 2.6. Schematic diagram of computing procedure according to Schmidt's method.

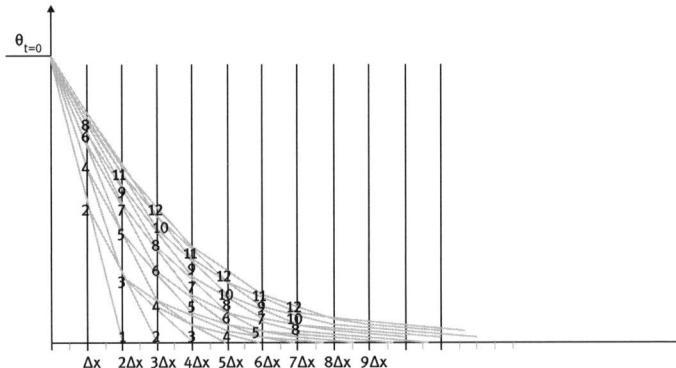

Fig. 2.7. Graphical representation of Schmidt's numerical methods for solving transport equations [59].

2.2.9 Flow regimes

There are no pure phenomena in nature such as pure diffusion or pure heat conduction. Fick had some problems ensuring conditions for pure diffusion, though molecular diffusion is the only method of mass transport in fluids. He could not eliminate spontaneous movement of fluid due to convection, which came from temperature gradients. Therefore, in order to obtain accurate results for pure diffusion, it is better to watch the phenomenon of pure diffusion in gels rather than in pure liquids. This is similar to the case of heat which is transferred by the principle of conduction. It is not possible to observe pure heat conduction in stagnant fluids, however. Convection of fluids is always more visible than diffusion or conduction.

When kinetic energy increases, the fluid or heat transfer rate of the mass also increases. However, after a certain point has been exceeded we observe a discontinuity in the rate of increase of transport and the heat and mass transfer becomes even more intense, thus making the effect of turbulence more visible. In all cases the transport changes of mass, heat and momentum (pressure drop) are carried out in a discontinuous manner under the influence of the increase of kinetic energy of fluid flow. The influence of hydrodynamics, i.e., flow rate or mixing of the fluid, intensifies the kinetics of heat and mass transport processes which are applied in engineering processes. But the cost of energy should not be forgotten and it must not be wasted. Therefore, the main objective is to minimize process costs by optimizing them so as to increase efficiency and yield. Turbulence is one of these problems, because on the one hand it results in an increase of the kinetics of transport (up to several orders of magnitude), but it also causes unwanted increase in hydraulic resistance and loss of energy. An additional problem is the formal change in the field of mathematical modeling in both areas. While the general transport equation can be solved analytically only in laminar area (of course after measuring the appropriate boundary conditions), only numerical solutions are attainable in turbulent areas. Therefore, it is of utmost importance that process engineers focus on detailed analysis of turbulence.

2.2.9.1 Laminar flow

During the laminar flow of fluid over a surface, the mass transfer at the boundary layer takes place by way of transport, which is molecular diffusion or heat transport via conduction. The thickness of the boundary layers under these conditions is dependent on the effects of shear stresses, which are determined by the velocity profile.

The solution to this system of equations allows determination of the residence time distribution of viscous fluid elements caused by velocity distribution depending on the distance from the walls.

Equations should take at least two dimensions into account because fluid flow is tangential to the surface between the phases, and the direction of transport (of mass or heat) is perpendicular to the direction of fluid flow. The following continuity equations can be written as follows for such two-dimensional (flat flow):

(a) the continuity equation for the whole system can be written as

$$\frac{\partial \rho}{\partial t} + \frac{\partial (\rho u_x)}{\partial x} + \frac{\partial (\rho u_y)}{\partial y} = 0; \tag{2.92}$$

(b) the equation of continuity of component "i" in a binary mixture is

$$\frac{\partial C_i}{\partial t} + \frac{\partial N_{ix}}{\partial x} + \frac{\partial N_{iy}}{\partial y} = 0; \tag{2.93}$$

(c) the momentum balance equation for laminar flow is

$$\rho \frac{\partial u_x}{\partial t} + \rho u_x \cdot \frac{\partial u_x}{\partial x} + \rho u_y \cdot \frac{\partial u_y}{\partial y} + \frac{\partial \tau_{xx}}{\partial x} + \frac{\partial \tau_{yx}}{\partial y} + \frac{\partial P}{\partial x} - \rho \cdot g_x = 0. \tag{2.94}$$

2.2.9.2 Turbulent flow

During turbulent flow, the transport of mass and momentum is performed by molecular motions (superscript m), and the random movements of the entire fluid portion known as eddies or turbulences (superscript t) which occur in different scales. Turbulence causes fluctuations in speed (u), the mass flow elements (N) as well as shear stress (τ). All these values vary locally in time, although the average values are constant (in steady-state processes), using the general laws of conservation of momentum and energy.

(a) The continuity equation for the component (i) of a binary mixture and turbulent flow is presented as follows:

$$\frac{\partial C_i}{\partial t} + \frac{\partial N_{ix}^m}{\partial x} + \frac{\partial N_{iy}^m}{\partial y} + \frac{\partial N_{ix}^t}{\partial x} + \frac{\partial N_{iy}^t}{\partial y} = 0. \tag{2.95}$$

(b) The momentum balance equation for turbulent flow:

$$\rho\frac{\partial u_x}{\partial t} + \rho u_x \cdot \frac{\partial u_x}{\partial x} + \rho u_y \cdot \frac{\partial u_y}{\partial y} + \frac{\partial \tau_{xx}^m}{\partial x} + \frac{\partial \tau_{yx}^m}{\partial y} + \frac{\partial \tau_{xx}^t}{\partial x} + \frac{\partial \tau_{yx}^t}{\partial y} + \frac{\partial P}{\partial x} - \rho \cdot g_x = 0. \tag{2.96}$$

The basic problem of obtaining analytical solutions of these equations is how to determine the components of turbulent flows, such as mass fluxes (N), and shear stresses (τ). Therefore, the description of the transport of mass and momentum is mainly based on empirical correlations with contribution of dimensional analysis.

2.2.9.3 The Einstein solution of the equation for momentum transport

The phenomenon of turbulence occurring in the flow of fluids has a stochastic nature. Vortices are formed at different scales, in random places and times. Some simplifications can be helpful for the estimation of the intensity of turbulences. For example, by assuming the constant mean size of vortexes, one can obtain their average frequency depending on the physical properties and the process conditions. Dimensional analysis is often used for the description of turbulence, whereby the following dimensionless variables are specified, such as distance from the wall, and frequency of eddies. Dimensionless distance from the wall of flow channel is defined as

$$y^+ = \frac{y}{\nu \cdot \sqrt{\frac{\rho}{\tau_w}}}, \tag{2.97}$$

where τ_w is the shear stress at the wall. Dimensionless frequency of the vortices is defined by the formula

$$n^+ = \frac{n \cdot \mu}{\tau_w}. \tag{2.98}$$

Einstein and Li [53] and Hanratty [54] determined the frequency of turbulence based on the momentum balance equation. They assumed that turbulence is associated with fluid eddies, which penetrate into the boundary layer from the bulk. The movement of the fluid elements towards a wall is nonsteady, because of the changes of their velocity depending on the distance. Assuming that the fluid is Newtonian, incompressible, and omitting the gravitational effects, the balance of momentum for this case can be described by the Navier-Stokes equation in simplified form, as follows:

$$\frac{\partial u}{\partial t} = v \cdot \frac{\partial^2 u}{\partial y^2}. \tag{2.99}$$

This equation can be solved provided the three limiting conditions are defined. One initial condition specifies the beginning of the history of the vortex. It was assumed that at the beginning (see Eq. (2.100a)), the speed of the fluid in the x-direction is constant and equal to u_0:

$$u_x = u_0 \quad \text{for} \quad t = 0. \tag{2.100a}$$

The following two boundary conditions (see Eqs. (2.100b) and (2.100c)) must be imposed on the two sites in the area and determine the speed at these sites:

$$u_x = 0 \quad \text{for} \quad y = 0, \tag{2.100b}$$

$$u_x = u_0 \quad \text{for} \quad y = \infty. \tag{2.100c}$$

The solution to the momentum balance equation, (under these conditions) is to determine the relation of velocity in a given place and time,

$$u(t, y) = u_0 \cdot \text{erf}\left(\frac{y}{2 \cdot \sqrt{v \cdot t}}\right). \tag{2.101}$$

Based on this relationship, the changes in shear stress in the direction of motion of the vortexes towards the wall in nonstationary conditions were calculated from Newton's equation:

$$\tau_w(t, 0) = \mu \cdot \frac{\partial u(t, 0)}{\partial y}. \tag{2.102}$$

If the average contact time of the vortex and the wall is known as t_{av}, it is possible on this basis to calculate the average value of the shear stress:

$$\overline{\tau_w} = \frac{1}{t_{av}} \int_0^{t_{av}} \tau_w \cdot dt, \tag{2.103}$$

$$\overline{\tau_w} = 2 \cdot u_0 \cdot \rho \cdot \sqrt{\frac{v}{\pi \cdot t_{av}}}. \tag{2.104}$$

On this basis, the dimensionless frequency of vortices may be determined as follows:

$$n^+ = \frac{\sqrt{\dfrac{v}{\pi \cdot t_{av}}}}{2 \cdot u_0}, \tag{2.105}$$

where kinematic viscosity (v) may be determined based on density (ρ) and dynamic viscosity (μ) as follows:

$$v = \frac{\mu}{\rho}. \tag{2.106}$$

Given that the journey of the vortex starts in the fluid bulk, Einstein and Li received a very simple formula which allows determination of the frequency of vortices on the basis of the coefficient of friction (f – Fanning factor):

$$n^+ = 0.392 \cdot f. \tag{2.107}$$

Shaw and Hanratty [55] performed tests using an electrochemical method which enabled their visualization in the form of fine bubbles, and led to the following relationship:

$$n^+ = 0.0022 \cdot Sc^{-0.41}. \tag{2.108}$$

Further experiments by Hanratty showed that the vortices are derived from y = 30, which allowed the calculation of their frequency:

$$n^+ = 4.3 \cdot 10^{-3}. \tag{2.109}$$

2.2.9.4 Definition and physical importance of the Reynolds number

The Reynolds number is a good criterion for determining the nature of flow, whether it is laminar or turbulent, and is defined as the dimensionless invariant, as follows:

$$Re = \frac{l \cdot \rho \cdot u}{\mu}, \tag{2.110}$$

where ρ – fluid density, u – fluid velocity, μ – fluid viscosity, and l – characteristic linear dimension, which is mostly hydraulic diameter. The Reynolds number is one of the criteria for similarity (in this case, hydrodynamic), which are discussed in the section on the theory of similarity. The Reynolds number has its own physical meaning. Generation of vortices is associated with the exchange of momentum (quantity of motion) between neighboring elements in motion and the stationary boundary layer (see Fig. 2.8). Velocity fluctuations are generated in turbulent flow, when the inertial forces cannot be damped by viscous forces.

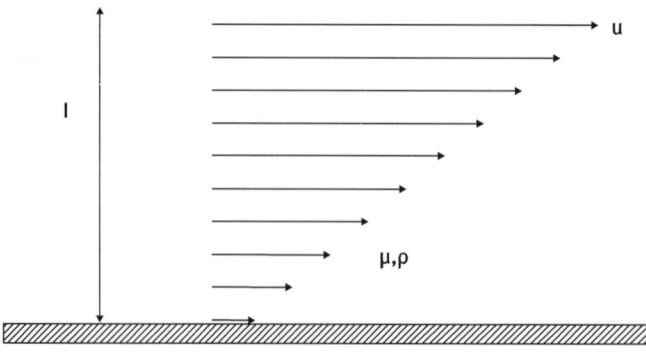

Fig. 2.8. Distribution of fluid velocity near the wall.

The physical significance of the Reynolds number is based on comparison of inertial forces and viscous forces. Inertial forces can be determined from Newton's second law as a change in momentum over time:

$$F_{inertail} = \frac{\Delta\,(m \cdot u)}{\Delta t} = \frac{\Delta\,(V \cdot \rho \cdot u)}{\Delta t} = \frac{\Delta\,(A \cdot l \cdot \rho \cdot u)}{\Delta t} = (A \cdot \rho \cdot u) \cdot \frac{\Delta l}{\Delta t} = A \cdot \rho \cdot u^2. \quad (2.111)$$

Viscous forces are calculated as the product of shear stress (R) and the surface (A), through which the transport of momentum is

$$F_{viscous} = R \cdot A = \mu \cdot \frac{u}{l} \cdot A. \quad (2.112)$$

In cases where the ratio (u/l) is linearly dependent on the shear rate, it is the Newtonian fluid. Therefore, comparison of these two forces leads eventually to the definition of the Reynolds number:

$$\frac{F_{inertial}}{F_{viscous}} = \frac{A \cdot \rho \cdot u^2}{\mu \cdot \frac{u}{l} \cdot A} = \frac{l \cdot \rho \cdot u}{\mu} = Re. \quad (2.113)$$

2.2.10 Residence time distribution

The time of each process has an impact on its economy. Regardless of the type of process, streams in the residence time of the device executing the process should be as short as possible. The kinetics for each process is the basic criterion for the assessment of the possibility of implementation. Therefore, the kinetics of processes is very important from a practical point of view. Generally speaking, in the same way process kinetic is the speed of all processes [56]. Some processes are very thoroughly described in terms of kinetic processes such as chemical, enzymatic, microbial, sorption, crystallization [57] and heat (thermodynamics). The generalization (in the same manner) of the concept of kinetics is very important for all unit processes including diffusion, thermal and mechanical processes. The industrial technologies are combined from different unit processes which form large systems. In any case the slowest process determines the speed of the entire system. Generally speaking, any of the properties changes that occur during the unit processes can be called conversion. Conversion efficiency (η) is the ratio between the useful output of an energy conversion and the input, in energy terms. The useful output may be electric power, mechanical work, or heat (see Fig. 2.10).

In any unit process with the aim of purposeful change properties, conversion efficiency can be: the ratio of the original amount (m_{in}) of the stream, whose property has been changed, to the amount (m_{out}) of the substance that has been converted:

$$\eta_E = \frac{E_{out}}{E_{in}} \quad \text{or} \quad \eta_m = \frac{m_{out}}{m_{in}}. \quad (2.114)$$

Laminar or viscous flow of fluids	Turbulent flow of fluids
Fluid flow is orderly flow lines are parallel, there is no velocity components outside the main flow direction x	There are fluctuations in the speed (vortexes) in the direction y perpendicular to the x direction, but the resultant is equal to zero
 $u_x \neq 0, u_y = 0$	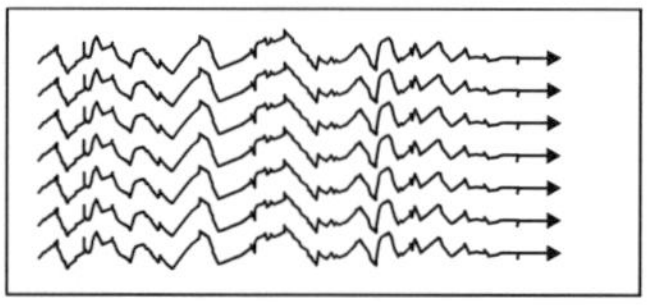 $u_x \neq 0, u_y \neq 0$
Distribution of velocities in pipes describe parabolas $$\frac{u_x(s)}{u_{max}} = \left[1 - \left(\frac{s}{r}\right)^2\right]$$ 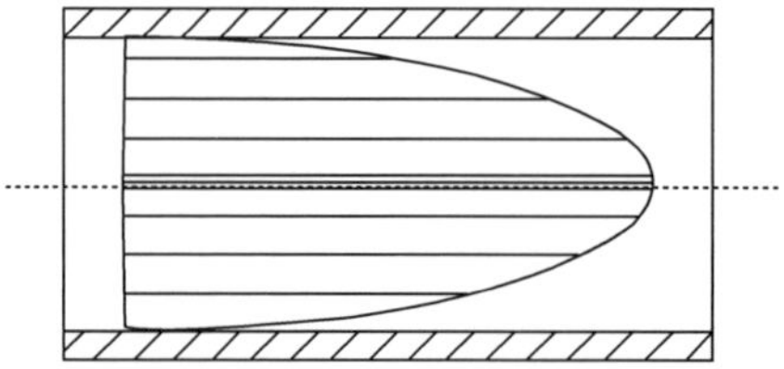	Distribution of velocities function is described compounded $$\frac{u_x(s)}{u_{max}} = \left[1 - \left(\frac{s}{r}\right)^{1/7}\right]$$ 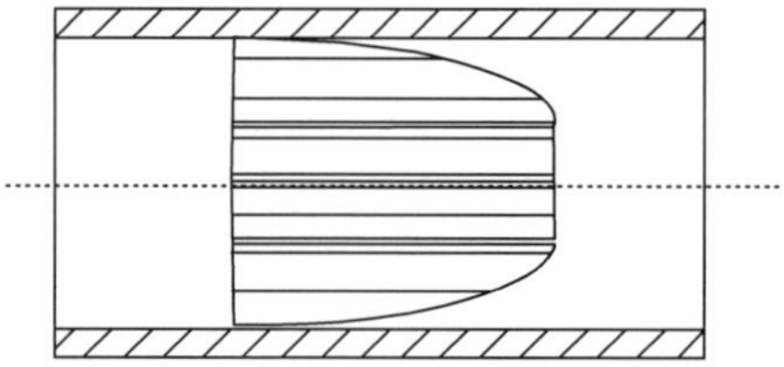
mean velocity $\bar{u} = 0.5 \cdot u^{max}$	mean velocity $\bar{u} = 0.82 \cdot u^{max}$
Shear stresses are proportional to the viscosity of the fluid (Newton's law) μ $$R_y = -\mu \cdot \frac{du_x}{dy}$$	Viscosity apparently increases, on the component, known as eddy viscosity $$R_y = -(\mu + \mu_w) \cdot \frac{du_x}{dy}$$
Mass transport under a concentration gradient follows a Fick's law, by pure diffusion $$N_y = -D \cdot \frac{dC}{dy}$$	Diffusivity apparently increases, on the component, known as eddy diffusivity $$N_y = -(D + D_w) \cdot \frac{dC}{dy}$$

Fig. 2.9. The comparison of laminar and turbulent flow regimes.

It is obvious that the efficiency of conversion of the substance in any process must depend on the duration of the process.

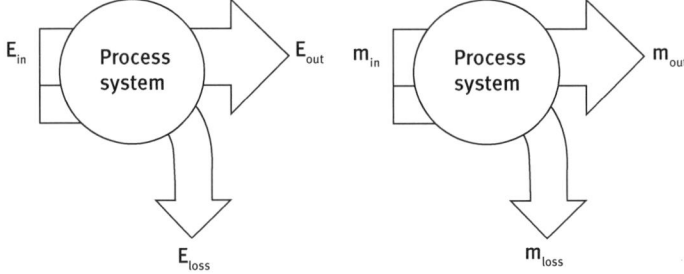

Fig. 2.10. Conversion of mass or heat in processes.

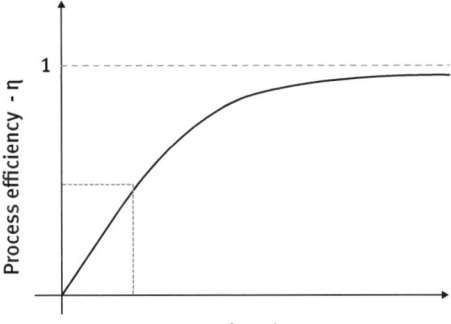

Process duration - t

Fig. 2.11. Process conversion efficiency.

A possible example of such a relationship is the speed of enzymatic processes which are often described by the Michaelis–Menten model. The rate of chemical reactions in respect to the concentration of reactants describes a kinetic equation in each case, with a respective power of the reaction, which is a power (n) at a suitable concentration. The rate of all equilibrium processes where the driving force is reduced during the process (for various reasons) (see Fig. 2.11). We can write the equation of the reaction rate of the form thus:

$$\frac{dC}{dt} = -k \cdot C^n \cdot t. \tag{2.115}$$

For a first order reaction, the solution of this differential equation (with the initial condition $C(0) = C_0$) leads to the exponential formula

$$C(t) = C_0 \cdot e^{-kt}. \tag{2.116}$$

This works for any flow system in stationary conditions, such as a reactor.

The mass balance may be written as

$$v \cdot C_{in} - v \cdot C_{out} - r \cdot V = 0. \tag{2.117}$$

In this mass balance v denotes volumetric flow rate, whereas V denotes volume of the flow reactor, C_{in} and C_{out} are concentrations at the inlet and outlet correspondingly.

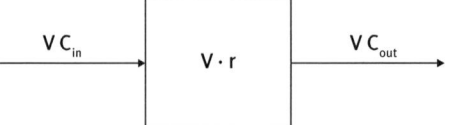

Fig. 2.12. The law of mass conservation (continuity) in the processes.

After rearrangement of this we get the formula for the residence time:

$$\frac{C_{in} - C_{out}}{r} = \frac{V}{v},$$ (2.118)

since, by definition, residence time is the volume divided by volumetric flow rate as follows:

$$\tau = \frac{V}{v}.$$ (2.119)

Thus, as shown in Figure 2.13, the axes represent correspondingly: horizontal – the appropriate concentrations, and vertical – the inverse of the reaction rate. Thus, the area under the curve is the residence time. In the case of the ideal mixing, the concentration immediately drops from the initial concentration to the final value. Then the area of the rectangle determines the mean residence time. In the case of plug flow, the concentration gradually decreases in a continuous manner along the flow path, and in accordance with the changing rate of the reaction, as described by the curve. Then, the area under the curve is a representation of the residence time and the plug flow is smaller, which is more effective.

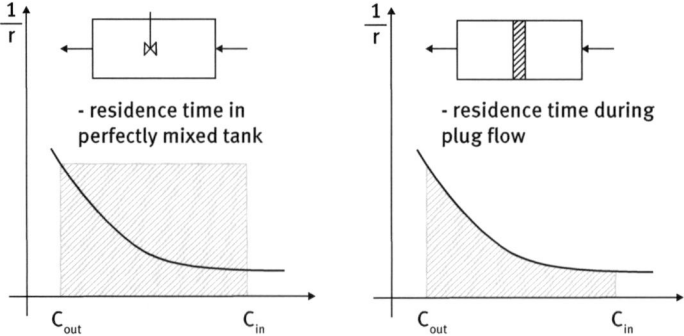

Fig. 2.13. Graphical representation of residence time in the two limiting cases, i.e. perfect mixing and plug flow.

In real cases, we have to determine the degree of mixing which is between extreme cases. This is due to the statistical nature of the process in a flow reactor, where there is a population of fluid elements which are present in the reactor at different times. The most effective way of determining the residence time distribution is a stimulus-response method (see Fig. 2.14).

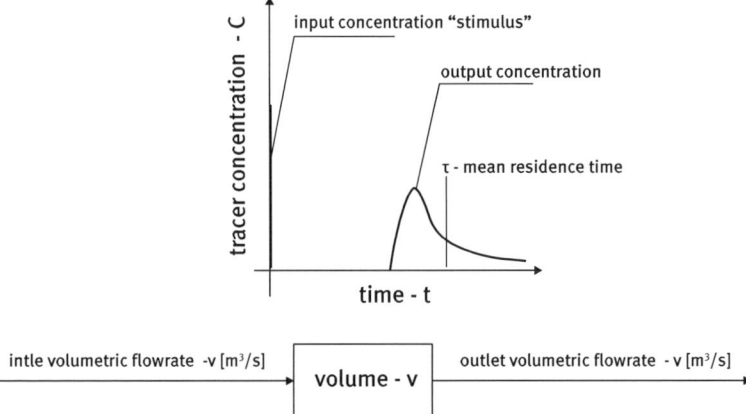

Fig. 2.14. Graphical representation of the stimulus-response method.

The tracer may be any substance which fulfills two conditions in order to be representative of the test fluid. Firstly, the elements of the tracer behave in the same way as the elements of the test fluid. Secondly, the tracer must exhibit some properties enabling observation and monitoring, such as a color, radioactivity or electrical conductivity. The tracer is then introduced to the tank in a certain way and observed at at least two points, i.e. at the inlet and outlet. The most spectacular case is if the tracer is introduced instantly in form of Dirac' delta function. The output signal then shows the residence time distribution directly (see Fig. 2.14).

2.3 Dynamic processes

2.3.1 Flow of fluids

Daniel Bernoulli (1700–1782) was a physicist who solved the problem of motion of stationary ideal fluid and derived a well-known equation, relating potential and dynamic flow energies, known as Bernoulli's equation (1743):

$$\Delta P + \rho \cdot g \cdot \Delta z + \frac{\rho \cdot \Delta u^2}{2} = \text{const.} \tag{2.120}$$

Although the Bernoulli equation is based on the unrealistic assumption that fluid is inviscid, it explains many phenomena and the behavior of many devices used widely in process engineering, such as the formation of a vacuum, i.e. negative pressure in jets, ejector action, propellers, wings and many others.

In 1822 Claude-Louis Navier and George Gabriel Stokes, described the motion of viscous fluids known as the Navier-Stokes equation. This is the basis for modern fluid mechanics, and makes the use of computational fluid dynamics (CFD) possible, simulation and visualization of flows and processes:

$$\rho \cdot \frac{du_x}{dt} + \frac{\partial P_x}{\partial x} - \mu \cdot \left[\frac{\partial^2 u_x}{\partial y^2} + \frac{\partial^2 u_x}{\partial z^2} \right] + \rho \cdot g_x = 0. \tag{2.121}$$

The main cost incurred in different technologies (involving flowing fluids) is the cost of pumping from place to place, i.e. between tanks and other parts of the system. These costs are related to energy losses which occur during the flow of real fluids through pipes and equipment. These losses arise mainly from the chaotic motion of the fluid elements, and thus conversion of kinetic energy into heat, which is returned to the environment. Thus, differential energy balance in fluid flow can be written as follows:

$$u \cdot du + g \cdot dz + v \cdot dP + \delta W + \delta F = 0. \tag{2.122}$$

If the fluid velocity du = 0 in the energy balance is assumed to be constant, (dz = 0), and the failure to perform external work (dW = 0), then the energy loss (F) reveals itself in the form of a pressure drop along the observed section of the flow channel

$$\Delta P = \rho \cdot \Delta F. \tag{2.123}$$

The formula is obtained by multiplying both sides of the equation by the flow rate (V), which shows that the amount of energy lost during the fluid flow is equivalent to the pressure drop (losses), along the flow channel. Flow resistance can occur over a length of pipe and locally:

$$\Delta E_{str} = \Delta P_{str} \cdot V. \tag{2.124}$$

It should be noted that the major cost associated with the flow is related to the energy loss, which results from hydraulic resistance and flow rate. This is a very important equation, which allows the calculation of the power of devices to pump liquids. Hydraulic resistance in pipes during fluid flow is expressed as pressure difference in any two cross-sections. Pressure loss occurs not only as a result of the friction on the wall of the flow channel. Friction loss in a horizontal pipe without changing the speed between sections 1 and 2 can be expressed by the general energy balance using the formula

$$F_{1-2} = -v \cdot (P_1 - P_2). \tag{2.125}$$

The pressure drop (ΔP) is usually experimentally determined, on the basis of flow in pipes (see Fig. 2.15). On the basis of these experiments, the most important variables of "hydraulic resistance" are determined as follows:

$$\Delta P = \Phi(u, \rho, \mu, l, d_h, e), \tag{2.126}$$

where
u – flow velocity,
ρ – fluid density,
μ – fluid viscosity,

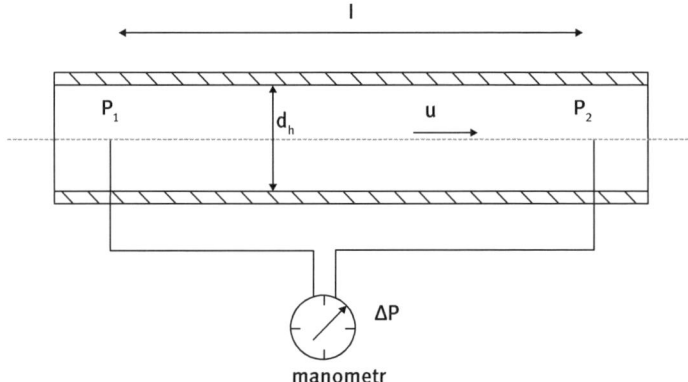

Fig. 2.15. Pressure drop during the flow of a fluid in a straight section of the tube.

l – the length of the path where the pressure drop (flow resistance) was measured

d_h – "hydraulic diameter" of the flow channel,

e – roughness of the flow channel wall.

The dimensionless relationship between values which are relevant to flow resistance can be determined using dimensional analysis (Buckingham Theorem Π):

$$\frac{\Delta P}{\rho \cdot u^2} = f\left(\frac{u \cdot d_h \cdot \rho}{\mu}, \frac{l}{d_h}, \frac{e}{d_h}\right). \tag{2.127}$$

Stokes first introduced the dimensionless criterion of flow regime which was further developed by Reynolds. Empirical relationships describing hydraulic resistance expressed with regard to the number of Euler and Reynolds criterion are as follows:

$$Eu = f\left(Re, \frac{l}{d_h}, \frac{e}{d_h}\right). \tag{2.128}$$

Research conducted by Panel and Stanton [59] allowed determination of this relationship (see identification models). It was found that the hydraulic resistance of the flow depends on the nature or degree of turbulence described by the Reynolds Number. It was found that the change in value of the Reynolds Number induces not only quantitative, but also qualitative changes in formulas determining hydraulic resistance. Studies show that during fluid flow in pipes of different roughness the coefficient of hydraulic resistance (friction coefficient) varies depending on the Reynolds number. The graph in Figure 2.16 shows quite different behavior of fluid in laminar and turbulent flow.

An example of such empirical relationships describing hydrodynamic flow resistance is the Darcy–Weissbach formula:

$$\Delta P = \lambda \cdot \frac{l}{d_h} \cdot \frac{\rho \cdot u^2}{2}, \tag{2.129}$$

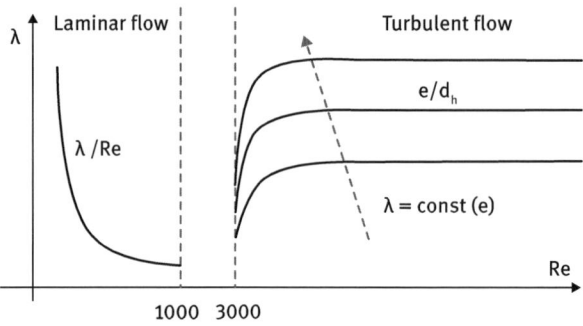

Fig. 2.16. The effect of turbulence on friction coefficient.

where l – length of flow channel, d_h – hydraulic diameter, ρ – fluid density, and u – mean velocity in the section of the flow channel. In the laminar flow region, the λ friction coefficient is independent of the roughness and varies inversely with the value of the Reynolds number according to the formula

$$\lambda = \frac{C}{Re}.$$

(2.130)

The constant (C) is characterized by the shape of the cross section of the flow channel as shown in Tab. 2.5.

Table 2.5. Hydraulic diameter and constant (C) in the formula.

Configuration	Hydraulic diameter	Constant (C)
Circle of diameter (d)	d	64
Square of a	a	57
Equilateral triangle of side a	0.58a	53
Ring with width a	2a	96

We obtain the relation of the pressure drop in the pipeline for laminar flow, based on the Darcy–Weissbach formula, as shown in the following transformation:

$$\Delta P = C\frac{\mu \cdot l \cdot u}{2 \cdot d_h^2} \quad \text{for} \quad \lambda = \frac{C}{Re} \quad \text{i.e. laminar flow,}$$

(2.131)

$$\Delta P = \lambda \cdot \frac{1}{d_h} \cdot \frac{\rho \cdot u^2}{2} = \frac{C}{Re} \cdot \frac{1}{d_h} \cdot \frac{\rho \cdot u^2}{2} = \frac{C \cdot \mu}{u \cdot d_h \cdot \rho} \cdot \frac{1}{d_h} \cdot \frac{\rho \cdot u^2}{2} = C\frac{\mu \cdot l \cdot u}{2 \cdot d_h^2}.$$

(2.132)

This relationship shows that the viscosity of the liquid is crucial in laminar flow but not for density at all. In this case, the roughness of the walls is negligible too. Moreover, the influence of the velocity (linear relationship) on the pressure drop is lower. In contrast to the previous case, during turbulent flow the effect of density appears while viscosity impact disappears. The influence of the velocity (velocity second power) is stronger at the turbulenct flow (see Fig. 2.17), and the influence of wall roughness is also revealed by the pressure drop

$$\Delta P = \lambda \cdot \frac{1}{d_h} \cdot \frac{\rho \cdot u^2}{2} \quad \text{for} \quad \lambda = \text{const}(e/d_n) \quad \text{i.e. turbulent flow.} \tag{2.133}$$

The disappearance of the roughness effect in laminar flow can be explained by the fact that the roughness of the walls is filled with "stagnant" fluid (see Fig. 2.17).

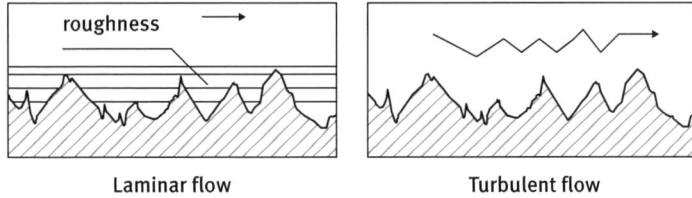

Laminar flow   Turbulent flow

Fig. 2.17. Influence of roughness of the wall on the flow.

Darcy (1803–1858) and Weissbach (1806–1871) provided the basis for the well-known Darcy-Weissbach equation, which is a purely empirical equation valid for turbulent flow, where the friction factor λ = const and dependent only on the relative roughness (e/d). For transient flow regime between laminar and turbulent flow ($3 \cdot 10^3 < \text{Re} < 10^5$) the friction coefficient can be calculated as [59]

$$\lambda = \frac{0.3164}{\sqrt[4]{\text{Re}}}. \tag{2.134}$$

For the same transient flow regime, when roughness is to be taken into account the Coolebrook–White equation may be used:

$$\frac{1}{\sqrt{\lambda}} = -2\lg\left(\frac{2.5}{\text{Re} \cdot \sqrt{\lambda}} + \frac{e}{3.7 \cdot d}\right), \tag{2.135}$$

or the purely empirical correlations such as

$$\lambda = 0.0032 + \frac{0.0221}{\text{Re}^{0.237}}. \tag{2.136}$$

For laminar flow regime, when the flow rate is introduced instead of velocity the so called Hagen–Poiseuille equation for calculation of pressure drop is obtained during viscous flow:

$$\Delta P = \frac{128 \cdot \mu \cdot l \cdot V}{\pi \cdot d^4}, \tag{2.137}$$

where the flow rate is related to flow velocity by means of the stream continuity law which is the kind of conservation law for volumetric flow of incompressible fluids:

$$V = u \cdot F = u \cdot \frac{\pi \cdot d_h^2}{4}. \tag{2.138}$$

It should be noted that in this way two correlations for two different areas of flow have been received, one for turbulent and another for laminar flow. The type of flow is determined by the value of the Reynolds number. This discontinuity in the correlation between flow resistance and flow rate is also found in other processes such as heat and mass transport, where the criterion of the turbulence is the Reynolds number. Then based on the similarity theory initiated by Newton, many physicists have studied the so-called similarity invariants for different phenomena. They introduced "criterial equations" and "criterial numbers", which are so important to Process Engineering. The names of these numbers were taken from the names of famous scientists and they are capitalized, for example the number Nu (Nusselt), Sh (Sherwood), Sc (Schmidt), Pe (Peclet), Eu (Euler), Gr (Grashoff) etc. These "odd" numbers initiated the detachment of process engineering from pure physics. Criterial equations allow for a quantitative description of the phenomena which cannot be described strictly physically or using exact mathematical models based on the equations of balance of mass, momentum and energy. Whenever necessary, the empirical approach is used to identify the criterial equations based on the experiment. Identification of the empirical model is based on the calculation of the model parameters based on regression analysis. Process engineering was a practical field from the beginning and its base is made up of representatives of various fields. In the implementation of conventional unit processes turbulent flow is more important than laminar because of the greater intensity of the process. However, strictly physical solutions of equations and models then have very little credibility. Therefore there has long been a big discrepancy between theoretical descriptions and their usefulness in practice.

Similarity theory is important for process engineering because most of the practical processes carry beyond the strictly physical area. In concluding this section, it is clear that process engineering has served since ancient times, solving real problems with technology. The largest of these problems were viscosity, compressibility and flow type. Therefore, process engineering" has diverged from physics wherever it was possible to find instant solutions based on experiments rather than on theoretical considerations.

2.3.2 Fluid flow through a fixed bed

This flow covers the following processes and devices: filtration, dryers, and furnaces. The same occurs in reactors, filled or packed columns during distillation, rectification, absorption, and adsorption and desorption, extraction and membranes. After crossing the sedimentation velocity of particles, the layer starts to move, if it is not immobilized, which is called fluidization, or pneumatic transport.

The flow rate of fluid (v_0) through the porous layer is proportional to the pressure difference on both sides of the layer and inversely proportional to the viscosity of the

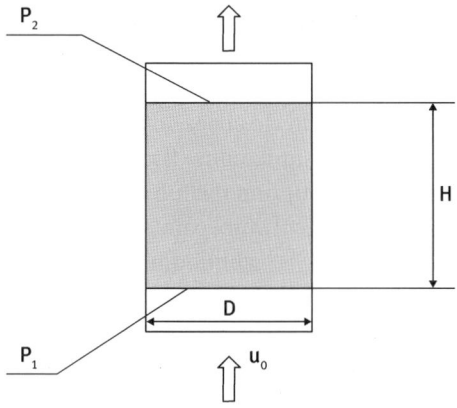

Fig. 2.18. Fluid flow through the porous fixed bed.

fluid and the thickness of the layer (see Fig. 2.18):

$$v_0 = \frac{V}{A} = K \cdot \frac{\Delta P}{\mu \cdot 1}.$$ (2.139)

The flow takes place at irregular capillaries. However, to simplify calculation they are assumed to be regular and cylindrical and to have average values. The permeability of the bed can be calculated from the properties of the grains making up the bed, such as porosity and surface area. In order to determine the permeability of the bed under conditions of laminar flow the Darcy-Weissbach equation may be used:

$$\Delta P = \lambda \cdot \left(\frac{1}{d_h}\right) \cdot \frac{\rho \cdot u^2}{2} \quad \text{where} \quad \lambda = \frac{64}{Re}.$$ (2.140)

After taking the friction coefficient based on the Reynolds number into account we get the formula which describes the pressure drop in laminar flow:

$$\Delta P = \frac{64 \cdot \mu}{u \cdot d_h \cdot \rho} \cdot \left(\frac{1}{d_h}\right) \cdot \frac{\rho \cdot u^2}{2} = \frac{32 \cdot \mu \cdot 1 \cdot u}{d_h^2}.$$ (2.141)

The impact of parameters such as flow velocity (of the first power) of the diameter (of the minus second power), furthermore impact of the density disappears, however, the viscosity effect appears, which is quite different to that in turbulent flow. After transforming this equation we obtain an important formula which shows that permeability of the bed is strongly dependent on the diameter of capillaries in the second power:

$$u = \left(\frac{d_h^2}{32}\right) \frac{\Delta P}{\mu \cdot 1}.$$ (2.142)

It can be assumed that the mean diameter of the capillary is equal to the hydraulic diameter, based on the known parameters characteristic of the layer, such as specific surface area, porosity and pore tortuosity. Starting from the definition of the hydraulic diameter, for any flow channel as 4× the cross-sectional area, divided by the wetted perimeter:

$$d_h = \frac{4 \cdot A}{O} = \frac{4 \times \text{cross-flow area}}{\text{``wetted perimeter''}}. \tag{2.143}$$

It is then convenient to multiply the numerator and denominator by the capillary length (l_k) to obtain a size which can easily be inserted:

$$d_k = d_h = \frac{4 \cdot A \cdot l_k}{O \cdot l_k} = \frac{4 \cdot V_k}{A_k} = \frac{4 \cdot V \cdot \epsilon}{a \cdot V \cdot (1 - \epsilon)} = \frac{4 \cdot \epsilon}{a \cdot (1 - \epsilon)}. \tag{2.144}$$

We take into account that the actual speed of the fluid flowing in the capillary depends on the speed calculated for the entire cross section of the flow channel:

$$u_0 = \epsilon \cdot u. \tag{2.145}$$

Besides, the length of the capillary is greater than the height of the layer, due to the pore tortuosity:

$$l_c = \tau \cdot H. \tag{2.146}$$

Finally, we obtain the formula determining volume stream of fluid flowing through a layer under the influence of the pressure difference:

$$u_0 = \left[\frac{\epsilon^3}{2 \cdot a^2 \cdot (1 - \epsilon)^2 \cdot \tau} \right] \cdot \frac{\Delta P}{\mu \cdot H}. \tag{2.147}$$

Comparing the contents of the square bracket with the Hagen–Poiseuille formula, the permeability (K) of the layer can be determined, which is characterized by parameters such as: the porosity of the layer, tortuosity of pores and the specific surface area. Note that the formula for permeability is true only after the assumption of laminar flow through a fluid bed. In other cases, empirical cases are used, approximations such as Ergun or Black-Kozeny, to be checked in each individual case. In general, the permeability of each layer depends strongly on the hydraulic diameter squared. This should be taken into account by decreasing the solid particles dimension to increase the surface ($a \cong 1/d$), because of the power losses $E \cong 1/d^2$ (and relevant costs) increased much faster than benefits from surface increase.

2.3.3 Rising or falling of particles of one phase in the second phase

The phenomena of sedimentation of solid particles or rising of liquid droplets are present in such processes as: sedimentation, decantation, hydraulic classification, fluidization, drying (especially spray) and extraction. Instead of gravitational forces the centrifugal forces acting on the dispersive elements may be involved in such equipment as centrifuges, cyclones, hydrocyclones can also be included in this group. Also, the movement of gas bubbles in a liquid is subject to the same rules. Bubbling is widely

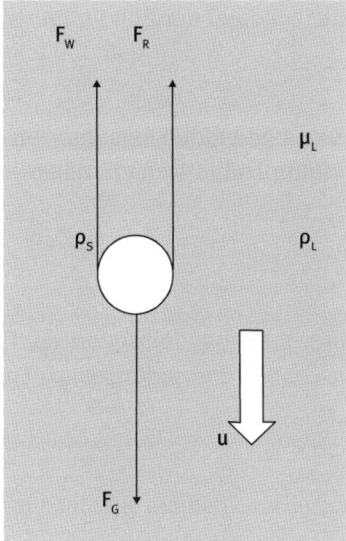

Fig. 2.19. Undistorted fall of the single particle in fluid.

used in the processes of absorption and rectification, distillation, and reactors and bioreactors. The same physical mechanism applies to all multi-phase systems such as suspensions of solid, liquid-liquid emulsions, and the bubbles.

The rate of falling or raising, (ascent or descent depending on the density of the fluid and particles) shows the same balance of forces acting on the elements of the dispersion, gas bubbles, liquid droplets or solid grains. In steady state conditions, all three forces must be balanced against each other (see Fig. 2.19):

$$F_G - F_R - F_W = 0, \tag{2.148}$$

where the gravitational force:

$$F_G = \frac{\pi \cdot d^3}{6} \cdot \rho_S \cdot g, \tag{2.149}$$

Archimedean buoyancy force:

$$F_W = \frac{-\pi \cdot d^3}{6} \cdot \rho_L \cdot g, \tag{2.150}$$

the resistance of fluid flow around a spherical particle:

$$F_W = \lambda \cdot A_0 \cdot \frac{\rho \cdot u^2}{2}; \tag{2.151}$$

the friction coefficient of spherical particle in laminar flow:

$$\lambda = \frac{24}{Re}. \tag{2.152}$$

Under these conditions, the resistance force flow around the model can be expressed as

$$F_R = 3\pi \cdot \mu \cdot d. \tag{2.153}$$

After adding up all the forces and after simplification we obtain the Stokes equation

$$u = \frac{d^2 \cdot (\rho_S - \rho_L) \cdot g}{18 \cdot \mu}. \tag{2.154}$$

Sedimentation velocity, which is expressed by the Stokes equation, relates to a single grain. In fact, we are dealing with a huge numbers of particles falling at the same time. Deviations from Stokes' formula are caused by interactions between solid particles, including pseudo-hydrostatic effects on increasing the buoyancy, due to the apparent increase in density of the suspension and increased suspension viscosity.

2.3.4 Bubble flow

"Barbotage", or bubble flow, is used in a number of unit processes between liquid and gas phases, such as rectification, distillation, absorption etc. Contact between the phases is obtained during the movement of the gas bubbles through the liquid layer during boiling, in reactors, in bubble columns, (plate, or packed) and in mechanical scrubbers (see Fig. 2.20). The surface and the contact time depend on the size of the bubbles.

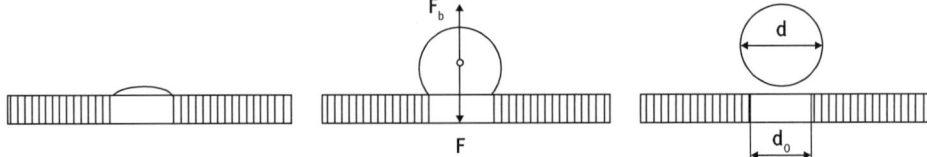

Fig. 2.20. The formation of bubbles during barbotage.

We are going to calculate the specific surface created by the bubbles (a) and the time of contact between the two phases. The time it takes for the bubble takes to rise through the liquid layer can be calculated from Stokes' formula. Both of these values depend on the diameter of bubbles (d), which is a key parameter. It is assumed that the bubble stops growing and detaches when the buoyant force (taking its weight into account) balances the surface tension force on the edge of the opening. The specific surface is defined as:

$$a = \frac{\text{total surface of bubbles}}{\text{foam volume}} = \frac{A_p}{V}. \tag{2.155}$$

Assuming uniform bubble diameter (d), we calculate the total area by the number of total bubbles:

$$A_p = n \cdot \pi \cdot d^2. \tag{2.156}$$

The number of bubbles in a foam volume (V) in which part (hold-up) the gas ? (g) can be determined with the formula

$$n = \frac{V_g}{V_p} = \frac{\Phi_g \cdot V}{\dfrac{\pi \cdot d^3}{6}}. \tag{2.157}$$

Thus, the specific surface area ultimately depends on the diameter of the foam bubbles in inverse proportion:

$$a = \frac{\dfrac{6 \cdot \Phi_g \cdot V}{\pi \cdot d^3} \cdot \pi \cdot d^2}{V} = \frac{6 \cdot \Phi_g}{d}. \tag{2.158}$$

Calculate the diameter of the bubble from the comparison of two forces on the bubble at the time of detachment:

$$F_w = F_\sigma. \tag{2.159}$$

The buoyancy force, taking into account the weight of the bubble, is as follows:

$$F_w = \frac{\pi \cdot d^3}{6} \cdot (\rho_L - \rho_G) \cdot g. \tag{2.160}$$

The force of the surface tension of the bubble still holds the edge of the hole with a diameter (d_0):

$$F_\sigma = \pi \cdot d_0 \cdot \sigma. \tag{2.161}$$

From the comparison of the forces acting on the gas bubble at the time of its detachment from the edge of the hole, its diameter may be calculated as

$$d = \sqrt[3]{\frac{6 \cdot d_0 \cdot \sigma}{(\rho_L - \rho_G.) \cdot g}}. \tag{2.162}$$

So the time of the bubble passage up through the liquid layer of height (H), can be calculated using the Stokes equation:

$$t = \frac{H}{u} = \frac{18 \cdot \mu \cdot H \cdot}{d^2 \cdot (\rho_L - \rho_G) \cdot g}. \tag{2.163}$$

2.4 Heat transfer processes

2.4.1 Basics of heat transfer

Heat transport is frequently accompanied by a number of natural phenomena, as well as industrial processes and operations. Heat transport is very important because of the high energy costs associated with it. But almost all types of heat transfer can be demonstrated using the everyday example of an ordinary pot of boiling liquid (see Fig. 2.21).

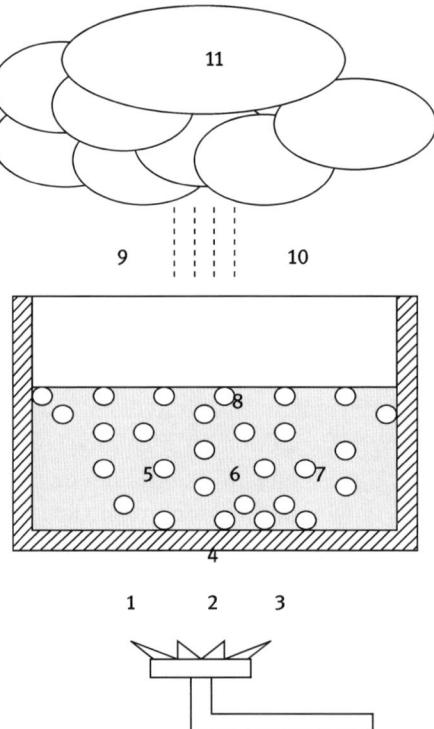

Fig. 2.21. Different methods of heat transfer in a pan of boiling liquid: (1) radiation (flame); (2) natural convection heat (in air); (3) heat conduction (in air); (4) heat conduction (through the vessel wall); (5) boiling (in the volume of fluid); (6) conduction (through the layer of liquid); (7) natural convection heat (in the liquid layer); (8) natural convection heat (steam and bubbles); (9) natural convection heat (steam); (10) heat conduction (by air).

Heat conduction is performed through the medium (i.e. material), but without moving the material. Thermal energy is propagated in the material through vibrations and collisions of molecules. Within the crystalline structure of some materials (such as metals), the greater portion of heat conduction takes place throughout the free electrons, which explains the observed interrelation between electrical and thermal conductivity. Thermal conductivity is a mechanism which occurs in all states of matter, i.e., gases, liquids and solids. Convection of heat (general) – this is a medium heat transport caused by the movement of fluids. The forced convection occurs when motion is forced by the application of mechanical energy supplied during mixing or pumping of fluid. Natural convection takes place when fluid movement occurs spontaneously as a result of thermal expansion and the associated change in fluid density, accompanied by the formation of buoyancy force. Radiation of heat is a type of transport during the propagation of electromagnetic waves. Thus, this mode of transport does not require any media in any form. Two other modes of transmission of heat such as condensation and evaporation are integrally combined with simultaneous mass transport.

2.4.2 Heat conduction through the flat plate in steady-state conditions

Heat may be transferred in the field of the temperature, only when a temperature gradient exists, which is a driving force of heat transport. The heat then flows in the opposite direction to the gradient. The value of heat flux in the direction x is expressed by Fourier's phenomenological equation. It should be noted that the heat flux and gradient of temperature are the vectors, since they have a value and direction (see Fig. 2.22):

$$\vec{q} = -\lambda \cdot \frac{\overrightarrow{dT}}{dx} \ [J/m^2 s] \, . \tag{2.164}$$

The heat flux

$$\vec{q} \ \left[\frac{J}{m^2 s} \right] \tag{2.165}$$

is a measure of the intensity of the heat flow, which refers to a unit of area A $[m^2]$, and the unit of time t $[s]$.

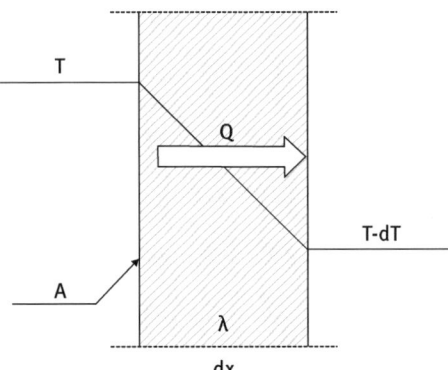

Fig. 2.22. Heat conduction through the flat plate

The rate of flow of the heat flowing through the total surface area is the product of the stream and the surface:

$$Q = q \cdot A \ \left[\frac{J}{s} = W \right] \, . \tag{2.166}$$

In practice, heat transfer is considered on a section of length $l = x_2 - x_1$, and therefore the Fourier equation must be integrated to obtain the numerical value of the heat flux q:

$$q = \lambda \cdot \frac{T_2 - T_1}{x_2 - x_1}.$$

(2.167)

Thus, the heat flow rate through the barrier is calculated on the basis of the so called "integral formula":

$$Q = \frac{\lambda}{l} \cdot A \cdot (T_1 - T_2).$$

(2.168)

2.4.3 Heat conduction through the multi-layer plate in steady-state conditions

Consider the multi-layer barrier in which each layer has a different thermal conductivity (λ) and different thicknesses (x) (see Fig. 2.23). We need to find a formula which describes the equivalent conductivity for the entire cascade (sandwich) through which heat flows. The intensity of the heat flow through each layer is the same:

$$Q = Q_1 = Q_2 = \cdots = Q_n.$$

(2.169)

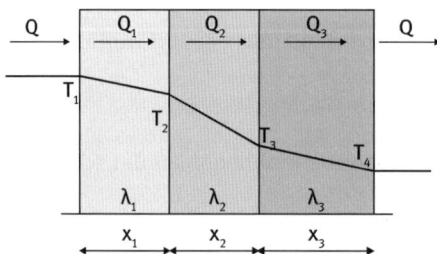

Fig. 2.23. Heat conduction through the multi-layer plate.

It is possible to write an equation for the heat flow rate for any i-th layer, which can then determine the appropriate temperature difference in the layer:

$$Q_i = \frac{\lambda_i}{x_i} \cdot A \cdot (T_i - T_{i+1}) \Rightarrow \Delta T_i = T_i - T_{i+1} = \frac{Q}{A} \cdot \frac{x_i}{\lambda_i} \quad \text{for} \quad i = 1, 2, \ldots, n.$$

(2.170)

By adding up the temperature difference across all n layers, we obtain the maximum difference in temperature between the first and last n-th layer as follows:

$$T_1 - T_n = \frac{Q}{A} \cdot \sum_{i=1}^{n} \frac{x_i}{\lambda_i}.$$

(2.171)

After rearranging, we get

$$Q = \frac{\lambda_z}{g} \cdot A \cdot (T_n - T_1),$$

(2.172)

where the equivalent coefficient of thermal conductivity can be calculated, as follows:

$$\lambda_z = \frac{\sum\limits_{i=1}^{n} x_i}{\sum\limits_{i=1}^{n} \frac{x_i}{\lambda_i}} = \frac{g}{\sum\limits_{i=1}^{n} \frac{x_i}{\lambda_i}}.$$

(2.173)

2.4.4 Heat conduction through a cylinder

As in the case of the multi-layer barrier, we must be able to calculate the heat entering the inner wall of the thick-walled cylinder (see Fig. 2.24). The problem in this case is the permanent changing of the surface along which the heat flows.

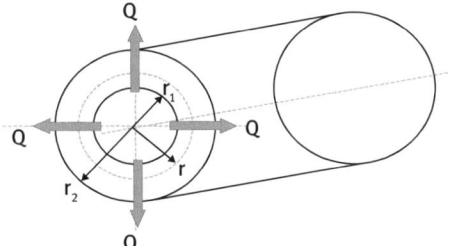

Fig. 2.24. Heat conduction through the wall of the cylinder.

The heat flow rate must in this case be represented by a differential formula:

$$Q = -\lambda \cdot A\,(r) \cdot \frac{dT}{dr}.$$

(2.174)

We now need to express the value of the surface, depending on the distance from the axis (r):

$$A\,(r) = 2\pi \cdot r \cdot l.$$

(2.175)

Reassuming the heat rate may be expressed by means of the following differential equation:

$$Q = -\lambda \cdot 2 \cdot \pi \cdot r \cdot l \cdot \frac{dT}{dr}.$$

(2.176)

This is then integrated over the range of space:

$$Q \cdot \int_{r_1}^{r_2} \frac{dr}{r} = -2\pi \cdot l \cdot \lambda \int_{T_1}^{T_2} dT.$$

(2.177)

The formula for the calculation of the heat flow rate can be represented as follows:

$$Q = \frac{\lambda}{r_2 - r_1} \cdot 2\pi \cdot r_m \cdot l \cdot (T_1 - T_2),$$

(2.178)

where the radius is calculated as the logarithmic mean of the inner and outer radius of the cylinder:

$$r_m = \frac{r_2 - r_1}{\ln\left(\frac{r_2}{r_1}\right)}. \tag{2.179}$$

2.4.5 Heat conduction through a sphere

Heat flows from the interior of a sphere radially in all directions (see Fig. 2.25). In this case (as in the preceding), the heat exchange surface varies along the flow path of heat.

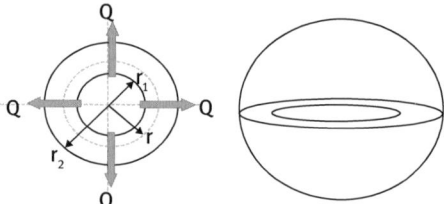

Fig. 2.25. Heat conduction through the wall of the sphere.

The heat flow rate is given by the differential equation of the form

$$Q = -\lambda \cdot A(r) \cdot \frac{dT}{dr}, \tag{2.180}$$

where the radius dependent surface of the heat exchange is

$$A(r) = 4\pi \cdot r^2. \tag{2.181}$$

After substituting we can obtain the differential equation

$$Q = -\lambda \cdot 4\pi \cdot r^2 \cdot \frac{dT}{dr}, \tag{2.182}$$

which gives the form

$$Q \cdot \int_{r_1}^{r_2} \frac{dr}{r^2} = -4\pi \cdot \lambda \cdot \int_{T_1}^{T_2} dT. \tag{2.183}$$

After integration we finally arrive at the form

$$Q = \frac{\lambda}{r_2 - r_1} \cdot 4\pi \cdot r_1 \cdot r_2 \cdot (T_1 - T_2). \tag{2.184}$$

2.4.6 Heat convection in one phase during fluid flow

Convection is the transfer of thermal energy from one place to another by the movement of fluids or gases (see Fig. 2.26). Although often discussed as a distinct method of heat transfer, convection describes the combined effects of conduction and fluid flow or mass exchange. That is why the convection of heat (convection or mass) cannot be described with purely physical models. Practically, however, convection is the most useful mechanism of heat transfer between different fluids (gases and liquids) because it concerns the so-called heat exchangers for use in various types of engineering. It is known for sure that when convection heat is transferred from a warmer fluid to a cooler fluid, (as in heat conductivity), the heat flow rate is proportional to the temperature difference between the two centers and their contact surfaces:

$$Q = \alpha \cdot A \cdot \Delta T. \tag{2.185}$$

Mostly, the contact area relates to the wall of the heat exchanger (solid) and fluid (gas or liquid). Moreover, this fluid is mostly in motion (flow). Between the wall and the liquid is a transition zone (the so-called boundary layer), the thickness of which it is difficult to measure, or even impossible to measure experimentally. We only know that the thickness of the boundary layers changes, depending on the flow rate.

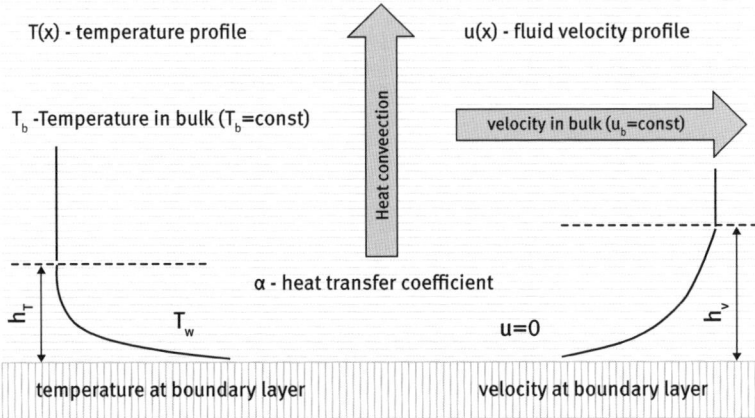

Fig. 2.26. Heat convection during fluid flow.

It is possible that as many as three different boundary layers exist independently, i.e. the laminar boundary layer, the temperature boundary layer (which is associated with heat transport) and the concentration boundary layer (which is related to mass transport). Therefore, the transfer coefficients are determined by using the so-called "black box"method, which is based on an approximation of purely empirical relations between the heat transfer coefficients and the significant independent variables:

$$\alpha \left(u, d_h, \mu, \rho, \lambda, \dots\right). \tag{2.186}$$

The values of the heat transfer coefficient are determined indirectly, by measuring the amount of heat at a specific time and relating it to the wall surface at a given temperature difference between the wall and the bulk of the fluid:

$$\alpha = \frac{Q}{A \cdot \Delta T} \left[\frac{W}{m^2 \cdot °K} \right]. \tag{2.187}$$

A convenient way of solving such problems is to use engineering dimensional analysis combined with the theory of similarity. The mass transfer coefficients in both phases must be determined based on the adequate criterial Nusselt numbers:

$$Nu = \frac{\alpha \cdot d_h}{\lambda}. \tag{2.188}$$

The Nusselt numbers for both phases should be calculated from the criterial equation taking the corresponding flow regime into account. For the laminar flow the general Leveque formula may be applied:

$$Nu = 1.62 \cdot \left(Re \cdot Pr \cdot \frac{d}{l} \right)^{0.33}, \tag{2.189}$$

or the similar but more accurate Sieder and Tate formula [58] which takes viscosity changes accompanying the temperature changes into account. This formula is as follows:

$$Nu = 1.86 \cdot \left(Re \cdot Pr \cdot \frac{d}{l} \right)^{0.33} \cdot \left(\frac{\mu_b}{\mu_w} \right)^{0.14}. \tag{2.190}$$

For turbulent flow the so-called Dittus and Boelter formula [59]:

$$Nu = 0.023 \cdot Re^{0.8} \overset{n}{Pr},$$

where $n = 0.4$ for heating and $n = 0.3$ for cooling. The dimensionless Reynolds number is

$$Re = \frac{u \cdot l \cdot \rho}{\mu}, \tag{2.191}$$

and the Prandtl number is

$$Pr = \frac{c_p \cdot \mu}{\lambda}. \tag{2.192}$$

It must be acknowledged that the formulas presented here are the best known, and for the specific types of heat exchangers. Data may be selected from the copious literature which best fits the given configuration of the heat exchanger. The temperature difference causes heat flow, but only where the fluids are stationary. When fluids flow in heat exchangers the temperature difference changes and the direction of fluid flow must then be taken into account when calculating the average temperature difference. In heat exchangers two fluids are contacted by the diaphragm, flowing in opposite directions (i.e. counter-current flow) or in the same direction (i.e. co-current flow). We use co-current flow when we want to heat or cool a fluid quickly. We use counter-current flow when we want to keep the uniform temperature difference between the fluids, which is the "driving force" of heat transport. In any case the average logarithmic temperature difference between the fluids should be determined as follows:

$$\Delta T_m = \frac{\Delta T_2 - \Delta T_1}{\ln \frac{\Delta T_2}{\Delta T_1}}. \tag{2.193}$$

2.4.7 Heat radiation

Heat radiation requires no medium because it is the transport of energy in the form of electromagnetic waves. Heat radiation is universal because all objects with a temperature above absolute zero radiate in all wavelengths. The total amount of energy that each black body emits can be calculated from the Boltzmann law:

$$q = \sigma \cdot T^4, \tag{2.194}$$

where the Boltzmann constant is

$$\sigma = 5.67 \cdot 10^{-8} \frac{W}{m^2 \cdot {}^\circ K^4}. \tag{2.195}$$

The black body is an object which absorbs the total amount of energy that falls on it. The radiation energy which reaches the surface of the real body can be divided into four parts (see Fig. 2.27): (1) the part which can be absorbed and converted into heat; (2) some of the energy conducted penetrates the body, as it is partially transparent to radiation; (3) some of the energy is reflected from the surface body (partly mirrored); and finally (4) a certain amount of energy will always be radiated in accordance with the Boltzmann law due to its emissivity.

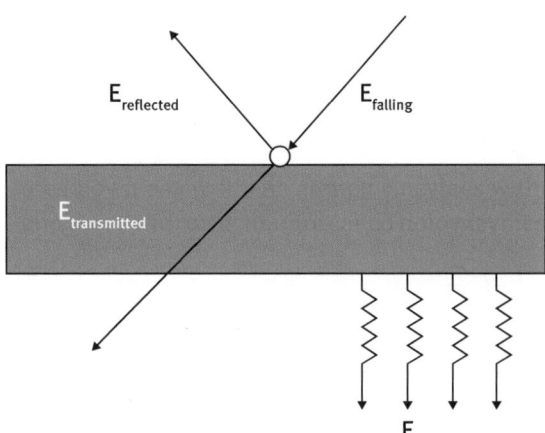

Fig. 2.27. Fate of energy radiation which falls on a surface.

Energy falling on the body's surface:

$$E_{falling} = J \cdot A \cdot t, \tag{2.196}$$

J – the intensity of the radiation source [$J/(m^2 s = W/m^2$)];
A – surface area [m^2];
t – time of exposition to radiation [s].

Energy absorbed on the body's surface:

$$E_{absorbed} = a \cdot E_{falling},$$ (2.197)

a – absorbance [–].

Energy emitted by the body's surface:

$$E_{emitted} = e \cdot q_{black} \cdot A \cdot t,$$ (2.198)

e – emittance. [–].

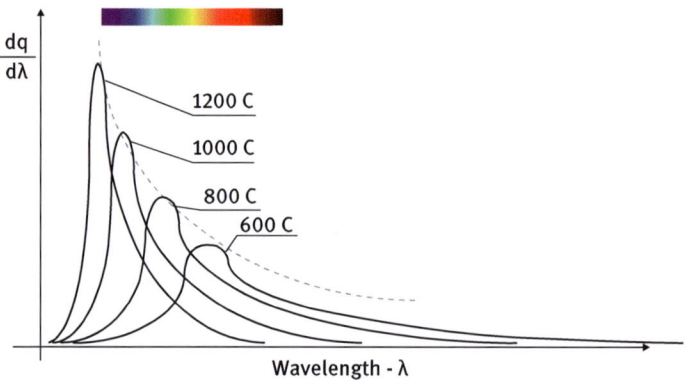

Fig. 2.28. The spectrum of wavelengths of the electromagnetic radiation of black body.

While for each spectrum, which corresponds to a given temperature (see Fig. 2,.28), the whole range of wavelength from zero to infinity is present:

$$q = \int_0^{\infty} \frac{dq}{d\lambda} \cdot d\lambda.$$ (2.199)

It shows a clear trend in the distribution of modal values ($\lambda_{mod\,al}$) dependant on the temperature (see dotted line in Fig. 1). This relationship is described by Wien's law as follows:

$$\lambda_m = \frac{0.00288}{T} \, [m] \,.$$ (2.200)

Kirchhoff's law states that absorptivity and emissivity are equal:

$$a = e.$$ (2.201)

Since all bodies with a temperature above zero always emit radiant energy, the transport of heat in one direction is equal to the net difference between the energies emitted by both bodies (see Fig. 2.29).

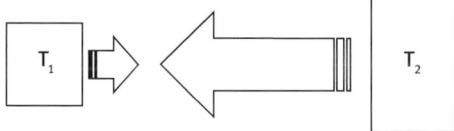

Fig. 2.29. Heat transfer between two objects with different temperatures during radiation.

The heat flux transported by radiation between two bodies at temperatures T_1 and T_2 can be determined using the Boltzmann formula

$$q = e \cdot \sigma \cdot \left(T_2^4 - T_1^4\right).$$ (2.202)

Using the definition of heat transfer coefficient, we can now determine this factor for the heat transfer due to radiation between two objects with known temperatures:

$$\alpha_{\text{net radiation}} = \frac{e \cdot \sigma \cdot \left(T_2^4 - T_1^4\right)}{\Delta T} \left[\frac{W}{m^2 \cdot {}^\circ K}\right].$$ (2.203)

2.4.8 Overall heat transfer between fluids in the heat exchanger

Heat transfer through the wall of the heat exchanger is composed of three stages (see Fig. 2.30). The first stage involves the penetration of heat from the hotter fluid (Phase 1) to the interface between the fluid and the wall of the heat exchanger. The next stage consists of heat conduction inside the wall of the heat exchanger. The third stage concerns the transport of heat from the surface of the wall to the fluid cooler (Phase 2).

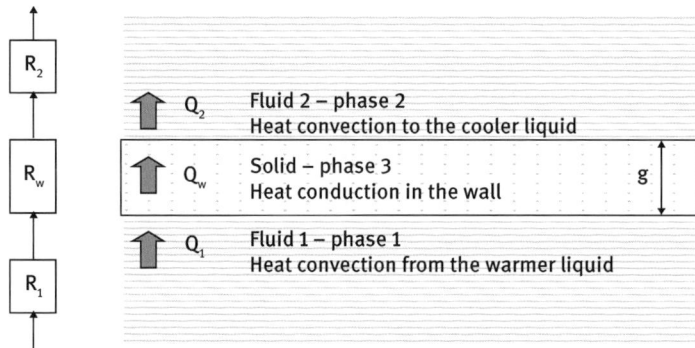

Fig. 2.30. The three stages of heat transport.

All three stages should be included when calculating the overall heat transfer coefficient:

$$Q_1 = \alpha_1 \cdot A \cdot \Delta T_1 \left[\frac{J}{s} \right] \tag{2.204}$$

$$Q_w = \frac{\lambda}{g} \cdot A \cdot \Delta T_w \left[\frac{J}{s} \right] \tag{2.205}$$

$$Q_2 = \alpha_2 \cdot A \cdot \Delta T_2 \left[\frac{J}{s} \right]. \tag{2.206}$$

After calculating the transfer coefficients for each phase, i.e. α_1 and α_2, we can calculate the overall heat transfer coefficient (k). Assuming "resistance in series" model, the total thermal resistance of all layers is equal to the sum of resistance in each layer:

$$R_{entire} = R_1 + R_W + R_2 + \cdots, \tag{2.207}$$

where the resistance is equal to the inverse transfer coefficients

$$R_i = \frac{1}{\alpha_i} \quad \text{where} \quad i = 1, 2 \quad \text{is the number of phases,} \tag{2.208}$$

thus arriving at the final formula

$$\frac{1}{k} = \frac{1}{\alpha_1} + \frac{g}{\lambda} + \frac{1}{\alpha_2}. \tag{2.209}$$

2.5 Mass transport processes

2.5.1 Diffusive mass transfer processes

Diffusion plays a dominant role in mass transport processes, which are perhaps the most important group of unit processes used to separate the different components at the molecular level. The practical application of these processes is often the use of a variety of simplifying assumptions and the appropriate models. Purely physical diffusion does not occur in equipment for processes such as distillation, extraction, absorption, adsorption and in the majority of membrane processes. Moreover these processes are intensified in many ways, in comparison to pure diffusion. Therefore, the physical interpretation of the phenomena which determine the kinetics of mass transport is important for the practical application of the relevant unit processes.

Knudsen diffusion occurs when the scale length of a system is comparable to or smaller than the mean free path of the particles involved. The new type diffusion has been extensively studied in recent times. This is single file diffusion, which is diffusion of some molecules without any impact [60]. The term file dynamics is the unidirectional motion of many particles in narrow channels [61]. This type of diffusion may occur in the following cases [62]:

1. molecular sieves (zeolites);
2. carbon nanotubes;
3. ionic transport through membranes;
4. reptation in polymer melts;
5. microfluid devices.

Molecular diffusion is the motion of all molecules in a temperature above absolute zero, taking collisions into account. The speed of this movement is a function of temperature, fluid viscosity and particle size. If a concentration gradient exists, then diffusion results in the gradual equalization of the concentration, which leads to a state of dynamic equilibrium of the molecules. The mass flow (expressed in moles per surface and time) of these molecules can then be determined at any point in space on the basis of mass balance (conservation law) (see Fig. 2.31).

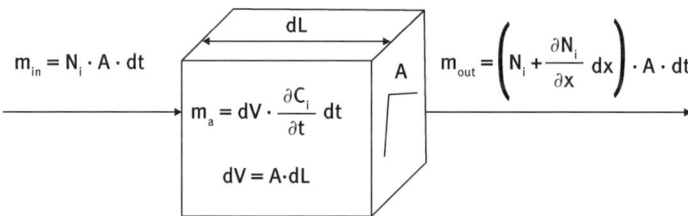

Fig. 2.31. Mass balance for the diffusing component at any point, i.e. infinitely small element of control in space.

Based on a simple mass balance including accumulation, we arrive at the general equation which is the basis for Fick's laws:

$$\frac{\partial N_i}{\partial x} = -\frac{\partial C_i}{\partial t}. \qquad (2.210)$$

From this general balance, in the absence of mass accumulation of the component

$$\frac{\partial C_i}{\partial t} = 0, \qquad (2.211)$$

we arrive at Fick's first law, which describes stationary diffusion:

$$N_i = -D_i \cdot \frac{\partial C_i}{\partial x}. \qquad (2.212)$$

The mass flux of the diffusing component (i) is then a constant on the way of the diffusion:

$$\frac{\partial N_i}{\partial x} = 0, \qquad (2.213)$$

where there is mass accumulation of the component "i" along the path of diffusion:

$$\frac{\partial C_i}{\partial t} \neq 0. \qquad (2.214)$$

The mass flux of the component is changed in accordance with Fick's second law of diffusion:

$$D_i \frac{\partial^2 C_i}{\partial x^2} = \frac{\partial C_i}{\partial t}.$$

(2.215)

The following types of diffusion can be distinguished: (i) an equimolar counter-current diffusion, (ii) diffusion of one component by the inert component, (iii) the diffusion of one component through the mixture of inerts, and finally (iv) multi-component diffusion. The mathematical models which describe these classic types of diffusion will now be presented. A particular type is that occurring in the membrane processes and in the components rejected on the membrane surface, the so-called concentration polarization boundary layer. Another very interesting type is diffusion in cell walls. This is a recently discovered specific type, the so-called "single-file" diffusion, i.e. transport in the ionic channel. It is currently a very intensively explored type of transportation which could revolutionize the process of desalination, solving the problem of water scarcity.

2.5.2 Counter-current equimolar diffusion

This type of diffusion occurs only in the process of distillation, when through the condensation of one mole of the less volatile component exactly the same amount of heat is released, which allows evaporation of one mole of a more volatile component (see Fig. 2.32).

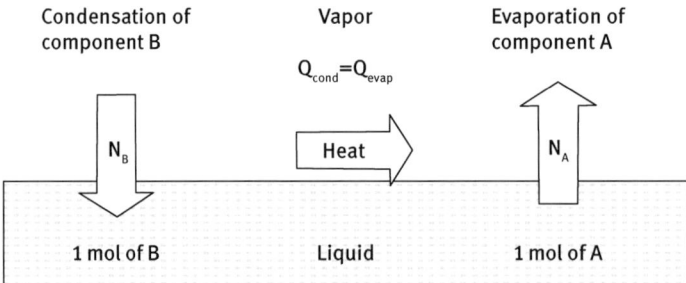

Fig. 2.32. Diagram of equimolar countercurrent diffusion.

The equal molar heat of vaporization and condensation of various components is known as the Trouton rule, which states that in the condition close to normal pressure, the molar heat of evaporation of liquids is about 85 J/mol K (typically from 84 to 92 J/mol K) [63]. During the distillation of a binary mixture, involving a more volatile component A and the less volatile component B, the vapor pressure gradients of the two components compensate each other:

$$\frac{\partial P_A}{\partial x} + \frac{\partial P_B}{\partial x} = 0. \tag{2.216}$$

The equation of state for gases can substitute the partial pressure with the concentration of components in the gas phase:

$$P_A = C_A \cdot R \cdot T. \tag{2.217}$$

Hence

$$\frac{\partial C_A}{\partial x} + \frac{\partial C_B}{\partial x} = 0. \tag{2.218}$$

From the assumption of equal diffusion coefficients DAB = DBA, comes the equivalence and opposite directions of molar fluxes of A and B:

$$N_A = -D_{AB} \cdot \frac{\partial C_A}{\partial x} = -D_{AB} \cdot \left(-\frac{\partial C_B}{\partial x} \right) = -\left(-D_{BA} \cdot \frac{\partial C_B}{\partial x} \right) = -N_B. \tag{2.219}$$

An integral form of the flux by diffusion between the points X1 and X2 is as follows:

$$N_A = -D_{AB} \cdot \frac{C_{A1} - C_{A2}}{x_1 - x_2}. \tag{2.220}$$

2.5.3 Diffusion of the component A by inert component B

This is the type of diffusion which can be used for the greatest number of unit processes. We are dealing with a model of a selectively absorbing surface of only one component A, while component B is still in the mixture (see Fig. 2.33).

When component A vanishes on the absorbing surface this is accompanied by the formation of a concentration gradient and the mass transport then occurs via diffusion of the component in the direction of the surface (see Fig. 2.33, diffusive flux of component A). A secondary effect is the effect of convective transport: the two components A and B in proportion to their concentrations, which is formed on the surface in accordance with the law of conservation of mass (continuity equation). The third effect is the return of inert component B, which cannot be absorbed by the surface. Thus, due to convective flux, component B constantly reaches the surface and generates a concentration gradient. This gradient is the reason for diffusive flux, i.e., in the opposite direction to the absorbing surface. For component B, convective and diffusive flows must be equal in stationary conditions (where no accumulation occurs).Since the net stream, is equal to the sum of all participating streams, as follows:

$$N_{DA} + N_{DB} + N_{KA} + N_{KB} = 0. \tag{2.221}$$

After reduction of the convective and diffusive flow, which are identical for the inert component B, and the arithmetical simplifications we obtain the following formula:

$$N_A = -D_{AB} \cdot \frac{dC_B}{dx} \cdot \frac{C_T}{C_{Bm}}. \tag{2.222}$$

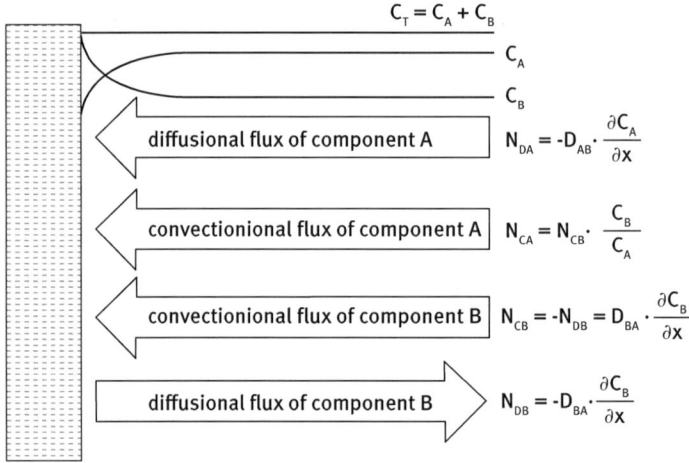

$$C_T = C_A + C_B$$

$$C_A$$

$$C_B$$

diffusional flux of component A | $N_{DA} = -D_{AB} \cdot \dfrac{\partial C_A}{\partial x}$

convectionional flux of component A | $N_{CA} = N_{CB} \cdot \dfrac{C_B}{C_A}$

convectionional flux of component B | $N_{CB} = -N_{DB} = D_{BA} \cdot \dfrac{\partial C_B}{\partial x}$

diffusional flux of component B | $N_{DB} = -D_{BA} \cdot \dfrac{\partial C_B}{\partial x}$

Fig. 2.33. The surface absorbing only the component A.

This is called Stefan's law, which describes the diffusion of a single inert ingredient A by B. In the case of low concentrations of the diffusing component A, Stefan's law coincides with Fick's law. The integral form leads to the formula for average flux in the path between the two positions X_1 and X_2:

$$N_A = -D_{AB} \cdot \frac{C_{B1} - C_{B2}}{x_1 - x_2} \cdot \frac{C_T}{C_{Bm}}, \tag{2.223}$$

where mean inert concentration of the inert B between the positions of X_1 and X_2 can be calculated using the formula for logarithmic mean value:

$$C_{Bm} = \frac{C_{B1} - C_{B2}}{\ln\left(\dfrac{C_{B1}}{C_{B2}}\right)} \tag{2.224}$$

2.5.4 Statistical-mechanistic model of mass transport

The diffusion of one component of the mixture is described by Maxwell [64] and gas concern. Later it was extended to liquids, and applies the same approach as the mechanistic-statistical theory of diffusion above.

If one component (A) diffuses through the mixture consisting of a number of other components (B, C, D, … etc.), the end result is the superposition of the effects of the partial diffusion of the component A by each component separately (See Fig. 2.34). Diffusion in gases can be described as the partial pressure gradients, which express the resistance of component A through all the other inert ingredients (B, C, D, etc.):

$$\frac{dP_A}{dx} = \frac{\partial P_{AB}}{\partial x} + \frac{\partial P_{AC}}{\partial x} + \frac{\partial P_{AC}}{\partial x} + \cdots = \sum_{i=B}^{N} \frac{\partial P_{Ai}}{\partial x}. \tag{2.225}$$

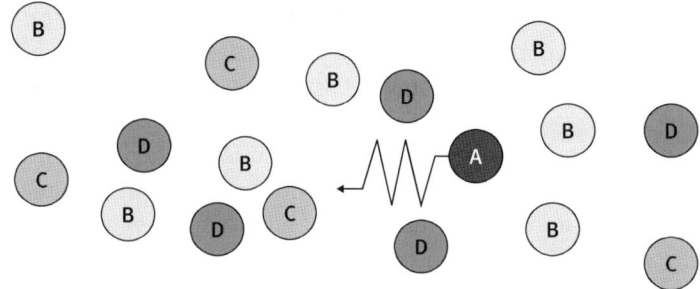

Fig. 2.34. Exemplification of diffusion of one active component A, through the mixture of inert components (B, C, D... etc.).

The same approach in relation to liquid is a certain simplification:

$$\frac{dC_A}{dx} = \frac{\partial C_{AB}}{\partial x} + \frac{\partial C_{AC}}{\partial x} + \frac{\partial C_{AC}}{\partial x} + \cdots = \sum_{i=B}^{N} \frac{\partial C_{Ai}}{\partial x}. \tag{2.226}$$

The partial gradient (for the binary system), according to Maxwell's assumptions, depends on the product of concentrations and the difference in velocity components:

$$\frac{\partial C_{Ai}}{\partial x} = -\frac{C_A \cdot C_i}{D_{Ai} \cdot C_T} \cdot (u_A - u_i). \tag{2.227}$$

Thus, the overall diffusion resistance of the component A, the mixture of inerts (i) can be calculated from the equation

$$-\frac{dC_A}{dx} = \sum_{i=B}^{N} \frac{C_A \cdot C_i}{D_{Ai} \cdot C_T} \cdot (u_A - u_i). \tag{2.228}$$

If all inertial components have (by definition) a zero rate of diffusion ($u_i = 0$ for $i = 1 - n$), and the average (statistically) of the diffusion rate of component A can be expressed with a stream of the N_A and the concentration of C_A:

$$u_A = \frac{N_A}{C_A}. \tag{2.229}$$

Then the gradient of the total (determining resistance to diffusion) can be calculated using the formula

$$\frac{dC_A}{dx} = -\sum \frac{C_A \cdot C_i}{D_{Ai} \cdot C_T} \cdot u_A = -\sum \frac{C_A \cdot C_i}{D_{Ai} \cdot C_T} \cdot \frac{N_A}{C_A}. \tag{2.230}$$

Thus, after a simple transformation, we obtain a formula similar to Fick's law:

$$N_A = D_e \cdot \frac{dC_A}{dx}, \tag{2.231}$$

where D_e is an effective diffusion coefficient of the component A, the mixture of any number (n) of inert components (B, C, D, ..., n), provided that their concentrations are known, and the binary diffusion coefficients:

$$D_e = \frac{C_T}{\displaystyle\sum_{i=B}^{N} \frac{C_{Ai}}{D_{Ai}}} = \frac{1}{\displaystyle\sum_{i=B}^{N} \frac{y_i}{D_{Ai}}}. \tag{2.232}$$

This is a very useful formula for calculating the effective diffusion coefficients of multicomponent mixtures. It should be emphasized that the previously discussed cases can be derived from the Maxwell model, such as Fick's law of diffusion, which is the equimolar counter-current, and Stefan's law, i.e., diffusion through the inert component. James Clerk Maxwell (1831–1879) was a Scottish mathematical physicist. His most prominent achievement was formulating classical electromagnetic theory. However, in process engineering the Maxwell–Stefan theory of gas diffusion, which describes these transport processes, was developed independently and simultaneously by him for dilute gases and Josef Stefan for fluids. This is a model which describes diffusion in multicomponent systems. Maxwell is considered by many physicists to be the 19th-century scientist who had the greatest influence on 20th-century physics. His contributions to the science are considered by many to be of the same magnitude as those of Isaac Newton and Albert Einstein. The Maxwell–Stefan equation is

$$\frac{\nabla \mu_i}{RT} = \nabla \ln a_i = \sum_{j=1}^{n} \frac{x_i x_j}{D_{ij}} \left(v_j - v_i\right) = \sum_{j=1}^{n} \frac{c_i c_j}{c^2 D_{ij}} \left(\frac{J_j}{c_j} - \frac{J_i}{c_i}\right), \tag{2.233}$$

∇ – vector differential operator,
x – mole fraction,
μ – chemical potential,
a – activity,
i, j – indexes for component i and j,
n – number of components,
D – Maxwell–Stefan-diffusion coefficient,
v – diffusion velocity of component i,
c_j and c_i – molar concentration of component i,
c – total molar concentration,
J_i – flux of component i.

2.5.5 Statistical-mechanical theory of membrane transport

This is the most general mass transport model, which describes the transport of multiple components of average velocity (u_i) under various driving forces [65]. The general equation describes the balance between the driving forces and the resistances associated with multicomponent transport:

$$\sum_{j=1}^{N} \frac{C_j}{C \cdot D_{ij}} \left(u_i - u_j\right) + \frac{u_i}{D_{iM}} = -\frac{1}{RT}\left(\nabla_T \mu_i - F_i\right) - \frac{\alpha_i \cdot B_0}{\eta \cdot D_{iM}} \left(\nabla P - C \cdot F\right) - \sum_{j=1}^{N} \frac{C_j}{C \cdot D_{ij}} \cdot D_{ij}^T \nabla \ln T.$$

(2.234)

The components of the equation on the left correspond to mass transport resistances and are related to the average diffusion velocities of components. These include the average relative velocity $(u_i - u_j)$ of i-th component through the rest j-th components, and velocity (u_i) of this i-th component within the membrane. The concentration of j-th component is C_j and total concentration is C. The diffusion coefficient in the binary system is D_{ij} and it is D_{iM} in the membrane.

The driving forces are listed on the righthand side of the equation. The first component refers to three different kinds of isothermal mass transport, i.e. diffusion enforced by chemical potential gradient (μ) and any given external forces (F_i). The second component on the righthand side corresponds to the driving force of viscous flow. The constant (B) characterizes membrane permeability. For cylindrical shaped pores this is $B = r^2/8$, according to the Poiseuille model. The different separation effects of the components in the membranes are expressed by dimensionless separation coefficients. The membrane permeability is characterized by the factor B_0. Kinematical viscosity is η, and F is the net external force acting on all of the components. The pressure gradient is denoted as ∇P. The last component on the right of the equation corresponds to thermodiffusion, which may be expressed by dimensionless temperature gradient and coefficient. The following designations were used in this equation:

$u_i = \dfrac{J_i}{c_i}$ – rate of i-th component,

$F = \sum \dfrac{c_i F_i}{c}$ – the net external force,

$c = \sum c_i$ – total concentration,

$\nabla_T \mu_i$ – gradient of chemical potential,

∇P – pressure gradient,

$\nabla \ln T = \dfrac{\nabla T}{T}$ – temperature gradient

D_{ij} D_{Mi} – diffusion coefficient in fluid and membrane respectively,

D_{ij}^T – thermodiffusion coefficient,

B_0 – permeability factor for viscous flow in membrane pores,

α_i – separation factor,

η_i – viscosity,

N – number of components,

N(N + 2) – total number of data parameters.

2.5.6 Diffusional methods of separation

The following (Fig. 2.35) is a graphical representation of almost all known unit processes involving mass transport. In almost all conventional unit processes, mass transfer occurs between the two phases. The criteria of belonging to particular unit processes are the kind of phase and the direction of mass transport. Membrane processes can be included for more recent unit processes, in which there are three phases. One of these phases also includes the same membrane, and may also be the liquid and gas phases.

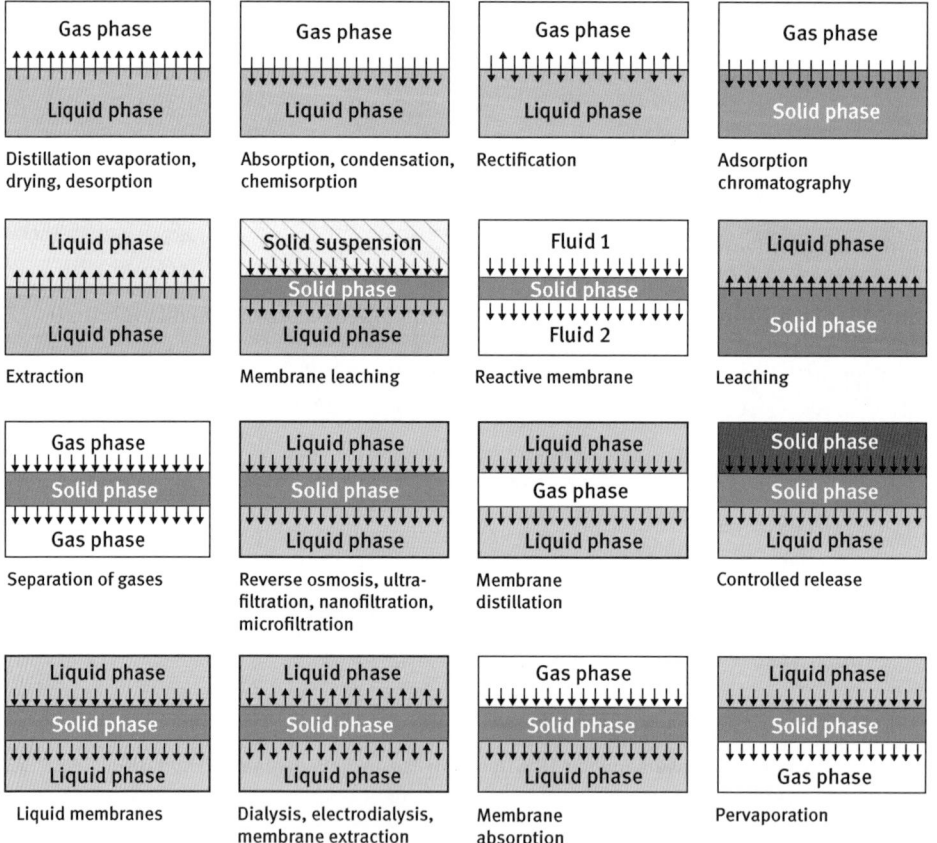

Fig. 2.35. Diffusional methods of separation.

2.5.7 Kinetics of mass transport processes

The kinetics of the transport of all processes depends on diffusivity as well as on the hydrodynamics phenomena which take place near the surfaces between phases. The so-called boundary layer is the area in the vicinity of the phase surface, in which a relevant gradient of concentration, temperature, or velocity occurs. For example, the transport of momentum, which is accompanied by energy loss due to friction, is present in the film or laminar viscous flow channel or device. Heat transfer is related to the boundary layer temperature, and mass transport is associated with an interfacial layer of concentration.

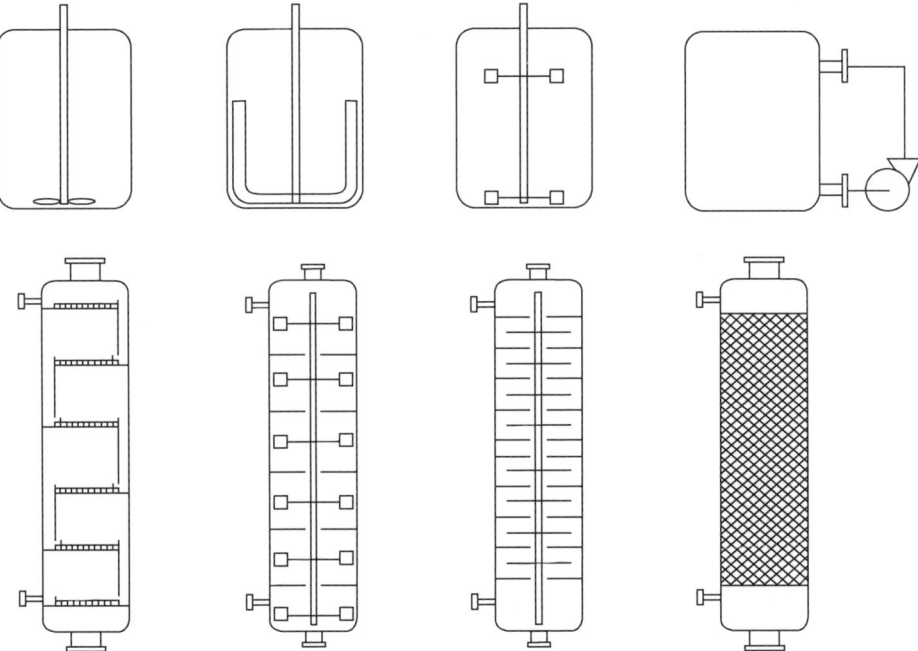

Fig. 2.36. Methods of reduction of the boundary layer in tanks and columns.

The means for controlling the boundary layer (Fig 2.37) are different methods such as the delivery of kinetic energy for mixing during the flows of phases, in such devices as various mixers and columns (see Fig. 2.36).

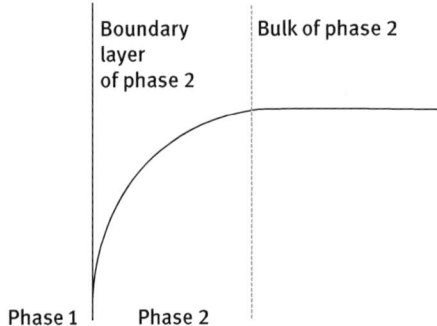

Fig. 2.37. Graphical representation of the boundary layer.

2.5.7.1 Nernst film model

Nernst (1904) formulated the model of mass transport in the boundary layer based on the following assumptions (see Fig. 2.38):

1. Total mass transport resistance is focused in the boundary layer.
2. Mass transport occurs by diffusion in the layer.
3. The concentration gradient within the layer is constant.
4. Adjacent to the layer is situated the so called "bulk region", in which concentration is constant.
5. Boundary layer thickness (δ) depends on the hydrodynamic conditions in the bulk region.
6. Velocity is constant in the entire bulk, but zero on the surface.
7. Shear stresses are constant in the entire boundary layer.

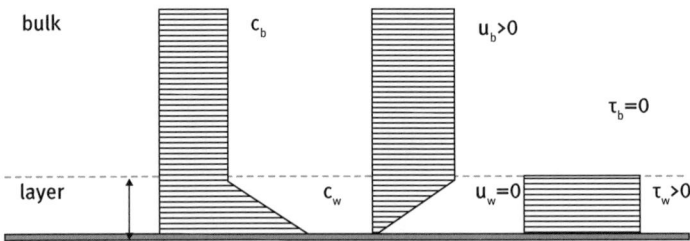

Fig. 2.38. Illustration of Nernst model of mass transport in the boundary layer.

The tickness of the boundary layer can be derived independently from two phenomenological equations:

(a) From the simplified version of Newton's phenomenological equation for the transport of momentum:

$$\tau_w = \mu \cdot \frac{du}{dy} \cong \mu \cdot \frac{u_b}{\delta} \quad \text{for} \quad \leq \delta. \tag{2.235}$$

After rearrangement the thickness of the hydrodynamic boundary layer can be derived with the formula

$$\delta_h = \mu \cdot \frac{u}{\tau}. \tag{2.236}$$

After substituting the shear stress to the above equation,

$$\tau = \lambda \cdot \frac{\rho u^2}{2}, \tag{2.237}$$

one can obtain the thickness of hydrodynamic boundary layer:

$$\delta_h = \mu \cdot \frac{u}{\lambda \cdot \frac{\rho u^2}{2}}. \tag{2.238}$$

(b) From the simplified Fick's phenomenological equation for diffusional mass transport:

$$N = D \cdot \frac{dC}{dy} \cong \frac{D}{\delta}(C_b - C_w) \quad \text{for} \quad y \leq \delta. \tag{2.239}$$

After simple rearrangements we can derive the thickness of the concentration boundary layer:

$$\delta_C = \frac{D}{N}(C_b - C_w). \tag{2.240}$$

The mass flux N may then be substituted,

$$N_i = \beta_i \cdot (C_{wi} - C_{bi}) \tag{2.241}$$

to obtain

$$\delta_C = \frac{D}{\beta}. \tag{2.242}$$

Assuming that these two layers i.e., of concentration and of hydrodynamic have the same thickness:

$$\delta_h = \delta_c. \tag{2.243}$$

The simple relation for mass transfer coefficient may therefore be obtained thus:

$$\beta_i = 0.5 \cdot \lambda \cdot u_b \cdot \frac{D \cdot \rho}{\mu}. \tag{2.244}$$

After further transformations, the criterial equation is presented with the numbers that are known from the theory of similarity:

$$Sh = 0.5 \cdot \lambda \cdot Re. \tag{2.245}$$

The Nernst model is purely theoretical assuming the predominant role of diffusion in the boundary layers. However, the equation resulting from this model can now be compared with the other models, even the purely empirical ones.

2.5.7.2 Reynolds penetration model of boundary layer

The Reynolds model assumes that mass transfer in the boundary layer is solely the result of turbulence, i.e. without any contribution from diffusion (see Fig. 2.39). The intensity of turbulence may be characterized by a number of vortices (n) per unit area and time. It should be noted that this is a simplification compared to reality. Several studies have confirmed the fact that the vortices are present in a wide range of dimensions.

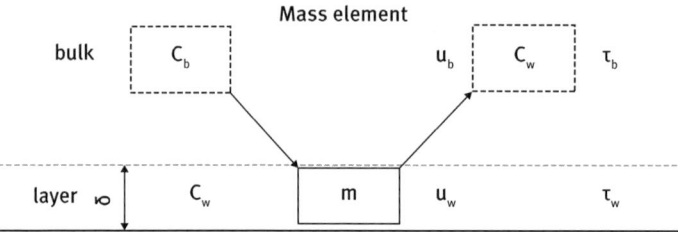

Fig. 2.39. Illustration of mass transport according to the Reynolds model.

Each eddy element carries the mass (m) from the core to the boundary layer. If the fluid is incompressible, then the same amount of fluid has to be supplanted from the boundary layer to the fluid core. Mass exchange between the interface and the core is the result of the different concentrations at these places:

$$N_i = n \cdot m \cdot \frac{(C_{bi} - C_{wi})}{\rho}. \tag{2.246}$$

Now the mass flow during such penetration can also be specified from the definition of mass transfer coefficient:

$$N_i = \beta \cdot (C_{bi} - C_{wi}). \tag{2.247}$$

A comparison of these equations shows the formula for determining the product of the intensity of the vortices, based on the mass transfer coefficient:

$$m \cdot n = \beta \cdot \rho. \tag{2.248}$$

At this expression, we may take the opportunity to show the physical meaning of mass transfer coefficient, as the average volume of fluid that is transported between the interface and the core. Momentum is exchanged simultaneously with the exchange mass between the fast elements coming from the bulk and the slow elements at the surface, resulting in a shear stress. The amount of momentum to the walls and transported per unit of the surface determines the equation

$$\tau_w = n \cdot m \cdot (u_b - u_w). \tag{2.249}$$

And just as in the previous case, but on the basis of hydrodynamics, we take the intensity of vortices. Previously, however, we needed to determine the value of shear stress, using the well-known Darcy–Weisbach formula

$$\tau_w = \lambda \cdot \frac{\rho u^2}{2}.$$ (2.250)

After substituting the shear stress, we obtain the intensity of the vortices (mn), as follows:

$$n \cdot m = \lambda \cdot \frac{\rho u^2}{2}.$$ (2.251)

Now, comparing the intensity of the vortexes obtained on the mass transfer (Eq. (2.254)) and the momentum exchange basis (Eq. (2.257)), we can obtain the formula for the mass transfer coefficient, based on the Reynolds' model:

$$\beta_i = 0.5 \cdot \lambda \cdot u_b.$$ (2.252)

It should be noted that the Reynolds model does not take the effect of molecular diffusion into account at all, only the hydrodynamic effects. Eventually, to calculate the mass transfer coefficient, the correlation in dimensionless form may be used:

$$Sh = 0.5 \cdot \lambda \cdot Re \cdot Sc.$$ (2.253)

2.5.7.3 The Highbie penetration model

Penetration models are based on the Reynolds analogy, assuming that mass and momentum transport take place due to eddies. The differences between the various models are focused on how the fluid elements are transferred between the contact surface and the bulk. Under the Reynolds model mass transport is instantaneous. According to other penetration models, mass transport takes place during the contact of the element with the interface by diffusion and then the element is mixed with the fluid in the bulk (see Fig. 2.40).

Highbie's penetration model describes mass transfer between phases, which takes place through contact with the small elements of liquid on the boundary layer. Elements adhere to the infinitely small surface but their length is unlimited. The identical time of exposure is unrealistic but was caused by the specific application of the Highbie model for gas absorption in the column plates. When the elements are disconnected from the surface they are instantly mixed with the bulk. The basic assumption of the model is the identical exposure time for all elements on the surface. The shorter the exposure time, the greater the value of the average flux. This is because the instantaneous flux decreases when elements have been saturated with time according to Fick's law (2nd), when assuming constant physical properties and the lack of a chemical reaction. This can be written thus:

$$\frac{\partial C}{\partial t} - D \frac{\partial^2 C}{\partial x^2} = 0.$$ (2.254)

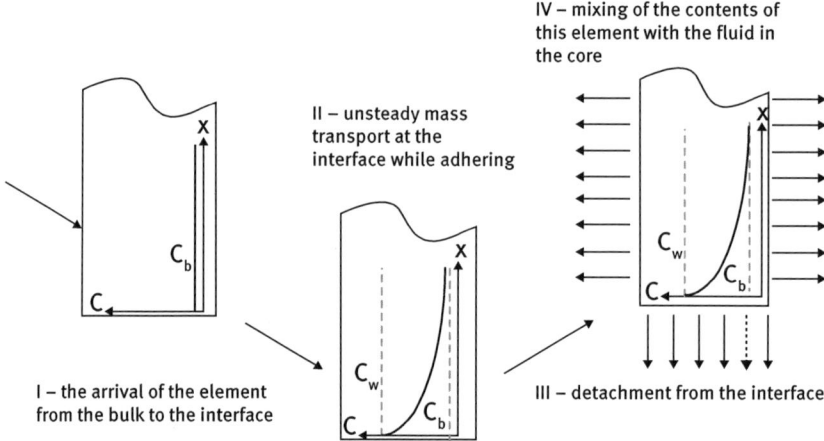

Fig. 2.40. Subsequent cycles of mass transport according to the Highbie model.

For the solution of unsteady mass transport equation (Fick's second law), two conditions must be established for space and one for time:

$$C = C_w \quad \text{for} \quad y = 0, \tag{2.255}$$

$$C = C_b \quad \text{for} \quad y = \infty, \tag{2.256}$$

$$C = C_b \quad \text{for} \quad t = 0. \tag{2.257}$$

This allows determination of the changes of the concentration profile within penetration element $C(x, t)$:

$$C(x, t) = (C_b - C_w) \cdot \text{erf}\left(\frac{x}{2\sqrt{Dt}}\right). \tag{2.258}$$

On the basis of the concentration $C(x, t)$, its gradient is calculated on the surface $(x = 0)$ and the instantaneous flux from Fick's law:

$$N(t) = -D \cdot \frac{\partial C(0, t)}{\partial x}. \tag{2.259}$$

The entire surface is covered with the elements to be detached from it after the same exposure time (te). In this period of time (from 0 to t), instantaneous flux changes and its value determines the average INTEGRAL:

$$\bar{N} = \frac{1}{t_e} \int_0^{t_e} N(t)dt, \tag{2.260}$$

from which, after integration we obtain a simple formula for the mass flux:

$$\bar{N} = 2\sqrt{\frac{D}{\pi \cdot t_e}} \cdot (C_w - C_b). \tag{2.261}$$

Please note that in this model, in addition to diffusion the effects of convective mass transport are also taken into account. The more frequently the exchange elements that are shorter exposure times run, the larger the average flux is. It is easy to extract the mass transfer coefficient from this equation as follows:

$$\beta = 2\sqrt{\frac{D}{\pi \cdot t_e}}.$$

(2.262)

Exposure time can be calculated as the inverse of the frequency of vortices which can be evaluated from analysis of turbulence. On this basis, it was found that the Schmidt number values have crucial importance, and therefore the criterial equation shows the ranges for the different Schmidt numbers as follows:

$$Sh = 0.080\sqrt{\frac{\lambda}{2}} \cdot Re \cdot Sc^{0.414} \qquad \text{for} \quad 1 < Sc < 500,$$

(2.263)

$$Sh = 0.167\sqrt{\frac{\lambda}{2}} \cdot Re \cdot Sc^{0.295} \qquad \text{for} \quad Sc > 500.$$

(2.264)

We have no reason to believe that the elements on the interface know how long to stay on the surface before mixing with bulk. However, there is one exception, which is almost exactly the case of the Highbie model. This case is the shelves in the absorption columns with barbotage which have some liquid layer on them, known as holdup. Since all the bubbles that are passing through the same height of layer, they have to have the same residence time, e.g. exposure time in the Highbie's model (t_p).

2.5.7.4 Danckwerts surface renewal theory

Dankwerts has modified the Highbie model in such a way to avoid the assumption of equal exposure time for the all elements. Instead, he assumed a stochastic nature of the mass transport phenomenon on the interface. In other words, it is assumed that the probability of surface renewal of any element is the same for each location and independent of the process time.

It is assumed that the surface of phase contact is made up of a mosaic of elements of different ages. It is also assumed that f(t) is a function defining the probability density of the occurrence of the element with the age interval (t, t + dt) at the surface of the site (the function of age).

Since the probability distribution of the events, that elements are being renewed is uniform over the entire surface (Bernoulli distribution) the probability of "surface renewal" for any given fraction of age f(t) dt is the same and is proportional to the area occupied by elements in this age range. To determine the age distribution of elements on the surface Danckwerts adopted the following assumptions:

1. The probability of the phenomenon of surface renewal is independent of the age of the element and its location.

2. The rate of surface renewal (s) is the same for the fractions of the age.
3. The parameter s is defined as a fraction of the surface being renewed per unit of time:

$$s = \frac{f(t + dt)dt - f(t)dt}{[f(t)dt]\, dt}. \tag{2.265}$$

This is a form of differential equation which can be used to give the function of the age distribution of the elements. For this purpose, however, one must take the initial condition into account. Instead Dankwerts benefited from a condition, which must be fulfilled by any density function of probability distribution:

$$\int_{0}^{\infty} f(t)dt. \tag{2.266}$$

In this way the formula for age distribution was obtained, which is the Poisson exponential distribution function

$$f(t) = s \cdot e^{-st}. \tag{2.267}$$

Instantaneous flux in a single element has been appointed in the same way as in Highbie's model, i.e. on the basis of the nonstationary transport equation. Mass flux over the entire surface in the Danckwerts model is calculated as the mean of this distribution:

$$\bar{N} = \int_{0}^{\infty} N(t) \cdot f(t) \cdot dt. \tag{2.268}$$

After integration one arrives at a simple formula:

$$\bar{N} = \sqrt{D \cdot s}\,(C_w - C_b). \tag{2.269}$$

Many researchers have identified the rate of renewal of the surface with an average frequency of occurrence of turbulence in the boundary layer.

The empirical formulas used to calculate the frequency of the possible turbulent mass transfer coefficient in dimensionless form are as follows:

$$Sh = 0.0707 \cdot \sqrt{\frac{\lambda}{2}} \cdot u_b \cdot Sc^{0.414} \quad \text{for} \quad 1 < Sc < 500 \tag{2.270}$$

$$Sh = 0.148 \cdot \sqrt{\frac{\lambda}{2}} \cdot u_b \cdot Sc^{0.295} \quad \text{for} \quad Sc > 500 \tag{2.271}$$

2.5.7.5 Comparison of mass transport kinetic models

In practice, the many models of mass transfer are variations of the basic models described here. For example, the Torr–Marchello model assumes the possibility of penetration on different depths and at different frequencies. If all penetrations are equal to the thickness of the layer, the model is identified as a Highbie model, but if penetration is zero it is simply reduced to the to the film model. It should be noted that in the various models mentioned here, the number of parameters is the same, although they have different physical interpretations. As discussed above, this leads to a difference in the contributions of molecular diffusion and convection in the entire transport (see Tab. 2.6).

Table 2.6. Comparison of mass transfer models

Mass transport model	Mass flux	Transfer coefficient	Kinetic parameter
Nernst' film model	$N_i = \dfrac{D_i}{\delta} \cdot (C_{wi} - C_{bi})$	$\beta = \dfrac{D_i}{\delta}.$	Layer thickness – δ
Reynold's penetration model	$N_i = n \cdot m \cdot \dfrac{(C_{bi} - C_{wi})}{\rho}$	$\beta_i = 0.5 \cdot \lambda \cdot u_b$	Velocity – u_b
Highbie's penetration model	$\bar{N} = 2\sqrt{\dfrac{D_i}{\pi \cdot t_e}} \cdot (C_w - C_b)$	$\beta = 2\sqrt{\dfrac{D}{\pi \cdot t_e}}.$	Exposure time – t_e
Danckwerts surface renewal	$\bar{N} = \sqrt{D_i \cdot s}\,(C_w - C_b)$	$\beta = \sqrt{D \cdot s}$	Surface renewal rate – s

2.5.7.6 The Colburn method of mass exchange between two phases in columns

Mass transport processes, such as absorption, desorption, extraction, distillation, rectification, wetting and drying are carried out in a column. The concentrations of the two contacting phases flowing along the columns have changed accordingly. Therefore, as in the case of heat exchangers, determining the average driving forces in the process is problematic. The driving forces depend on relative direction of movement of the contacting phases (see Fig. 2.41).

The mass balance of the narrow slice of the column with a differential section (dH) allows the right formula necessary to calculate the assumed process to be found. Consider the case of a countercurrent flow of the two phases in the column (see Fig. 2.42).

Intensities of the mass flow are indicated with symbols: G [kg/s], referring to those components whose quantity does not change along the flow path of the phases. These constituents may be inerts of processes such as absorption or extraction. The total number of moles may also be used as the constant value in rectification. Concentrations of C [kg/m^3] represent the ratio of component "A" to the volume of inert ingredients. Index (0) refers to the core in each phase, and the "i" indices refers to the interface. The indices 1 and 2 refer to the appropriate phases.

We use two approaches to determine the same amount of mass of component shared between phases 1 and 2 in the segment of the column of height (dH), in the same way as for calculation of the average logarithmic temperature. This method takes

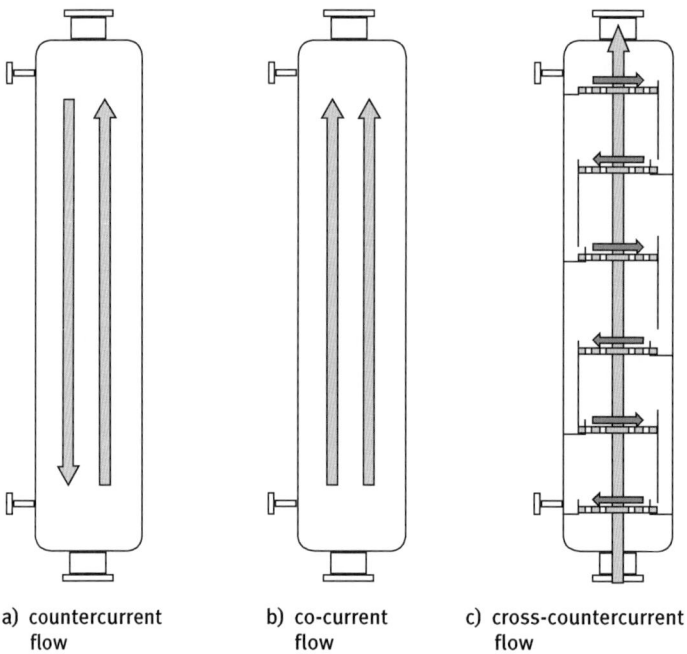

a) countercurrent
   flow

b) co-current
   flow

c) cross-countercurrent
   flow

Fig. 2.41. Schematic diagram of the stream flows in a column.

phase G_1 that is taking
the component A

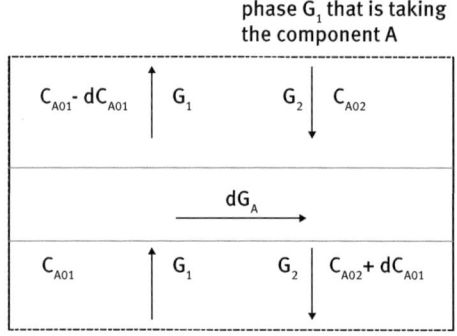

phase G_1 that is giving
back component A

Fig. 2.42. The differential balance sheet of the
local section of the columns.

the changing driving force of the process into account, which is the concentration difference between the core and the interfacial surface in each separate phase.

A stationary observer (see Fig. 2.42) records the differential amount of mass (dG_A) exchanged through the slice in the column of thickness dH between phases 1 and 2. In the control element, the surface area (dA) of contact between the phases is the same for each the phase.

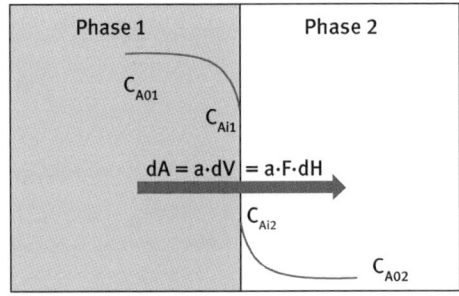

Fig. 2.43. Mass exchange between the two phases.

Each phase has its own transport rate expressed as mass transfer coefficient (β) and flow (G). Thus, the appropriate equations expressing the exchanged component (A) will have the form (see Fig. 2.43):

(a) the mass given from the core of phase 1 to the mirror of the interface 1–2:

$$dG_A = \beta_1 \cdot dA \cdot (C_{A01} - C_{Ai1});$$ (2.272)

(b) the mass taken from the interface 1–2 to the bulk of phase 2:

$$dG_A = \beta_2 \cdot dA \cdot (C_{Ai2} - C_{A02}).$$ (2.273)

The contact area between the phases is dependent on the design of the device. In the filled column, it will be of the packing surface, and in the "plate column", it will be the total surface area of bubbles or droplets. In each case it is possible to determine this surface as a parameter called the specific surface. The specific surface area of the interface expresses the volume referred to the volume of the filling, i.e. foam, bubbles, droplets etc. This surface, in the control element, is given by

$$dA = a \cdot dV = a \cdot F \cdot dH,$$ (2.274)

where F is the cross-section of the column. Thus, the mass flow rate is seen from the point of view of the stationary observer, and the relationship is expressed as follows:

$$dG_A = \beta_1 \cdot a \cdot F \cdot dH \cdot (C_{A01} - C_{Ai1}),$$ (2.275)

$$dG_A = \beta_2 \cdot a \cdot F \cdot dH \cdot (C_{Ai2} - C_{A02}).$$ (2.276)

The moving observer notices the changes in the concentrations during the passage through the infinitesimal path (dH) separately for each phase:

$$dG_A = V_1 \cdot dC_{A01},$$ (2.277)

$$dG_A = V_2 \cdot dC_{A02},$$ (2.278)

where V_1 and V_2 are the volumetric flow rates of both phases. Combining the relevant observations of the two observers we now get

$$V_1 \cdot dC_{A01} = \beta_1 \cdot a \cdot F \cdot dH \cdot (C_{A01} - C_{Ai1}),$$ (2.279)

$$V_2 \cdot dC_{A02} = \beta_2 \cdot a \cdot F \cdot dH \cdot (C_{Ai2} - C_{A02}).$$ (2.280)

The differential (dH) can be determined independently from each of these equations:

$$dH = \frac{V_1}{\beta_1 \cdot a \cdot F} \cdot \frac{dC_{A01}}{(C_{A01} - C_{Ai1})} = \frac{u_1}{\beta_1 \cdot a} \cdot \frac{dC_{A01}}{(C_{A01} - C_{Ai1})} \tag{2.281}$$

$$dH = \frac{V_2}{\beta_2 \cdot a \cdot F} \cdot \frac{dC_{A02}}{(C_{Ai2} - C_{A02})} = \frac{u_2}{\beta_2 \cdot a} \cdot \frac{dC_{A02}}{(C_{Ai2} - C_{A02})} \tag{2.282}$$

Here, u_1 and u_2 are the respective flow rates of phases related to the cross-section of the column (F) and are obtained after integration formulas for the total height of the column. Using the values for the first phase we calculate the height of the column of formulas:

$$H = \frac{V_1}{\beta_1 \cdot a \cdot F} \cdot \int_{C_{1\,inlet}}^{C_{1\,outlet}} \frac{dC_{A01}}{(C_{A01} - C_{Ai1})} = \frac{u_1}{\beta_1 \cdot a} \cdot \int_{C_{1\,inlet}}^{C_{1\,outlet}} \frac{dC_{A01}}{(C_{A01} - C_{Ai1})} \tag{2.283}$$

$$H = \frac{V_2}{\beta_2 \cdot a \cdot F} \cdot \int_{C_{2\,inlet}}^{C_{2\,outlet}} \frac{dC_{A02}}{(C_{Ai2} - C_{A02})} = \frac{u_2}{\beta_2 \cdot a} \cdot \int_{C_{2\,inlet}}^{C_{2\,outlet}} \frac{dC_{A02}}{(C_{Ai2} - C_{A02})}, \tag{2.284}$$

and after defining and substituting the term of unit height

$$h_1 = \frac{u_1}{\beta_1 \cdot a} \tag{2.285}$$

$$h_2 = \frac{u_2}{\beta_2 \cdot a} \tag{2.286}$$

and by defining the concept of the number of mass transfer units

$$N_1 = \int_{C_{1\,inlet}}^{C_{1\,outlet}} \frac{dC_{A01}}{(C_{A01} - C_{Ai1})} \tag{2.287}$$

$$N_2 = \int_{C_{2\,inlet}}^{C_{2\,outlet}} \frac{dC_{A02}}{(C_{Ai2} - C_{A02})} \tag{2.288}$$

we then arrive at a simple formula in the form of the products

$$H = h_1 \cdot N_1 \tag{2.289}$$

$$H = h_2 \cdot N_2. \tag{2.290}$$

Mass transfer between two phases is the transport of mass from the core of one phase to the core of the second phase

$$G_A = k \cdot a \cdot F \cdot H \cdot (m \cdot C_{A01} - C_{A02}). \tag{2.291}$$

The mass transfer coefficient (k) will thus be defined by the intensity of the mass transport through the two boundary layers. Thus the intensity of the mass transfer from the first phase bulk to the interface will be

$$G_A = \beta_1 \cdot a \cdot F \cdot H \cdot (C_{A01} - C_{Ai1,}) \tag{2.292}$$

and mass of transfer through the second phase, the interface to the core

$$G_A = \beta_2 \cdot a \cdot F \cdot H \cdot (C_{Ai2} - C_{A02}). \tag{2.293}$$

Taking the relationship between the equilibrium concentrations at the interface into consideration

$$C_{Ai2} = m \cdot C_{Ai1}, \tag{2.294}$$

and by expressing the respective differences in concentration by mass transfer equations, we obtain

$$C_{A01} - C_{Ai2} = \frac{G_A}{\cdot \beta_1 \cdot a \cdot F \cdot H} \tag{2.295}$$

$$C_{Ai2} - \frac{C_{A02}}{m} = \frac{G_A}{m \cdot \beta_2 \cdot a \cdot F \cdot H}, \tag{2.296}$$

and, after summing up the parties

$$C_{A01} - \frac{C_{A02}}{m} = \frac{G_A}{\beta_1 \cdot a \cdot F \cdot H} + \frac{G_A}{m \cdot \beta_2 \cdot a \cdot F \cdot H} = \frac{G_A}{a \cdot F \cdot H} \cdot \left(\frac{1}{\beta_1} + \frac{1}{m \cdot \beta_2}\right), \tag{2.297}$$

multiplication by m of both sides of the equation gives

$$m \cdot C_{A01} - C_{A02} = \frac{G_A}{a \cdot F \cdot H}\left(\frac{m}{\beta_1 \cdot} + \frac{1}{\beta_2}\right). \tag{2.298}$$

After further rearrangements we can get

$$G_A = \frac{1}{\left(\frac{m}{\beta_1 \cdot} + \frac{1}{\beta_2}\right)} \cdot a \cdot F \cdot H \cdot (m \cdot C_{A01} - C_{A02}). \tag{2.299}$$

Comparison of the mass flow by mass transfer equations allows determination of the overall mass transfer coefficient (k) on the basis of the values of mass transfer coefficients β_1 and β_2:

$$k = \frac{1}{\left(\frac{m}{\beta_1 \cdot} + \frac{1}{\beta_2}\right)}. \tag{2.300}$$

Similarly, after the definition of mass transfer units and unit height we get the formula for the number of overall mass transfer units (N_{og}) and the overall mass transfer unit height (h_{og}):

$$N_{og} = \frac{1}{\left(\frac{m}{N_1 \cdot} + \frac{1}{N_2}\right)} \tag{2.301}$$

$$h_{og} = \frac{1}{\left(\frac{1}{h_1 \cdot} + \frac{1}{m \cdot h_2}\right)}. \tag{2.302}$$

2.6 Kinetics of reactions

2.6.1 Chemical reactions

Chemical reactions are the processes leading to the conversion of the reactants, to quite different substances, which are called products, and which typically have different chemical properties [66]. The kinetics of chemical reactions is of practical importance because it determines all the realities regarding the possibility of the reaction under the circumstances, especially those that determine time and cost. In 1864, Peter Waage and Cato Gulberg formulated "The law of mass action", which states that the reaction rate is proportional to the number of reactive substances. The strength of chemical bonds differs and can vary considerably. Covalent or ionic bonds are considered "strong", whereas dipole–dipole interactions, the London dispersion, van der Waals forces and hydrogen bonds are rather "weak" (see Tab. 2.7). Generally, the intermolecular forces which attract neighboring atoms, molecules or ions are weaker than those which keep atoms together within a molecule. Experimental observations show that reactions are made up of a number of elementary reactions occurring simultaneously or sequentially. The best-known elementary reactions are synthesis reactions, dissociation or metathesis. The reaction mechanism describes the step-by-step sequence of elementary reactions and improves the amount of energy consumed during production [67].

The kinetics of a reaction can help to identify which step is rate-determining, but this step may also be rate of mass transport, heat transport and other physical phenomena. The concept of the rate-determining step is very important because it simplifies the calculation of the rate equation. Chemical reactions can often be described by chemical equations, which are symbolic representations where the reactants are given on the lefthand side and the productes on the right, whereas the mass of the reactants equals the mass of the products. Consider a typical chemical reaction.

$$\alpha \cdot A + \beta \cdot B \overset{r}{\rightleftharpoons} \gamma \cdot C + \delta \cdot D \tag{2.303}$$

The coefficients ($\alpha, \beta, \gamma, \delta$) next to the symbols and formulae of entities are the stoichiometric coefficients. For a closed system with no byproducts, the reaction rate can be written either as the rate of reagents vanishing (minus sign), or the rate of the formation of products:

$$r = -\frac{1}{\alpha} \cdot \frac{d[A]}{dt} = -\frac{1}{\beta} \cdot \frac{d[B]}{dt} = \frac{1}{\gamma} \cdot \frac{d[C]}{dt} = \frac{1}{\delta} \cdot \frac{d[D]}{dt} \tag{2.304}$$

It should be noted that symbols with square brackets indicate the chemical activities of components, which are the reactants or the products of a chemical reaction. The chemical activity is the ability of a substance to undergo a chemical reaction. Thus activity is a measure of the "effective concentration" of a species in a mixture. The activity of a substance is defined as

$$a_i = \exp\left(\frac{\mu_i - \mu_i^0}{RT}\right),\qquad(2.305)$$

where R is the gas constant, T is the absolute temperature, μ_i is the chemical potential of the components in the given conditions, and μ_0 is the chemical potential of the components in standard conditions. Activity can be represented as a quantity proportional to the measured concentration and the known activity coefficient:

$$a_i = \gamma_{x,i} \cdot x_i.\qquad(2.306)$$

The chemical activity depends on each variable affecting the chemical potential such as temperature, pressure, chemical environment, etc. The rates of chemical reactions can be calculated from the following equations:

$$r_1 = k_1 \cdot [A]^m \cdot [B]^n\qquad(2.307)$$

$$r_2 = k_2 \cdot [C]^o \cdot [D]^p,\qquad(2.308)$$

when the forward reaction rate is equal to the reverse reaction rate,

$$r_1 = r_2\qquad(2.309)$$

$$k_1 \cdot [A]^m \cdot [B]^n = k_2 \cdot [C]^o \cdot [D]^p,\qquad(2.310)$$

chemical equilibrium has then been reached. The balance is associated with the same rate in both directions. In that case, the ratio of the rate constants (k_1 and k_2) is also a constant, now known as an equilibrium constant:

$$K = \frac{k_1}{k_2} = \frac{[C]^o \cdot [D]^p}{[A]^m \cdot [B]^n}.\qquad(2.311)$$

The values m, n, o and p are known as the reaction order and they must be experimentally determined. Orders of complex reactions may or may not be due to their stoichiometric coefficients. The coefficient of the reaction rate (k) can be expressed by the Arrhenius formula as follows:

$$k = A \cdot e^{-\frac{E_a}{k_B \cdot T}},\qquad(2.312)$$

where E_a is the activation energy and the coefficient (A) depends on the frequency of collisions. Activation energy is the amount of energy which must be supplied to initiate a chemical reaction. Catalysts allow such reactions which would otherwise be blocked or slowed down by this kinetic barrier. Since at temperature (T) the molecules have energies according to the Boltzmann distribution, A is the pre-exponential factor or so-called frequency factor. The Arrhenius equation is a remarkably accurate formula for the calculation of the temperature dependence of reaction rates. The equation was proposed in 1884. Five years later, in 1889, the Dutch chemist JH van't Hoff provided its physical interpretation and justification.

The Eyring–Polanyi equation is equivalent to the empirical Arrhenius equation. This equation results from the transition state theory can be obtained from statistical thermodynamics and kinetic theory of gases. The general form of the Eyring–Polanyi equation is as follows:

$$k = \frac{k_B}{h} \cdot e^{-\frac{\Delta G}{RT}}, \tag{2.313}$$

where ΔG is the Gibbs energy of activation, k_B Boltzmann's constant ($k_B = 1.3806503 \times 10^{-23}$ [m^2 kgs^{-2} K^{-1}]), and h is Planck's constant (h = 6.62606957 (29) $\times 10^{-34}$ [Js]). The Boltzmann constant (k_B) is a physical constant relating energy to temperature at the individual particle level. It is the gas constant R = 8.3144621 (75) J $mol^{-1}K^{-1}$ divided by the Avogadro constant $N_A = 6.02214129$ (27) $\times 1023$ mol^{-1}:

$$k_B = \frac{R}{N_A}. \tag{2.314}$$

During thermodynamic change from a well-defined initial state to a well-defined end state, the Gibbs free energy (ΔG) is equal to the work exchanged by the system with the environment, less work pressure forces, during a reversible transformation of the system from the same initial state to the same final state. Gibbs energy (also referred to as ΔG) is also the chemical potential which is minimized when a system reaches equilibrium at constant pressure and temperature. The optimum conditions for a chemical reaction can be determined based on the following variables:

- *Physical state (solid, liquid, or gas):* When reactants are in the same phase, such as in aqueous solution, then thermal motion brings them into contact. However, when they are in different phases the reaction is limited solely to the interface between the reactants. This means that the more finely divided a solid or liquid reactant, the greater its specific surface area and therefore the faster the reaction.
- *Concentration:* The reaction proceeds more rapidly when there is a great number of collisions per unit of time, and this depends on the likelihood of the occurrence of particles in space, i.e. on their concentrations. Thus, an increase in the concentrations of the reactants will result in a corresponding increase in the reaction rate, while a decrease in the concentrations will have the opposite effect.
- *Temperature.* The higher the temperature, the greater the amount of energy supplied to the reaction, which increases the number of collisions. However, the activation energy has an even greater effect on reaction, resulting in an effective increase in the number of collisions which cause chemical bonds.
- *Pressure:* Increasing the pressure of the gas phase reaction of reactants increases the number of collisions, thus increasing the reaction rate. This is because the activity of gas is directly proportional to its partial pressure. This is similar to the effect of increasing the concentration of the solution

$$P_i = c_i \cdot RT \tag{2.315}$$

- *Electromagnetic radiation* is a form of energy, and therefore can accelerate reactions, or even spontaneously lead to breaking intramolecular bonds.

– *Agitating or mixing* of the reactants of a chemical reaction accelerates the reaction rate, since this increases the number of collisions.
– The *ionic strength of the solvent* is a measure of the concentration of ions in solution. The total concentration of ions present in the solvent may affect the solubility or dissociation constant of the various salts.

$$I = \frac{1}{2} \sum_{i=1}^{n} c_i \cdot z_i^2, \tag{2.316}$$

where c_i is the concentration of ions [$mol \cdot dm^{-3}$], z is the number of charges of these ions, and the sum takes all the ions in the solution into account. For electrolyte monovalent ions, the ionic strength is equal to the concentration.

2.6.2 Types of chemical bonds

Various attracting forces exist between atoms forming molecules of chemical compounds [68]. As a result of chemical reactions, these bonds between the atoms are changed, that is, the existing bonds can be decomposed or created entirely new. The bindings are caused by electrostatic forces between opposite charges, either between electrons and nuclei, or as the result of a dipole attraction (see Tab. 2.7).

Covalent chemical bonds is the bringing together of pairs of electrons between atoms, in order to achieve a stable balance of attractive and repulsive forces. In many molecules each atom is filled with an outer shell, which corresponds to a stable electronic configuration. Covalence is greatest between atoms of similar electronegativities [69]. In chemistry, binding energy – dissociation (BDE) or D0, is a measure of the strength of the chemical bond.

Hydrogen bonding interactions are electromagnetic polar hydrogen atoms of electronegative in the molecule. This type of attraction occurs between inorganic particles such as molecules of water and organic molecules such as proteins, DNA. Hydrogen bonds are responsible for the high boiling point of water.

Ionic bonding is a type of chemical bond formed as a result of the attraction of oppositely charged ions. Static electricity is generated when the total number of electrons is not equal to the total number of protons, giving the atom a positive or negative electric charge. These forces are described by Coulomb's law. The ions are formed between the cation, which is usually made of metal, and the anion, which is usually nonmetallic.

Metallic bonding involves electrostatic forces of attraction between delocalized electrons and positively charged metal ions. It is defined as the sharing of "free" electrons between the crystal lattice of positively charged ions (cations). Metallic bonding is sometimes compared to molten salt, however, this simplified view refers to very few metals.

Aromatic bonding occurs in when certain configurations of electrons and orbitals provide extra stability to a molecule. In benzene, the prototypical aromatic compound, the bonding electrons bind 6 carbon atoms together to form a planar ring structure.

Intermolecular forces which attract generally adjacent atoms, molecules or ions are weaker intramolecular forces keeping the atoms together within the molecule. There are four types of intermolecular forces: attractive dipole-force, ion-dipole forces, dipole-dipole or Debye forces, the force of attraction of the instantaneous dipole-dipole or London dispersion forces.

Debye force occurs in any polar molecule combined with a nonpolar and symmetrical molecule. This is much weaker than the dipole-dipole, but greater than the London dispersion force.

London dispersion force is formed by the instantaneous formation of dipoles in the neighboring atoms. The negative charge of the electron is not constantly uniform around the entire atom. This small charge induces a corresponding dipole in the vicinity of the molecule, causing mutual attraction. London dispersion forces are the forces acting between the kinds of atoms and molecules [70] and are part of the van der Waals forces are not present between the atoms in liquid form.

Van der Waals forces are relatively weak compared to covalent bonds but play an important role in sorption processes as well as in fields as diverse as supramolecular chemistry, structural biology, polymer science, nanotechnology, surface science, and condensed matter physics.

Table 2.7. The strength of chemical bonds

Bond type	Dissociation energy [J/mol]
Covalent	1675
Hydrogen	50–67
Dipole-dipole	2–8
London and van der Waals	< 4

2.6.3 Catalysts

The use of catalytic rather than stoichiometric reactions is one of the twelve postulates of green chemistry because of the smaller number of waste and byproducts. In contrast to other substrates, the enzyme is not consumed in the reaction itself.

Catalysts are of great importance in the global economy. It has been calculated that catalytic processes led to production of goods with a total value of more than $ 900 billion in 2005. Catalysts do not alter the course of the reaction and have no effect on the thermodynamic equilibrium, because they accelerate the reactions simultaneously in two opposite directions.

Catalysts which delay reaction rates are called inhibitors. Substances that enhance the activity of catalysts are called promoters and those that inactivate or reduce the activity of catalysts are called poisons. In general catalysts reduce the energy required, which means that the reaction is performed faster at the same temperature. The rates of catalytic reaction can be determined based on the weight of the catalyst (g mol^{-1} s^{-1}) or the surface on which it occurs (mol m^{-2} s^{-1}) or the renewal rate, also known as turn-over rate (s^{-1}).

There are two main types of catalysts, i.e. hetero- and homogeneous. Heterogeneous catalysts operate in a different phase to the substrates present. Most heterogeneous catalysts are usually immobilized in dispersed form which increases their efficiency. The most common catalyst is the proton H$^+$, and the rare earth metals.

Homogeneous catalysts operate in the same phase in which the reactants are present.

The presence of the catalyst opens different reaction pathways with lower activation energy. However, the catalyst does not change the end result. Proteins which act as catalysts in biochemical reactions are known as enzymes. Michaelis-Menten kinetics describes the simple rate of enzymatic reactions.

3 Mathematical methods in design

3.1 Dimensional analysis

Dimensional analysis is a mathematical method of determining the functional relationships between physical quantities based on their dimensions. The basic principles of dimensional analysis were formulated by Isaac Newton in 1686, who wrote about them in the book "Great Principle of Similitude" [71]. James Clerk Maxwell also played an important role in the use of dimensional analysis by highlighting the fundamental units of mass, length, and time. In the 19th century, French mathematician Joseph Fourier [72] wrote: "Beginning apparently with Maxwell, mass, length and time began to be interpreted as having a privileged fundamental character and all other quantities as derivative, not merely with respect to measurement, but with respect to their physical status as well." This led to the conclusion that the law must be invariant and uniform in relation to the different units of measure. These demands are formulated in the form of the famous "PI" theorem by Edgar Buckingham (1867–1940) [73, 74]. Edgar Buckingham was a physicist. He graduated from Harvard with a bachelor's degree in physics in 1887. He did additional graduate work at the University of Strasbourg and the University of Leipzig, where he received a PhD in 1893.

Dimensional analysis has become a convenient tool for mathematical modeling of "difficult cases", especially when the experimental data available show evident relationships but a strictly physical model is not yet available.

Dimensional variables are those whose values depend on the adopted system of units. The fundamental dimensional units are the physical quantities which are always dimensionally-independent and can generate all the other dimensional elements. The elements of dimensional space are dimensionally-independent if from the equation

$$A_1^{a_1} \cdot A_2^{a_2} \cdot A_3^{a_3} \cdot \dots \cdot \prod_{i=1}^{m} A_i = \alpha_i, \tag{3.1}$$

where α – dimensionless quantity, and a_i – real numbers, indicates that all of them are zero:

$$a_1 = a_2 = \dots = a_m = 0. \tag{3.2}$$

3.1.1 The general method for verifying dimensional independence

1. Select a number of the dimensional size A^i (index).
2. Assume a system of basic units, all of which, by definition, are dimensionally-independent.

3. Express dimensional variables A_i by means of basic units X_i in the selected system, as follows:

$$A_1 = \alpha_1 \cdot X_1^{a_{11}} \cdot X_2^{a_{12}} \cdots \cdots X_n^{a_{1n}} \tag{3.3}$$

$$A_2 = \alpha_2 \cdot X_1^{a_{21}} \cdot X_2^{a_{22}} \cdots \cdots X_n^{a_{2n}} \tag{3.4}$$

$$\cdots\cdots\cdots\cdots\cdots\cdots\cdots\cdots\cdots\cdots$$

$$A_m = \alpha_m \cdot X_1^{a_{m1}} \cdot X_1^{a_{m2}} \cdots \cdots X_1^{a_{mn}}, \tag{3.5}$$

where n is the number of base units, m is the number of dimensional variables, and the inequality m < n occurs.

4. Create a matrix with the corresponding exponents:

$$
\begin{matrix}
a_{11} & a_{12} & a_{13} & \cdots & a_{1n} \\
a_{21} & a_{22} & a_{23} & \cdots & a_{2n} \\
a_{31} & a_{32} & a_{33} & \cdots & a_{3n} \\
\cdots & \cdots & \cdots & \cdots & \cdots \\
a_{m1} & a_{3m} & a_{3m} & \cdots & a_{mn.}
\end{matrix}
\tag{3.6}
$$

5. The "minor" is the biggest determinant of the given rectangular matrix (here n x m). Check all minors of degree "m". If at least one of them is different from zero the dimensional variables being investigated are dimensionally-independent. A minor of a matrix A is the determinant of some smaller square matrix, cut down from A by removing one or more of its rows or columns. Minors are obtained by removing just one row and one column from square matrices.

3.1.2 Modeling of the functions with all dimensionally-independent arguments

Dimensional function is a function the arguments and values of which are dimensional values:

$$Z = \Phi\left(Z_1, Z_2, \ldots, Z_s\right). \tag{3.7}$$

Theorem 1 of dimensional analysis. If dimensionally-invariant and uniform dimensional function has dimensionally-independent arguments, then its form is determined by the product of the powers, as follows:

$$\Phi\left(A_1, A_2, \ldots, A_s\right) = \alpha \cdot \prod_{i=1}^{m} A_i^{a_i}, \tag{3.8}$$

where α – dimensionless coefficients and a_i – are real-exponents. It should be emphasized that the condition of dimensional independence of the arguments seriously limits the application of this theorem. However, keep in mind that the set dimensionally independent arguments can be made up of different combinations of physical variables, as long as they meet the requirement of dimensional independence. However, the number of independent dimensions of such arguments is always constant and refers to the dimension of the so-called dimensional space, i.e. the number of basic units describing the given physical phenomenon, such as kilogram, meter, second etc. The practical usefulness of this theorem can be checked using a simple example.

Example 1. Based on Theorem 1 of dimensional analysis, determine the dimensional function which will describe the period of vibration of the pendulum assuming that the arguments are the following values: m – mass of the pendulum, l – length of the pendulum, and g – the acceleration of gravity:

$$T = \Phi(m, l, g) \tag{3.9}$$

Solution:

1. Check the dimensional independence arguments: m, l, g to this end, must express established arguments (Z: m, l, g) in the assumed system of basic units (X: kg, m, s):

$$[m] = kg^1 m^0 s^0 \tag{3.10}$$

$$[l] = kg^0 m^1 s^0 \tag{3.11}$$

$$[g] = kg^0 m^1 s^{-2}. \tag{3.12}$$

2. Then create and calculate the determinant formed from the corresponding exponents. The determinant may be calculated by sequentially reducing its degree according to known rule as follows:

$$\begin{vmatrix} 1 & 0 & 0 \\ 0 & 1 & 0 \\ 0 & 1 & -2 \end{vmatrix} = 1 \cdot (-1)^{1+1} \begin{vmatrix} 1 & 0 \\ 1 & -2 \end{vmatrix} = -2 \neq 0. \tag{3.13}$$

3. Now you can see that the value of the determinant is nonzero and thus examined arguments are dimensionally-independent. Thus, by Theorem 1, the dimensional form of the function is the product of series of the powers

$$T = \alpha \cdot m^{a_1} \cdot l^{a_2} \cdot g^{a_3}. \tag{3.14}$$

4. Now we determine the exponents in the exponential expression based on the identity of the dimensions on both sides of the equation

$$\left(kg^0 m^0 s^1\right) \equiv \alpha \cdot \left(kg^1 m^0 s^0\right)^{a_1} \cdot \left(kg^0 m^1 s^0\right)^{a_2} \cdot \left(kg^0 m^1 s^{-2}\right)^{a_3}. \tag{3.15}$$

5. From this identity, the three equations result from the appropriate exponents of the basic units (in accordance with the rules of exponentiation):

$$\text{kg:} \quad 1 \cdot a_1 + 0 \cdot a_2 + 0 \cdot a_3 = 0 \tag{3.16}$$

$$\text{m:} \quad 0 \cdot a_1 + 1 \cdot a_2 + 1 \cdot a_3 = 0 \tag{3.17}$$

$$\text{s:} \quad 0 \cdot a_1 + 0 \cdot a_2 - 2 \cdot a_3 = 1 \tag{3.18}$$

Now, from the set of these three equations, the three missing exponents may be calculated

$$a_1 = 0$$
$$a_2 = -a_3 = \frac{1}{2}$$
$$a_3 = \frac{1}{2}.$$

Finally, the sought (by Theorem 1) dimensional function is

$$T = \alpha \cdot m^0 \cdot l^{\frac{1}{2}} \cdot g^{-\frac{1}{2}} = \alpha \cdot \sqrt{\frac{l}{g}}. \tag{3.19}$$

Here, the constant α, can be determined experimentally using the experimental identification procedure of the mathematical models, which is discussed later. Noteworthy is the elimination of the variable "m" as an argument of dimensional function determining T – the period of oscillation of the pendulum. In this case, dimensional analysis is enabled for the verification of the variables yet before performing the experiment, which is the additional role of dimensional analysis. The condition imposed on the number of arguments of dimensional functions always causes major difficulty during the process of mathematical modeling. Taking into account the nature of the physical phenomena described the arguments of dimensional functions are selected, rather than their numbers.

3.1.3 Modeling of dimensional function with the "dimensionally dependent" arguments

In most practical problems we are dealing with a numbers of arguments (s) which is greater than the number of dimensionally independent (m) units that is the number of base units that must be used to describe the phenomenon:

$$Z = \Phi(Z_1, Z_2, \ldots, Z_s). \tag{3.20}$$

Among all of the s-dimensional function arguments, we can extract the number (m = n) of arguments that are always dimensionally-independent (the dimensional base) and can be marked by the symbols A, and the remaining (r = s − m) arguments that are dimensionally-dependent, are labeled B:

$$Z = \Phi(A_1, A_2, \ldots, A_m, B_1, B_2, \ldots, B_r). \tag{3.21}$$

There is a choice between different dimensional bases, although the number of variables included in each of these bases is constant and equal to the size of the m-dimensional space (i.e., the number of base units which describe the phenomenon).

The problem of using dimensional analysis for mathematical modeling is to find a function of the dimension for an unlimited number of arguments that will meet the conditions of dimensional invariance and dimensional uniformity. The character of such a function is determined by the central theorem of dimensional analysis called as Buckingham's or "Π" theorem.

Theorem 2 (Buckingham's theorem, "Π" theorem). If a dimensionally-invariant function

$$Z = \Phi\left(A_1, A_2, \ldots, A_m, B_1, B_2, \ldots, B_r\right) = f\left(\pi_1, \pi_2, \ldots, \pi_r\right) \cdot \prod_{i=1}^{m} A_i^{a_i} \tag{3.22}$$

has m-dimensionally independent arguments (type A), and other arguments (type B) that are dimensionally-dependent (from previous ones), and which can be expressed by using a dimensionally-independent arguments:

$$B_j = \pi_j \prod_{i=1}^{m} A_i^{a_{ij}} \quad \text{for} \quad j = 1, 2, \ldots, r, \tag{3.23}$$

then this function can be expressed as

$$Z = \Phi\left(A_1, A_2, \ldots, A_m, B_1, B_2, \ldots, B_r\right) = f\left(\pi_1, \pi_2, \ldots, \pi_r\right) \cdot \prod_{i=1}^{m} A_i^{a_i}, \tag{3.24}$$

where $f(\pi_1, \pi_2, \ldots, \pi_r)$ is a real function of the dimensionless arguments π_j. The arguments π_j and a_i are real numbers, where $j = 1, 2, 3, \ldots, n$ and $i = 1, 2, 3, \ldots, m$. Dimensionless modules π_j are determined by expressing the dimensionally-dependent quantities B_y, by the dimensionally-independent A_i variables, according to the following formula:

$$\pi_j = \frac{B_j}{\prod\limits_{i=1}^{m} A_i^{a_{ij}}} \tag{3.25}$$

Each function of the form specified by the Buckingham theorem has the following properties: it is dimensionally homogeneous, and is dimensionally invariant.

Example 2. Determine the form of the dimensional function which determines the sedimentation rate of the solid particles in a liquid, if the observation is that the main physical quantities that determine the sedimentation rate are the following: μ – the viscosity of the liquid, $\Delta\rho$ – a difference of density between solid and fluid, d – particle size, and g – acceleration due to gravity.

Solution. In this case, the dimensional function (u) has four parameters (s = 4):

$$u = \Phi(\mu, d, \Delta\rho, g). \tag{3.26}$$

The dimension of dimensional space, which describes the phenomenon of gravitational sedimentation, is m = 3, since all physical quantities occurring as function arguments dimensional (u) can be expressed in the system of three basic units (X) here: weight [kg], linear dimension [m] and time [s]. The number of dimensionally-independent arguments is m = 3, total number of the arguments of dimensional function u is s = 4. Thus, one argument is dependent on the other dimensions (r = s − m = 4 − 3 = 1). The first step is to determine which arguments are dimensionally-independent. Select, and then check the dimensional independence such as the first three arguments (μ, d, Δρ). For this purpose, we express them in a system of the base units (kg, m, and s):

$$\mu = \alpha_1 \cdot kg^{a_{11}} \cdot m^{a_{21}} \cdot s^{a_{31}} = \alpha_1 \cdot kg^1 \cdot m^{-1} \cdot s^{-1} \qquad (3.27)$$

$$d = \alpha_2 \cdot kg^{a_{12}} \cdot m^{a_{22}} \cdot s^{a_{32}} = \alpha_2 \cdot kg^0 \cdot m^1 \cdot s^0 \qquad (3.28)$$

$$\Delta\rho = \alpha_3 \cdot kg^{a_{13}} \cdot m^{a_{23}} \cdot s^{a_{33}} = \alpha_3 \cdot kg^1 \cdot m^{-3} \cdot s^0. \qquad (3.29)$$

Now we can create and then check the value of the determinant formed from the corresponding exponents for the basic units. To this end, we reduce the degree of the determinant by expanding it of the second row and second column:

$$\begin{vmatrix} a_{11} & a_{12} & a_{13} \\ a_{21} & a_{22} & a_{23} \\ a_{31} & a_{32} & a_{33} \end{vmatrix} = \begin{vmatrix} 1 & -1 & -1 \\ 0 & 1 & 0 \\ 1 & -3 & 0 \end{vmatrix} = 1 \cdot (.1)^{2+2} \begin{vmatrix} 1 & -1 \\ 1 & 0 \end{vmatrix} = 1 \cdot (1\dot{0} - (-1) \cdot 1) = 1 \neq 0. \quad (3.30)$$

The value of the determinant is nonzero, which means that the selected variables are dimensionally-independent. Therefore, the dimensional form of the function describes the Buckingham theorem:

$$u = f(\pi_1) \cdot \mu^{a_1} \cdot d^{a_2} \cdot \Delta\rho^{a_3} \qquad (3.31)$$

The values of the exponents are calculated based on the identity of the dimensions on both sides of the equation:

$$\left(kg^0 \cdot m^1 \cdot s^{-1}\right) \equiv \left(kg^1 \cdot m^{-1} \cdot s^{-1}\right)^{a_1} \cdot \left(kg^0 \cdot m^1 \cdot s^0\right)^{a_2} \cdot \left(kg^1 \cdot m^{-3} \cdot s^0\right)^{a_3}.$$

Balancing of exponents at the appropriate basic units leads to three equations:

$$kg: \qquad 1 \cdot a_1 + 0 \cdot a_2 + 1 \cdot a_3 = 0 \qquad (3.32)$$

$$m: \qquad -1 \cdot a_1 + 1 \cdot a_2 - 3 \cdot a_3 = 1 \qquad (3.33)$$

$$s: \qquad -1 \cdot a_1 + 0 \cdot a_2 + 0 \cdot a_3 = -1. \qquad (3.34)$$

The values of exponents are calculated from these equations to obtain: $a_1 = 1$, $a_2 = -1$, $a_3 = -1$. Thus, using Buckingham's theorem, the rate of descent of solid particles in liquids can be described as a function of the form

$$u = f(\pi_1) \cdot \mu^1 \cdot d^{-1} \cdot \Delta\rho^{-1} = f(\pi_1) \cdot \frac{\mu}{d \cdot \Delta\rho}. \qquad (3.35)$$

The dimensionless module π_1 is determined on the basis of the equation expressing the variable (g) dependent on the arguments dimensionally independent:

$$g = \pi_1 \cdot \mu^{a_{11}} \cdot d^{a_{21}} \cdot \Delta\rho^{a_{31}}. \tag{3.36}$$

Again, for this purpose we use the following dimensional identity:

$$\left(kg^0 \cdot m^1 \cdot s^{-2}\right) \equiv \left(kg^1 \cdot m^{-1} \cdot s^{-1}\right)^{a_{11}} \cdot \left(kg^0 \cdot m^1 \cdot s^0\right)^{a_{21}} \cdot \left(kg^1 \cdot m^{-3} \cdot s^0\right)^{a_{31}}, \tag{3.37}$$

which leads to a system of three equations:

$$\text{kg:} \qquad 1 \cdot a_{11} + 0 \cdot a_{21} + 1 \cdot a_{31} = 0 \tag{3.38}$$

$$\text{m:} \qquad -1 \cdot a_{11} + 1 \cdot a_{21} - 3 \cdot a_{31} = 1 \tag{3.39}$$

$$\text{s:} \qquad -1 \cdot a_{11} + 0 \cdot a_{21} + 0 \cdot a_{31} = -2. \tag{3.40}$$

The values of the exponents in this case are as follows: $a_{11} = 2$, $a_{21} = -3$, $a_{31} = -2$. The dimensionally dependent variable g is expressed by the independent variables as

$$g = \pi_1 \cdot \mu^2 \cdot d^{-3} \cdot \Delta\rho^{-2}. \tag{3.41}$$

Hence the dimensionless unit (in this case the Galileo number) is given by:

$$\pi_1 = \frac{g \cdot d^3 \cdot \Delta\rho^2}{\mu^2} = Ga. \tag{3.42}$$

Thus, the sought rate of sedimentation, determined on the basis of Buckingham's theorem, can be presented in dimensional form as follows:

$$u = f\left(\frac{g \cdot d^3 \cdot \Delta\rho^2}{\mu^2}\right) \cdot \frac{\mu}{d \cdot \Delta\rho}, \tag{3.43}$$

or in the dimensionless form

$$\frac{u \cdot d \cdot \Delta\rho}{\mu} = f\left(\frac{g \cdot d^3 \cdot \Delta\rho^2}{\mu^2}\right), \tag{3.44}$$

and finally

$$Re = f(Ga). \tag{3.45}$$

The overall conclusion is that the dimensional analysis made it possible to replace the dimensional function (here rate of descent), by the ordinary numerical function f, which still at this point cannot be determined. It may be determined on the basis of experiments using one of the known approximation procedures. This may be a regression analysis, also known as the least squares method, which will be discussed in the next section.

3.2 Identification of mathematical models of processes

The relationships obtained using the pi theorem do not specify a finite form of the numerical function $f(\pi_1, \pi_2, \pi_3, \ldots, \pi_n)$, but suggest only the form of dimensionless modules instead of dimensional variables. However, the complete form of this function can be determined, based on the experiment using identification procedures (see Fig. 3.1). Identification of the function is determination of class, type, form, and estimation of the numerical values of parameters, which are based on the experiments. The tasks associated with the identification of the function, are the following:

1. definition of input and output variables object;
2. designation of function (model) class;
3. carrying out the experiments on the given subject;
4. selection of the algorithms associated with this criterion for identification;
5. determination of model parameters.

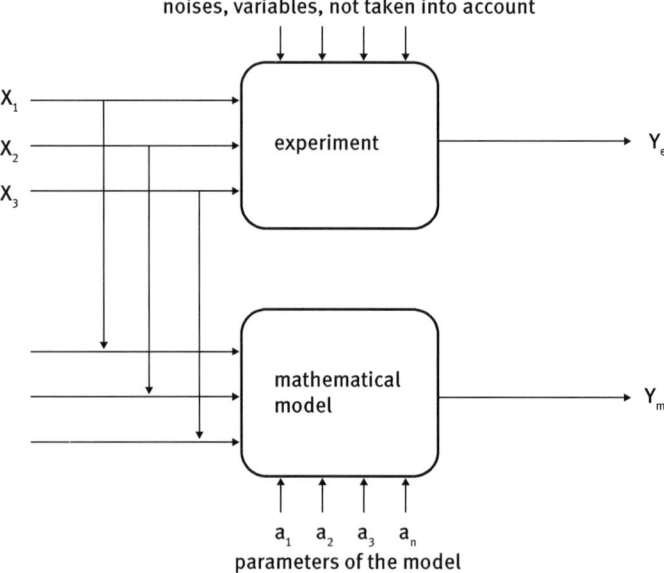

Fig. 3.1. The idea of identifying the mathematical model based on experimental results.

In the general case identified, function can have many input variables (arguments):

$$Y = f(X_1, X_2, \ldots, X_r). \tag{3.46}$$

Restrict further considerations to linear functions, i.e. suitable for linearization:

$$Y = a_0 + a_1 X_1 + a_2 X_2 + \cdots + a_r X_r. \tag{3.47}$$

Many functions can be linearized using the appropriate substitutions. In the experimental part, the values of the output variables (Y_i) of the object are determined for the vector of input variables ($X_{1i}, X_{2i}, \ldots, X_{RI}$) during subsequent realization of (i) the experiment, giving so called "experimental points" (see Tab. 3.1).

Table 3.1. Example of a table with experimental results

Number of experiment i \ number of variable	X_{1i}	X_{2i}	...	Y_{ei}	Y_{mi}
1			...		
2			...		
3			...		
...	...	...	...		
n			...		

Note that the experimental values are always subject to some error (see Fig. 3.2). Thus the mathematical model must be a compromise between the accuracy of the description, and the number of experiments made.

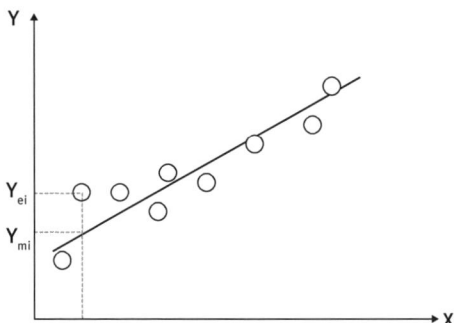

Fig. 3.2. The linear approximation of experimental results.

The criteria of model identification determine the "distance" between a set of experimental results and the results of model calculations which are carried out for the same input variable vector. The criteria are numerical measures of differences between "response" of the object and the model. The criteria are defined in such a way that the differences "in plus" and "in minus" could not be mutually reduced. Here are some examples that are used in practice as identification criteria:

$$d = \sum_{i=1}^{n} (Y_{mi} - Y_{ei})^2 \tag{3.48}$$

$$d = \max (Y_{mi} - Y_{ei}) \tag{3.49}$$

$$d = \sqrt{\sum_{i=1}^{n} (Y_{mi} - Y_{ei})^2}.$$

(3.50)

The first of these criteria is then a measure of the quality of approximation used in the regression analysis method known as the "least squares". The optimal vector of model parameters is the vector of the established parameters $\vec{a} = (a_0, a_1, a_2, \ldots, a_r)$, for which "identification criterion d", reaches its minimum, as follows:

$$\vec{a}_{opt} = (a_0, a_1, a_2, \ldots, a_r) = a : \{d = d(a_0, a_1, a_2, \ldots, a_r) = d_{min}\}.$$

(3.51)

If we want to use the "least squares method" of regression, we now need to incorporate the standard criterion according to this method:

$$d = \sum_{i=1}^{n} (Y_i^m - Y_i^e)^2.$$

(3.52)

Assuming linear regression, we got the form of the identified linear model:

$$Y_i^m = a_0 + a_1 \cdot X_1 + a_2 \cdot X_2 + \cdots + a_r \cdot X_r.$$

(3.53)

To calculate the value of Y_i^m, a parameter vector must be set $\vec{a}_{opt} = (a_0, a_1, a_2, \ldots, a_r) = a : \{d = d(a_0, a_1, a_2, \ldots, a_r) = d_{min}\}$, which at this stage we do not know. So we need to optimize our criterion "d", which is a function of "a" parameter experiments after Y_i^m values are known. After substituting Y_i^m to our criterion function d, we get the criterion for optimizing as follows:

$$d = \sum_{i=1}^{n} (a_0 + a_1 \cdot X_1 + a_2 \cdot X_2 + \cdots + a_r \cdot X_r - Y_i^e)^2.$$

(3.54)

The optimal vector of parameters (regression coefficients) of the linear model is determined in such way as the conditions for the minimum of a function of many variables, so throughout comparing successive partial derivatives to zero:

$$\frac{\partial d}{\partial a_0} = 0 = \frac{\partial}{\partial a_0} \left[\sum_{i=1}^{n} (a_0 + a_1 \cdot X_{1i} + a_2 \cdot X_{2i} + \cdots + a_r \cdot X_{ri} - Y_i^e)^2 \right]$$

(3.55)

$$\frac{\partial d}{\partial a_1} = 0 = \frac{\partial}{\partial a_1} \left[\sum_{i=1}^{n} (a_0 + a_1 \cdot X_{1i} + a_2 \cdot X_{2i} + \cdots + a_r \cdot X_{ri} - Y_i^e)^2 \right]$$

(3.56)

$$\cdots\cdots\cdots\cdots\cdots\cdots\cdots\cdots\cdots\cdots\cdots\cdots\cdots\cdots\cdots\cdots\cdots\cdots\cdots\cdots$$

$$\frac{\partial d}{\partial a_r} = 0 = \frac{\partial}{\partial a_r} \left[\sum_{i=1}^{n} (a_0 + a_1 \cdot X_{1i} + a_2 \cdot X_{2i} + \cdots + a_r \cdot X_{ri} - Y_i^e)^2 \right].$$

(3.57)

After differentiation of the criterial expression, rearrangement and entering the appropriate sums that are known after performing the experiment, the regression coefficients can now be calculated based on the following set of equations:

$$a_0 \cdot n + a_1 \cdot \sum_{i=1}^{n} X_{1i} + a_2 \cdot \sum_{i=1}^{n} X_{2i} + \cdots + a_r \cdot \sum_{i=1}^{n} X_{ri} = \sum_{i=1}^{n} Y_i^e \tag{3.58}$$

$$a_0 \cdot \sum_{i=1}^{n} X_{1i} + a_1 \cdot \sum_{i=1}^{n} (X_{1i})^2 + a_2 \cdot \sum_{i=1}^{n} (X_{1i} \cdot X_{2i}) + \cdots + a_r \cdot \sum_{i=1}^{n} (X_{1i} \cdot X_{ri}) = \sum_{i=1}^{n} (X_{1i} \cdot Y_i^e) \tag{3.59}$$

$$a_0 \cdot \sum_{i=1}^{n} X_{2i} + a_1 \cdot \sum_{i=1}^{n} (X_{1i} \cdot X_{2i}) + a_2 \cdot \sum_{i=1}^{n} (X_{2i})^2 + \cdots + a_r \cdot \sum_{i=1}^{n} (X_{2i} \cdot X_{ri}) = \sum_{i=1}^{n} (X_{2i} \cdot Y_i^e) \tag{3.60}$$

$$a_0 \cdot \sum_{i=1}^{n} X_{2i} + a_1 \cdot \sum_{i=1}^{n} (X_{1i} \cdot X_{2i}) + a_2 \cdot \sum_{i=1}^{n} (X_{2i} \cdot X_{ri}) + \cdots + a_r \cdot \sum_{i=1}^{n} (X_{ri}) = \sum_{i=1}^{n} (X_{ri} \cdot X_i^e). \tag{3.61}$$

Please note that the system of equations is always unambiguously determined in that the number of equations is always equal to the number of unknowns. In this equations system, the parameters of the model sought are factors a. The known are relevant sums, since combinations of the relevant empirical variables of X and Y under sum are definitely known after the experiments have been performed.

3.3 The theory of similarity

Testing on real objects is very expensive and often not feasible. These difficulties can be overcome by performing tests on physical models of real objects. Models are made in a different scale usually smaller (for example on a laboratory scale, technical, pilot, etc.) (see Fig. 3.3). Information about objects in real scale (target) can be obtained easily on the basis of this smaller scale, provided that criteria for the similarity of the model and the object have been applied. The model is marked by the lower dash under variable (see in Fig. 3.3).

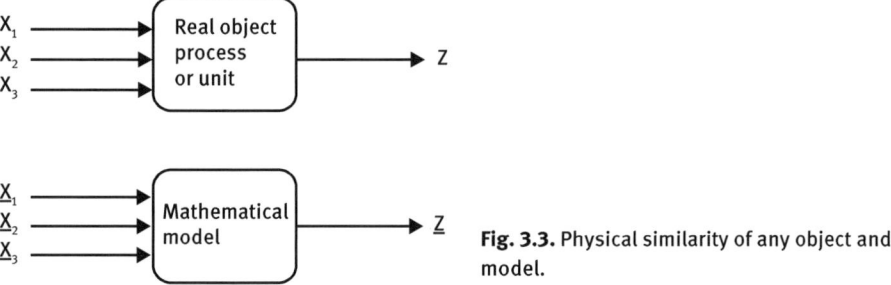

Fig. 3.3. Physical similarity of any object and model.

Dimensional analysis suggests that the answer to the question of how to determine the similarity criteria for different phenomena, or what conditions must be met to be able to determine the scale of this similarity. Let us consider the case in which the phenomenon is described in a variable that is a function of the assumed dimensional variables, wherein the variable A is dimensionally-independent but the variable B is dimensionally-dependent on A, as follows:

$$Z = \Phi\left(A_1, A_2, \ldots, A_m, B_1, B_2, \ldots, B_r\right). \tag{3.62}$$

The same conditions may be assumed for the model variables:

$$\underline{Z} = \Phi\left(\underline{A}_1, \underline{A}_2, \ldots, \underline{A}_m, \underline{B}_1, \underline{B}_2, \ldots, \underline{B}_r\right). \tag{3.63}$$

The ratios between relevant input variables are known as scales:

$$\frac{A_1}{\underline{A}_1} = \alpha_1, \ldots, \frac{A_m}{\underline{A}_m} = \alpha_m, \tag{3.64}$$

$$\frac{B_1}{\underline{B}_1} = \alpha_1, \ldots, \frac{B_m}{\underline{B}_m} = \beta_m. \tag{3.65}$$

Scales output variables are defined as the transition scales:

$$\frac{Z}{\underline{Z}} = \lambda. \tag{3.66}$$

Using the theorem of Buckingham we may define a scale of the transition:

$$\lambda = \frac{Z}{\underline{Z}} = \frac{\Phi\left(A_1, A_2, \ldots, A_m, B_1, B_2, \ldots, B_r\right)}{\Phi\left(\underline{A}_1, \underline{A}_2, \ldots, \underline{A}_m, \underline{B}_1, \underline{B}_2, \ldots, \underline{B}_r\right)} = \frac{f\left(\pi_1, \pi_2, \ldots, \pi_r\right) \cdot \prod_{i=1}^{m} A_i^{a_i}}{f\left(\underline{\pi}_1, \underline{\pi}_2, \ldots, \underline{\pi}_r\right) \cdot \prod_{i=1}^{m} \underline{A}_i^{a_i}}. \tag{3.67}$$

Thus, the scale of the transition can be expressed by means of scale arguments:

$$\lambda = \frac{f\left(\pi_1, \pi_2, \ldots, \pi_r\right)}{f\left(\underline{\pi}_1, \underline{\pi}_2, \ldots, \underline{\pi}_r\right)} \prod_{i=1}^{m} \alpha_i^{a_i}. \tag{3.68}$$

The form of the function f is not yet known, but if the arguments are equal then the scale of the transition can be easily calculated without knowing the function f, as follows:

$$\lambda = \prod_{i=1}^{m} \alpha_i^{a_i}. \tag{3.69}$$

Note, however, that it is a necessary condition for equality of the relevant dimensionless of modules, which we call in this case the similarity criteria.

$$\pi_j = \underline{\pi}_j \quad \text{for} \quad j = 1, 2, \ldots, r. \tag{3.70}$$

Criteria of similarity in this case can be written in a different form by reasserting the Buckingham theorem to express the arguments of numerical function f:

$$\frac{B_j}{\prod\limits_{i=1}^{m} A_i^{a_{ij}}} = \frac{\underline{B}_j}{\prod\limits_{i=1}^{m} \underline{B}_i^{a_{ij}}}. \tag{3.71}$$

Thus, as a result of equality π dimensionless of modules, we get the equation, which is the "criterial equation" of similarity. Newton's first theorem of similarity, which is used in the cases described by dimensional function, can thus be expressed as follows: if the criteria of similarity are met:

$$\beta_j = \prod_{i=1}^{m} \alpha_i^{a_{ij}}. \tag{3.72}$$

Then these phenomena are similar and the scale of the transition can be calculated as the products of the dimensionally independent arguments:

$$\lambda = \prod_{i=1}^{m} \alpha_i^{a_i} \quad \text{for} \quad j = 1, 2, \ldots, r \quad \text{and} \quad i = 1, 2, \ldots, m. \tag{3.73}$$

If the real phenomenon is described by equations (differential, integral or algebraic), whose solution is difficult or impossible, the criterial equation can be used to describe the phenomenon under the second theorem, as formulated by Newton. The equation or set of equations that are complete and homogeneous and describe the physical phenomenon can be replaced by criterial equations, in the form of a functional relationship between the dimensionless similarity criteria.

Example 3. The Navier–Stokes equation describes the flow of real fluids (viscous). It is very difficult to use manually, though it is a very useful equation that is lately being used in computational fluid dynamics software. However, based on dimensional analysis of the equation, we can generate a set of similarity criteria which are often used in the engineering process for empirical correlations:

$$\rho \cdot \frac{\partial u_x}{\partial \tau} + \rho \cdot \left(u_x \cdot \frac{\partial u_x}{\partial x} + u_y \cdot \frac{\partial u_x}{\partial y} + u_z \cdot \frac{\partial u_x}{\partial z} \right) - \rho \cdot g_x + \frac{\partial P}{\partial x} - \mu \cdot \left(\frac{\partial^2 u_x}{\partial x^2} + \frac{\partial^2 u_y}{\partial y^2} + \frac{\partial^2 u_z}{\partial z^2} \right) = 0 \tag{3.74}$$

Assuming that the scales of all dimensions of this equation are given (see Tab. 3.2 below), flow rate must be described with the identical equation wherein all parameters (geometric and physical characteristics) can have different values. The equation for the 'similar realization' (in the sense of the similarity theory) is written using underlined variables to distinguish from the primary realization.

$$\underline{\rho} \cdot \frac{\partial \underline{u_x}}{\partial \underline{\tau}} + \underline{\rho} \cdot \left(\underline{u_x} \cdot \frac{\partial \underline{u_x}}{\partial \underline{x}} + \underline{u_y} \cdot \frac{\partial \underline{u_x}}{\partial \underline{y}} + \underline{u_z} \cdot \frac{\partial \underline{u_x}}{\partial \underline{z}} \right) - \underline{\rho} \cdot \underline{g}_x + \frac{\partial \underline{P}}{\partial \underline{x}} - \underline{\mu} \cdot \left(\frac{\partial^2 \underline{u_x}}{\partial \underline{x}^2} + \frac{\partial^2 \underline{u_y}}{\partial \underline{y}^2} + \frac{\partial^2 \underline{u_z}}{\partial \underline{z}^2} \right) = 0 \tag{3.75}$$

We can now express the relationship between the corresponding values (i.e. between the first realization and its similar values) by using the corresponding scales:

$$\left[\frac{\alpha_\rho \cdot \alpha_u}{\alpha_\tau} \right] \cdot \rho \cdot \frac{\partial u_x}{\partial \tau} + \left[\frac{\alpha_\rho \cdot \alpha_u^2}{\alpha_l} \right] \cdot \rho \cdot \left(u_x \cdot \frac{\partial u_x}{\partial x} + u_y \cdot \frac{\partial u_x}{\partial y} + u_z \cdot \frac{\partial u_x}{\partial z} \right)$$

$$- \left[\alpha_\rho \cdot \alpha_g \right] \cdot \rho \cdot g_x + \left[\frac{\alpha_P}{\alpha_l} \right] \cdot \frac{\partial P}{\partial x} - \left[\frac{\alpha_\mu \cdot \alpha_u}{\alpha_l^2} \right] \cdot \mu \cdot \left(\frac{\partial^2 u_x}{\partial x^2} + \frac{\partial^2 u_y}{\partial y^2} + \frac{\partial^2 u_z}{\partial z^2} \right) = 0. \tag{3.76}$$

If this equation describes the flow in both realization (i.e. the first and the second, similar realization with underlined variables), the values in the square brackets must be identically equal to the unity. Conditions that must be met so that both implementations are similar are called similarity conditions (see Tab. 3.3):

$$\left[\frac{\alpha_p \cdot \alpha_u}{\alpha_\tau}\right] = \left[\frac{\alpha_p \cdot \alpha_u^2}{\alpha_l}\right] = \left[\alpha_p \cdot \alpha_g\right] = \left[\frac{\alpha_P}{\alpha_l}\right] = \left[\frac{\alpha_\mu \cdot \alpha_u}{\alpha_l^2}\right] = 1. \tag{3.77}$$

Table 3.2. Scales of similarities for various physical quantities

Scale of density	$\alpha_p = \dfrac{\rho}{\underline{\rho}}$
Scale of velocity	$\alpha_u = \dfrac{u_x}{\underline{u}_x} = \dfrac{u_y}{\underline{u}_y} = \dfrac{u_z}{\underline{u}_z}$
Scale of linear dimensions	$\alpha_l = \dfrac{x}{\underline{x}} = \dfrac{y}{\underline{y}} = \dfrac{z}{\underline{z}}$
Scale of acceleration	$\alpha_g = \dfrac{g_x}{\underline{g}_x}$
Scale of pressure	$\alpha_P = \dfrac{P}{\underline{P}}$
Scale of viscosity	$\alpha_\mu = \dfrac{\mu}{\underline{\mu}}$
Scale of time	$\alpha_\mu = \dfrac{\tau}{\underline{\tau}}$

Table 3.3. The method for constructing characteristic criterial numbers

$\left[\dfrac{\alpha_p \cdot \alpha_u^2}{\alpha_l}\right] = \left[\dfrac{\alpha_\mu \cdot \alpha_u}{\alpha_l^2}\right]$	$\dfrac{u \cdot l \cdot \rho}{\mu} = \dfrac{\underline{u} \cdot \underline{l} \cdot \underline{\rho}}{\underline{\mu}} = Re$	The Reynolds number is the ratio of inertial forces to viscous forces and is the criterion of turbulence.
$\left[\dfrac{\alpha_p \cdot \alpha_u^2}{\alpha_l}\right] = \left[\dfrac{\alpha_P}{\alpha_l}\right]$	$\dfrac{P}{\rho \cdot u^2} = \dfrac{\underline{P}}{\underline{\rho} \cdot \underline{u}^2} = Eu$	The Euler number is the ratio of pressure forces to inertial forces and is the criterion of flow resistance.
$\left[\dfrac{\alpha_p \cdot \alpha_u^2}{\alpha_l}\right] = \left[\alpha_p \cdot \alpha_g\right]$	$\dfrac{g \cdot l}{u^2} = \dfrac{\underline{g} \cdot \underline{l}}{\underline{u}^2} = Fr$	The Froud number is the ratio of gravitational forces to the forces of inertia; it is used as a criterion of mixing conditions.
$\left[\dfrac{\alpha_p \cdot \alpha_u}{\alpha_\tau}\right] = \left[\dfrac{\alpha_p \cdot \alpha_u^2}{\alpha_l}\right]$	$\dfrac{u \cdot \tau}{l} = \dfrac{\underline{u} \cdot \underline{\tau}}{\underline{l}} = St$	The Stokes number is the time-dependent criterion of process dynamics.

Table 3.4. Other notable criterial numbers

$Nu = \dfrac{\alpha \cdot l}{\lambda}$	Nuselt number	Ratio of heat convection to conduction
$Pr = \dfrac{c_p \cdot \mu}{\lambda}$	Prandtl number	Ratio of heat conduction to accumulation
$Sh = \dfrac{\beta \cdot l}{D}$	Sherwood number	Ratio of mass convection to diffusion in boundary layer, both driven by concentration gradient
$Sc = \dfrac{\mu}{\rho \cdot D}$	Schmidt number	Ratio of momentum diffusivity (viscosity) and mass diffusivity
$We = \dfrac{\rho \cdot u \cdot l}{\sigma}$	Weber number	Ratio of inertial to surface forces
$Pe = \dfrac{u \cdot l}{D}$	Peclet number	Ratio of advection of a mass by the flow to the rate of diffusion

By comparing the corresponding expressions in square brackets one can arrive at the so-called criterial number, well-known in process engineering (see Tab. 3.4). Based on other similar equations describing the transport phenomena of mass and energy, many other similarity invariants for specific physical phenomena are defined in the same way, where

c_p – specific heat [J/kg °C];

D – diffusion coefficient [m^2/s];

u – velocity [m/s];

l – linear dimension [m];

α – heat transfer coefficient [W/m^2];

β – mass transfer coefficient [m/s];

λ – thermal conductivity [W/m];

μ – kinematic viscosity [kg s/m];

ρ – density [kg/m^3];

σ – surface tension [N/m].

4 Nanoprocesses

4.1 Microreactors

Microreactors are the biggest innovation in process engineering of the past 5 years. They allow strict control of reactions that are very fast, exothermic and hazardous. Microreactors are particularly suitable for reactants with different viscosities, for changing viscosity, for hazardous reactions and intermediates. Generally microreactors:

1. are more efficient
2. can increase selectivity
3. minimize reagent consumption
4. allow optimization of conditions
5. are better for control
6. have excellent heat transfer
7. are made of more robust equipment
8. achieve an easy scale-up of the process by the increase in the number of reactors.

4.1.1 Structure and function of microreactors

The spectacular career of microreactors is due to a number of advantages and even led to the ermergence of a new branch of (micro-) process engineering. This is an area for carrying out different unit processes in microchannels. The idea of microreactors evolved in the nineties at the Nuclear Research Center Karlsruhe [75–77]. Microreactors are defined as a device consisting of a number of parallel microchannels with dimensions of less than 1mm and with precise temperature control for conducting continuous operation. The simplest and most obvious feature of such microstructures is the order of magnitude of the size of the characteristic in the range of 10 microns to 1000 microns, as shown in Figure 4.1, which systematises known chemical reactors due to their characteristic size.

More detailed definitions of microreactors for gas-liquid systems are also presented in the literature. Brauner and Maron [78] reported that the following relationship defines the hydraulic diameter of a microreactor:

$$\frac{\left(\rho_L - \rho_g\right) d^2 \, g}{\sigma} < (2\pi)^2 \, . \tag{4.1}$$

The left side of the equation (4.1) is the Eötvös number (Eo), which determines the ratio of the force of gravity to the force of surface tension. Triplett et al. [79] found that a hydraulic diameter specifying the microreactor size should fulfill the following

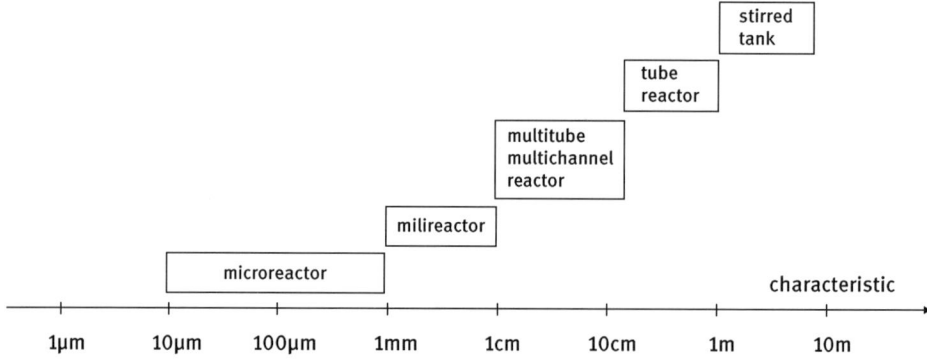

Fig. 4.1. Classification of chemical reactors, due to their characteristic size.

condition:

$$d < \sqrt{\frac{\sigma}{g\left(\rho_L - \rho_g\right)}}. \tag{4.2}$$

As can be seen, at the boundary of the air-water system, the maximum value of the hydraulic diameter is 17 mm respectively, according to relation (4.2). These values are about one order of magnitude too large as proposed in Figure 4.1 classification. Therefore the definition of the microreactor's size (as such) is rather vague. However, it roughly determines the essential property of these devices. So, during the flow of fluids in microreactor microchannels the forces of surface tension and adhesion were observed to be dominant over the forces of gravity and buoyancy (depending on the Eötvös number). As a result, differences in hydrodynamics and the processes of heat and mass transfer in microchannels, compared to the same processes occurring in larger diameter reactors, are also observed. Dziubiński and Prywer [80], comparing different definitions, recognized that features of microchannels are those in which the value of the Eötvös number is less than unity. This means that the hydraulic diameter is $d_h = 0.43$ mm for water-air systems. In view of these differences it is proposed that microreactors are explicitly defined by their characteristic size, regardless of the fluids (see Fig. 4.1).

However, the purpose of microreactors is not to miniaturize industrial installations, but to intensify production through improved reaction conditions and to reduce production costs. In addition to a network of microchannels, the microreactors also have micro-reservoirs, microsensors, microvalves, microactuators and micropumps. Micropumps can operate on different principles depending on the size of the microreactor. This may be a syringe pump, or one operating on electrokinetic, electrophoretics or electroosmosis principles. Micro-reservoirs may contain appropriate reagents which react in a specific place and time. Sometimes additional channels are applied in microreactors, in which cooling/heating fluids flow. In this way the temperature may change very rapidly when nonisothermal conditions are required.

4.1.2 Characteristics of microreactors

The small size of the flow channels in the microreactors causes good properties and excellent conditions for the reaction, which were previously unachievablee with conventional reactors. It was possible for several reasons, which are listed below. The small size of the flow channels in microreactors allows development of a very large "specific surface area", which is the surface area to volume ratio. It should be noted that the specific surface area is always inversely related to the size of the surface-forming elements, in this case to the size of the flow channels. Typically, specific surface area is very high, up to a = 20 000 m^2/m^3. This creates a very favorable environment for heat exchange and therefore allows maintainence of a constant temperature, even if the reactions are highly exothermic. In addition, the very small dimensions of microreactors reduce the distance for the transport of mass and heat transfer from the bulk to the wall of channels, and therefore the corresponding fluxes are very large (see Tab. 4.1). This is because of the large driving forces that are the relevant gradients of temperature and concentration.

The precise arrangement of the microchannels and micro-reservoirs, combined with the precisely controlled flow, allow for accurate determination of the place and time of a given reaction. This is particularly important when complex reactions composed of the chains of consecutive and competitive (i.e. parallel) reactions could occur. In contrast, this might be extremely difficult to control in conventional reactors.

The precise control of the residence time in milliseconds is very important, which results in very high conversion rates and no byproducts during the continuous production which may be achieved. Microreactors are normally operated continuously and very quickly. This rapid operation prevents the decomposition of precious intermediates, which always allows better selectivity. The minimal longitudinal mixing (which takes place in very narrow channels) continuously maintains adequate driving force for heat and mass transport on a high level, along the flow path. It should be noted that there are many reasons for reduction of the driving forces in conventional reactors such as the recycling streams, the by-passes and side streams. What's more, dead zones may lead to local overheating and sometimes even an explosion. The best situation would be a perfect mixing flow over the entire cross-section, but such mixing is completely lacking along the flow path. Many misconceptions result from the frequent use of the word *mixing* instead of *turbulence*, which means the convective transport, caused by increasing heat and mass transfer. Flow in channels is characterized by a Reynolds number 1 < Re < 100 and is fully laminar. It should be underlined that there can be no question of any turbulence in such capillaries, where only pure diffusion may take place, mostly with the contribution of the shear forces.

The very easy heat transport in microreactors is more efficient than in conventional reactors and even critical reactions such as nitration can be performed safely at high temperatures. Microreactors typically have heat exchange coefficients of at least 1 megawatt (MW) up to 500 MW m^{-3} K^{-1} vs. a few kilowatts in conventional glassware.

Table 4.1. Comparison of mass transfer parameters in different gas-liquid contactors

Type of gas liquid contactor	a (m²/m³)	$k_L a \times 10^2$ (s⁻¹)	Ref.
Gas–liquid microchannel contactor of this study	3400–9000	30–2100	[548]
Static mixers	100–1000	10–250	[549]
Impinging jet absorbers	90–2050	2.5–122	[550]
Packed columns, co-current	10–1700	0.04–102	[551]
Tube reactors, vertical	100–2000	2–100	[551]
Tube reactors, horizontal and coiled	50–700	0.5–70	[551]
Stirred tank	100–2000	3–40	[552]
Bubble columns	50–600	0.5–24	[551]
Couette–Taylor flow reactor	200–1200	3–21	[553]
Packed columns, countercurrent	10–350	0.04–7	[551]
Spray column	75–170	1.5–2.2	[552]

Heating and cooling a microreactor is also much quicker and operating temperatures can be as low as −100 °C. Microreactors are also feasible in cases of a large difference in viscosity between components as well as in cases when a quick increase in viscosity occurs during the reaction.

The values of the volumetric mass transfer coefficients in the microreactors range from 10 to as much as 100 times higher than those achieved in conventional gas-liquid contactors. It should be noted that this is related to the extremely large interfacial surface values (a), while the mass transfer coefficient on the side of the liquid (kL) is more typical and average. Of course, the intensity of the mass transport is related to the large values of the volumetric mass transfer coefficients, which are related to the miniaturization and increased surface to volume ratio.

The small size of the flow channels encourages the use of high pressures in microreactors. It would also increase the temperature above the boiling point of the solvent. This in turn helps to increase rate of the Arrhenius-type reactions, and should be considered as an advantage. Increasing the pressure may also be capable of dissolving the reactants in the gas flow stream.

Thus almost all of the advantages of microreactors result from the small dimensions of the flow channels, which makes precise control of all of the conditions influencing chemical reactions in the microchannels possible.

4.1.3 Microreactor design

The technologies used to produce microreactors, known as MT (microfabrication techniques), are the following:
1.   wet etching;
2.   dry etching;

3. micromachining: cutting, milling, turning, drilling, punching, embossing;
4. lithography, including laser lithography.

The basic manufacturing technique is photolithographic etching of microchannels in plates (which can be covered by appropriate masks) by using appropriate solutions. These plates are then placed in a pile and pressed for tightness. This technique is known as two-dimensional i.e. 2D. Depending on the needs of the microreactors they are made mainly of glass, metals, polymers, quartz, and silicon. For example, glass is etched with 1 % HF and 5 % NH_4F at a temperature of 65 degrees at a rate of 0.5 microns per minute. Other methods are also used such as embossing, injection molding or laser lithography. UK scientists have developed 3D printing technology for making miniaturized fluidic reaction ware devices that can be used for chemical syntheses in just a few hours [81].

From a user standpoint, i.e. operation and maintenance, the installations with microreactors are characterized by increased safety, reliability and simpler process control. The design of microreactors is also much simpler due to the scalability, or scaling-up, i.e. linear transfer from laboratory to production. This is due to the modular design of microreactors and the complete installations. Please note that the microreactor (in 2D structure) consists of identical plates (called wafers), and the whole system always consists of identical modules, whose parts can be replaced if necessary. This is not the ordinary scaling-up in process engineering. Here an increase in the number of microchannels, not their size, is proposed. This approach is known in the literature as "numbering-up". As a result, it is much easier to increase the productivity of the process merely by increasing the number of microchannels. Moreover, the results of one microchannel, obtained during laboratory investigations, may then be directly transferred into practice by multiplying the number of channels. This is very important because later generalizations and tedious working out of the laboratory results are not necessary. In Figure 4.2 a platform with several different microstructures is presented. Modularity increases the flexibility to adapt to current needs for performance, and also to the type of production (on-site and on-demand), and in the end offers a lot more opportunities for control and maintenance of the process quality of the desired product (see Fig. 4.2).

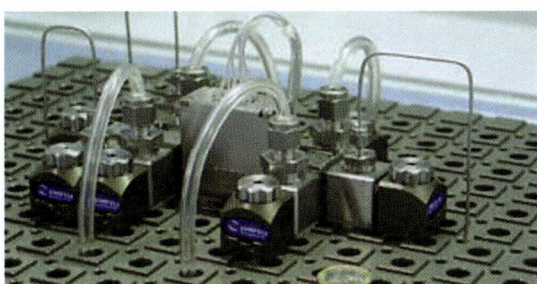

Fig. 4.2. Modular microstructure system (courtesy of of EHRFELD MICROTECHNIK BTS).

Finally, the mobility of the system (presented in Fig. 4.2 [82]) is much greater. The desired product can be produced at the destination, as might pertain to dangerousor toxic products.

Even though it is very promising, the development of microreactors is still in its infancy. Manufacturers of microreactors and micro-structural components are mostly partners for researchers seeking innovative synthesis technology. The close and extensive cooperation of scientists engaged in mathematical models of reaction using computational fluid dynamics with performers of all equipment related to microreactors is very important. The scale of production is usually from a few grams to a maximum of 100kg per day which can satisfy only certain types of industries such as fine chemicals, pharmaceuticals and cosmetics. However, the final success often rewards these efforts.

The convenient way to pump fluid in microreactors is by using electrical phenomena such as electrophoresis and electroosmosis. In this case, the liquid flow in microreactor tubules can be caused by the difference in electrical potential. Fluid velocity due to electroosmosis can then be calculated from the formula proposed by Paul [82]:

$$u_{eo} = \frac{E \cdot \epsilon \cdot \epsilon_0 \cdot \zeta}{\eta}, \tag{4.3}$$

(a)   (b)

Fig. 4.3. Scale up of microreactors by ordinary multiplication. On the left hand is shown a microreactor with a capacity of 1 kg/h, and on the right system 500 times larger and consisting of 500 microreactors with a capacity of 500 kg/h. (courtesy of DSM Pharmaceutical Products [83]).

where

u – fluid velocity;

E – electrical potential (typically hundreds V/cm);

ϵ_0 – dielecric constant of the free space;

ϵ – dielectric constant of the fluid;

ζ – zeta potential of the fluid-wall interface;

η – liquid viscosity.

For example in a glass microreactor the water solutions of electrically neutral molecules during such electrically driven flow the velocity was approximately 0.1–1 mm/s, at neutral pH = 7 and electric field of hundreds V/cm and zeta potential (–50 to –150 mV).

In the case of the charged molecules the additional velocity appears [82, 83] due to electrophoresis, which can be calculated as follows:

$$u_{ef} = \frac{z \cdot e \cdot E \cdot D}{k \cdot T}, \tag{4.4}$$

z – charge number on the species;

e – charge of electron ($1.602176565 \times 10^{-19}$ coulombs);

E – electrical potential (typically hundreds V/cm);

D – diffusion coefficient;

k – Boltzmann constant;

T – temperature.

Analysis of the work of the microreactor is usually performed by means of microscopic imaging of absorbance or direct fluorescence signals inside the microreactor. It is also possible to take microsamples and perform analysis outside the microreactors by means of HPLC or mass spectroscopy. The best way would be to use a miniaturized system for measurement and control in real time, all the chemical parameters. Such systems allow automatic download and preparation of samples, detection of constituents and processing of data. Such systems are known as micro-TAS [84] and are under investigation. They are the most advanced in biotechnology and biomedical applications for DNA resolution [85].

4.1.4 Applications of microreactors

Microreactors are used everywhere where accuracy and precision are more important than productivity, so in the pharmaceutical, cosmetics, polymers, fine chemicals industry and the production of analytical reagents. Microreactors are particularly feasible for the development of green organic chemistry [86]. Reactions in microreactors can be strictly controlled to allow for the implementation of particularly difficult cases such as highly exothermic reactions e.g. nitration reactions. They may be applied in combination with photochemistry, electrosynthesis, multicomponent reactions and polymerization (for example that of butyl acrylate). Synthesis may also be combined with online purification of the product. The reactions in microreactors run several times faster than in conventional reactors. For example, the condensation of 2-trimethylsilylethanol and p-nitrophenyl chloroformate to produce 2-(trimethylsilyl)ethyl 4-nitrophenyl carbonate requires 14 hours to complete in a conventional setup but only 18.4 minutes in a microreactor [87]. The disadvantages of conventional solutions are the difficulty of maintaining homogeneous reaction conditions, and the risk of complicated reaction sequences occurring or the outbreak of highly exothermic reactions.

Microreactors have already been used for the following reactions:
1. selective partial hydrogenation of a cyclic triene;
2. oxidative dehydrogenation of alcohols;
3. exchange between immiscible fluids;
4. partial oxidation of methane;
5. partial oxidation of propene to acrolein;
6. oxidation of ammonia with integrated catalyst;
7. gas-phase synthesis at high temperature;
8. synthesis of hazardous gases;
9. synthesis of ethylene oxide;
10. synthesis of vitamin precursor.

Nitration reactions are extremely exothermic and require perfect temperature control. Millions of tons of nitration products are produced each year for various applications. Doku et al. [88] presented the results of the nitration of benzene in a borosilicate glass microreactor. These were extremely dangerous conditions because nitrobenzene is explosive. Burns and Ramslaw [89] also investigated nitration of benzene in a microreactor and made very interesting observations on the linear increase of the conversion with temperature, and flow.

Wittig olefination is a chemical reaction of an aldehyde or ketone with a triphenyl phosphonium ylide (often called a Wittig reagent) to give an alkene and triphenylphosphine oxide. It is widely used in organic synthesis for the preparation of alkenes [90]. Skelton [91] reported carrying out this reaction in a borosilicate glass reactor, where it was found to attain a 10 % increase in efficiency compared with batch reactor. Moreover the typically used excess of reagent (here the aldehyde relative to the phosphonium salt) was entirely eliminated, demonstrating the usefulness of microreactors for green chemistry.

Enamines (universal intermediates) are unsaturated compounds derived from the condensation of an aldehyde or ketone with a secondary amine [92]. In a microreactor enamines are produced from cyclohexane and pyrrolidine. They are directly mixed in methanol solution and dicyclohexylcarbodiimide (DDC) is added to remove water [93].

Aldol, β-hydroxyaldehydes, aldehyde-alcohols – organic compounds containing an aldehyde and alcohol, obtained by an aldol condensation and discovered by Charles Adolph Wurtzs. The aldol reaction is a powerful means of forming carbon–carbon bonds in organic chemistry. The application of microreactors has shown 100 % conversion and significant acceleration of the chemical reaction (near one hundred times) as compared to a conventional batch system: from 24 h to 15 min [94].

An interesting example of multistage synthesis of peptides in a cascade of microreactors is presented by Watts et al. [95]. Several stages were needed due to the low degree of conversion in one step. The paper reports on the conversion of 93 % in 5 stages.

The reactions of synthesis of diazo dyes were performed of 4-nitrobenzene diazonium tetrafluoroborate with N, N-dimethylaniline, i.e. the red diazo compound was obtained as a result with 100 % conversion [96]. It should be mentioned that some azo dyes, such as dinitroaniline orange, ortho-nitroaniline orange or pigment orange 1, 2, and 5 have been found to be mutagenic [97] and carcinogenic [98]. Therefore manufacture and sale of azo dyes was recently prohibited in the European Union in September 2003.

The reaction of hydrogenation of 1,5-cyclooctadiene to cyclooctene was performed in a stainless steel microreactor [99] with palladium catalyst at 150 °C. The conversion was up to 99.5 % but increasing residence time decreased the selectivity. Also hydrogenation of c, t, t, 1, 5, 9-cyclododecatriene to cyclododecenes [100] under the same conditions was reported with 85–90 % selectivity which was much higher than in fixed bed reactor. The catalytic dehydrogenation of benzene to cyclohexene on ruthenium/zinc catalyst in microreactor was also described using palladium catalyst the cyclohexane was obtained. The advantage of the microreactor was ease of using elemental fluorine in gaseous form, which is rather problematic in organic synthesis.

In fluorination reactions which are exothermic and difficult to control, the benefit of microreactor use is in contrast safety and using the small amount of elemental fluorine in gaseous phase. Using the microreactor for fluorination reaction with nickel/copper catalyst where fluorine has been diluted with nitrogenhas been described in the literature [101, 102]. Another fluorination described in literature [103] used direct fluorine in a silicon microreactor with a nickel catalyst, almost impossible in a conventional reactor. The fluorination of toluene was achieved at room temperature with the use of methanol.

The oxidation of ammonia on platinum catalyst in silicon microreactors has been described in the literature [104]. The influence of temperature on the exhaust gas composition was investigated. The feasibility of microreactors for endodermic reaction was demonstrated during production of hydrogen cyanide [105], which is easily hydrolyzed to ammonia at low temperatures. A microheat exchanger was helpful in this case.

The use of microreactors employed in a two gas-liquid allows the use of very large interfacial area [106]. The basic and most extensive classification of gas-liquid microreactors divides them into microreactors with closed and open channels. Figure 4.4

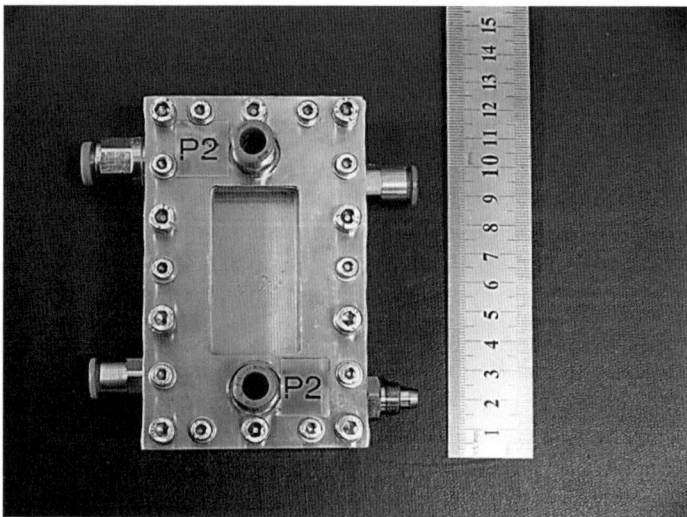

Fig. 4.4. Falling film microreactors.

shows a photograph of falling liquid microreactors in our laboratory. In the literature, reactors of this type are referred to by the acronym FFMR (falling film microreactor), and are produced by IMM Maitz. Gas-liquid microreactors are also increasingly being used to produce a specific product, as well as for analytical methods. Examples of these in the literature follow: synthesis of hydrogen peroxide [107], fluorination [102], photochemical reactions of gas-liquid systems [108] and the fast detection of solubility and diffusivity of gases in liquids [109, 110].

4.1.5 Catalytic reactions in microreactors

Many coupling reactions have found their way into the pharmaceutical industry [111] and into conjugated organic materials [112]. Microreactors can be used to synthesize and purify extremely reactive organometallic compounds for ALD and CVD applications, with improved safety in operations and higher purity of products. Microreactors are used to prepare an ultra-pure organometallic compound using a microchannel device for synthesis in reaction of a metal halide with an alkylating agent to produce an ultra-pure alkyl metal compound for processes such as chemical vapor deposition.

The Grignard reaction is an organometallic chemical reaction in which alkyl- or aryl-magnesium halides (Grignard reagents) are added to a carbonyl group in an aldehyde or ketone. The procedure uses transition metal catalysts, typically nickel or palladium, to couple a combination of two alkyl, aryl or vinyl groups. Robert Corriu and Makoto Kumada reported the reaction independently in 1972 [113, 114]. They used a polypropylene tubular pressure-driven microreactor with an insert of nickel catalyst on glass wool.

The Suzuki reaction is a chemical reaction between an aryl or vinyl boronic acid, a vinyl halide or aryl, which is catalyzed by palladium complexes in the zero oxidation state. The Suzuki reaction plays a significant role in organic synthesis. This reaction is most often used to build the carbon skeleton of particles of during the production of polyolefins, styrene derivatives and biphenyls. Greenway et al. optimized microreactor for the Suzuki reaction by applying semi-continuous operation mode in his microreactor and obtained up to 95 % without any byproducts.

4.1.6 Microfotoreactors

Photochemical reaction for benzylophenone production in a microreactor made of silicone/quartz wafers has been reported [115]. The benzophenone is used in the cosmetic industry, and is currently regulated in the cosmetics directive in annex VII, part 1 List of permitted UV filters which cosmetic products may contain [116]. A pinacol coupling reaction is an organic reaction in which a carbon–carbon covalent bond is formed between the carbonyl groups of an aldehyde or a ketone in the presence of an electron donor in a free radical process.

4.1.7 Manufacturers of microreactors

SIGMA-ALDRICH has its own microreactor system which works at a pressure of 6.5 bar within the broad temperature ranges of −70 to 150 °C. Production of methylenecyclopentane with 99 % conversion in microreactor enables raising the boiling point which is close to catalyst temperature 55 °C. 100 % conversion has been achieved during the production of retinol (vitamin A). The problem with retinol is that it is sensitive to air, light and temperature. With microreactor technology the product can be collected continuously. In conventional batch mode the product begins to decompose while being kept at reaction conditions of 60 °C for full conversion of the starting material.

The IMM, or Institut für Mikrotechnik Mainz GmbH [117] deals in the area of microprocess, engineering a variety of microstructured components, which are available for industrial analytics, biomedical diagnostics and environmental analysis.

Microglas Chemtech GmbH specializes in the production of micro-technological products made of glass with great success. Microreactors are an innovative alternative to the large-scale production in the chemical industry. Glass is an excellent material for components in microreaction technology because of its unique properties and advantages.

The Parr Instrument Company [118] manufactures innovative equipment such as customizable reactors and pressure vessels, which can be designed for up to 350 bar and temperatures of up to 500 °C, and supercritical fluid systems. Applications include extractions, nanoparticle and nano-structured film formation, supercritical drying, carbon capture and storage, as well as enhanced oil recovery.

Syrris is based in Royston (near Cambridge), UK. Syrris has subsidiary offices in the USA, Japan, India and Brazil and over 30 distributors worldwide. Their microreactors are equipped with a heat exchanger system for temperature control and can be heated up to 250 °C and cooled down to 0 °C.

LTF (little thing factory – Optic groups) is a German company but with large sales in Asia, America and Canada in the chemicals and pharmaceuticals industry as well as in process technology.

HiTEC ZANG is a company which deals with micro-reaction technology which enables exact control of complex or critical reactions in fine chemistry (special products are agricultural and pharmaceutical chemicals), which considerably increase the operational safety of the chemical production.

Ehrfeld Mikrotchnik BTS GmbH (Bayer) manufactures microreactors for the dairy industry.

4.2 Membranes and their unlimited opportunities

Widely recognized as the "father" of modern membrane processes Jean-Antoine Nollet (1700–1770), published the first work on the phenomenon of osmosis in 1748. Nollett was a French abbot and also a member of the Royal Society of London and the French Academy of Sciences. He was the first professor of experimental physics at the University of Paris. He is best known for his spectacular experience of the phenomenon of osmosis in natural membranes. It was not until many years later that Thomas Graham (1805–1869) studied gas separation on rubber membranes in 1866. His research on the diffusion of gases resulted in "Graham's Law", which states that the rate of diffusion of a gas is inversely proportional to the square root of its molar mass. His discovery of dialysis, which is used in many medical facilities today, was the result of his study of colloids. He was a precursor of the field known today as colloid chemistry. It so happens that all of the scientists who contributed greatly to the development of process engineering also worked and had success in various fields such as chemistry, physics, biology and mathematics. The greatest progress of a new era in the field of membranes is associated with the invention of asymmetric membranes by Loeb and Surirayan in 1963. The characteristics of asymmetric (or anisotropic) membranes are a combination of the two desired membrane characteristics, i.e. high separation selectivity with high yield. It should be added that these features were formerly considered to be contradictory.

Membrane processes and membrane-based hybrid processes are actually the most effective separation processes, and they are still in rapid development, creating new prospects of their applications. They serve practically unlimited selectivity of separation, which is essential for environmental engineering. In particular the separation of large streams of diluted mixtures (by heterogenic and homogenous contaminants) by means of membranes and membrane-based hybrid processes seem to be very effective, promising and profitable.

The opinion is often voiced that membranes should be defined depending on what they do rather than by what they are. In practice the four most common functions of the membranes have been established: i.e. separation, contact, immobilization and controlled release (see Fig. 4.5).

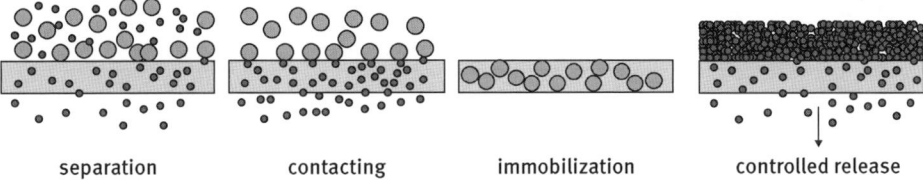

| separation | contacting | immobilization | controlled release |

Fig. 4.5. Membrane functions.

4.2.1 Membrane manufacture

A membrane is usually a semi-permeable surface, which may be artificial or natural. Separation membranes must be made as thin as possible in relation to permeate flux. However, in the case of very thin layers defects may be revealed which may affect the selectivity of the membrane. The result must be the usual trade between flow and separation selectivity. Synthetic membranes can be made from a single (integral) or several materials (composite). The most commonly used materials are polymers, ceramics and metals. It is believed that membranes can be porous or dense, although such classification seems to be very conventional without distinct border. It is much more accurate to distinguish symmetrical from asymmetrical membranes. Asymmetric membranes were developed by Loeb and Surrirayan in 1963 [119]. They have made a true breakthrough, thanks to their exceptional advantages. First of all, the problem of reconciling high performance with high separation selectivity has been solved. Formerly membranes were considered to be interesting but completely useless in practice, a laboratory curiosity. The porosity of symmetrical membranes changes perpendicular to the membrane surface. Therefore the layer which is responsible for the separation can be very thin, and still provide the enough mechanical strength without affecting a reduction in permeate stream. Asymmetrical structure is obtained when polymer membranes are cast. The membranes are fabricated from a casting solution of cellulose acetate and acetone, to which aqueous magnesium perchlorate solution is added.

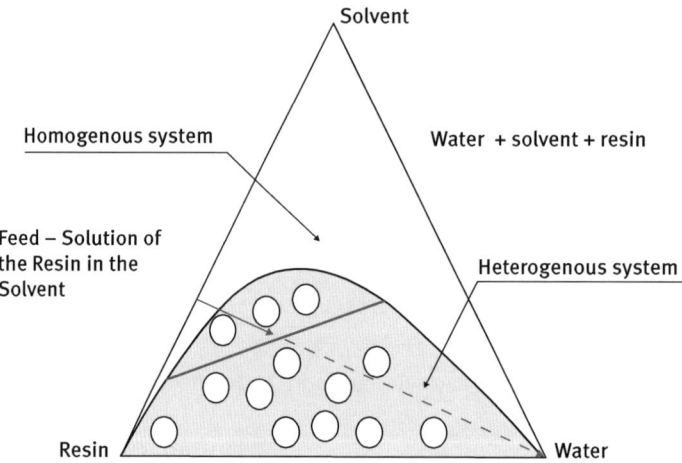

Fig. 4.6. Production of membranes by casting a polymer resin.

There are two competing mechanisms, i.e. coalescence (merging of small droplets into larger droplets) of the dispersed phase is precipitated, and the liquid phase changes to a solid by polymerization (see Fig. 4.6).

Another method of preparation of asymmetrical membranes is combining layers of different porosities, e.g. by sintering etc. The composite membranes consist of several layers. Hybrid membranes are made of different materials such as polymers and inorganic layers such as zeolites. In the photographs below (see Fig. 4.7 on the next page) various structures of membranes (symmetrical and asymmetrical) made of various materials are shown.

4.2.2 Membrane contactors

Membrane contactors are the novel use of membranes for unit processes such as extraction, absorption, chromatography and distillation (see Fig. 4.8). One can imagine that the further development of contactors will extend to other unit processes in the future such as a membrane adsorption (see Section 4.5) and membrane leaching (see Section 4.6). The main function of membrane contactors is effective contacting during both phases. In conventional unit processes the surface between phases is formed by dispersing gas bubbles in the liquid or the dispersion of one liquid in another insoluble liquid.

Dispersion of the fine particles causes an increase in surface area on the one hand, but creates a lot of disturbances on the other, such as mutual entrainment and mixing of the phases, longitudinal mixing, foaming etc. In unit processes for solid-liquid systems, dispersion of solid particles will also increase the flow resistance and energy

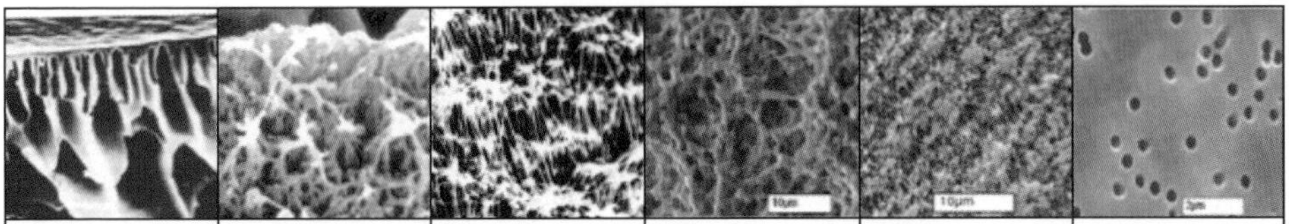

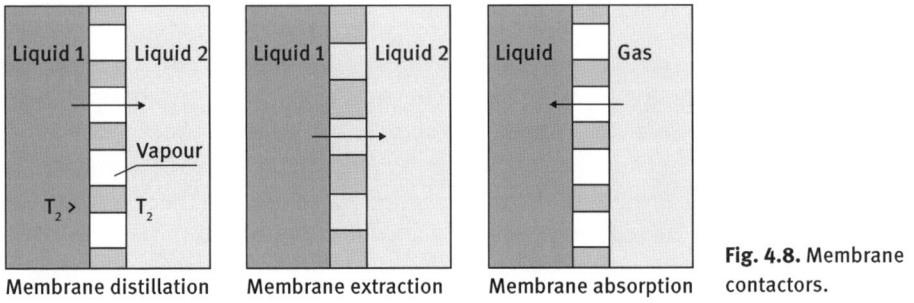

Polysulphone	Polysulphone	Polysulphone	PVDF	PTFE	Polycarbonate
Casting	Casting	Stretched	Stretched	Stretched	Track-etched
Assymetric	Symetric	Symetric	Symetric	Symetric	Symetric

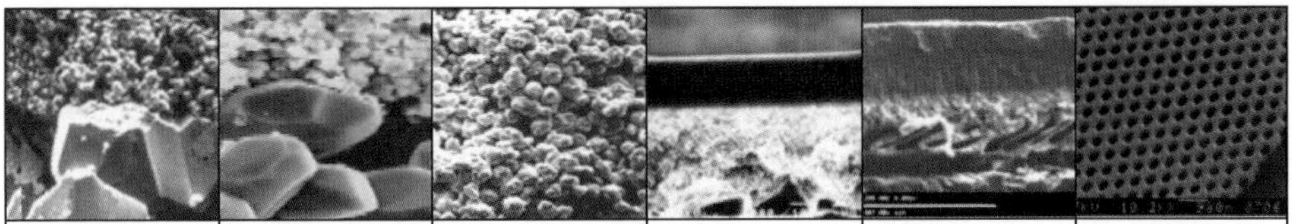

ZrO/α-Al₂O₃	TiO2/α-Al2O3	Glass	Zeolite on α-Al2O3	Zeolite on PTFE	Metal
Sintered	Sintered	Sintered	Hybrid	Hybrid	Laser lithography
Assymetric	Assymetric	Symetric	Assymetric	Assymetric	Symetric

Fig. 4.7. Membrane structures.

Membrane distillation — Liquid 1 | Liquid 2, Vapour, $T_2 > T_2$

Membrane extraction — Liquid 1 | Liquid 2

Membrane absorption — Liquid | Gas

Fig. 4.8. Membrane contactors.

losses. The use of membranes in these unit processes helps to eliminate these adverse effects. In later chapters of this book entirely new contactor processes such as membrane sorption and leaching membranes will be discussed.

4.2.3 Immobilization of species on membranes

Immobilization of various substances on the surfaces of the membranes enabled the creation of many different hybrid systems, capable of integrating functions of membrane separation with certain reactor functions (see Fig. 4.9). These membranes allow the immobilization of a variety of active substances such as:

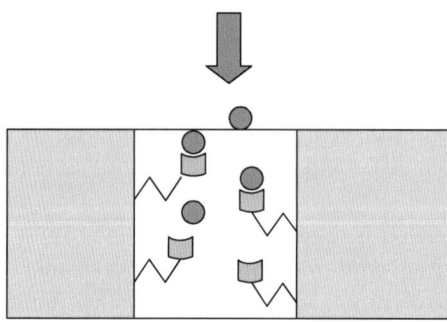

Fig. 4.9. Ligands immobilized in membrane pore.

1. catalyst and enzymes, photocatalyst;
2. cells;
3. tissues (culturing);
4. functional groups (affinity separation);
5. Electrical charges.

Catalysts immobilized on a variety of ligands can be: aminoacids, antigen and antibody ligands, dye ligands, metal affinity ligands, chelate adsorbents, ion exchange ligands, the tg19318 peptide, thiophilic adsorbents and metals such as Cu, Co, Mn, Cr, Fe, Pt, Pd, Ru, Mo, Pc, Y.

The catalytic membranes enable contact between phases. The pore size of the catalytic layer can be adjusted according to the needs of the reaction. There are different variants of this combination of separation of chemical reactions such as controlled delivery of substrates or receiving certain products. This allows full control of reaction conditions, including reaction to stop on a particular product, avoiding the production of waste, shift of the equilibrium conditions and acceleration of the reaction. Solid polyelectrolytes (SPE) are used as new electrocatalytics. They have a suitable composition and electronic properties, predetermined crystallographic orientation and order to achieve maximum catalytic activity.

They can be used for the following applications:
1. generation of hydrogen and oxygen;
2. production of ozone;
3. Separation of isotopes: hydrogen deuterium;
4. hydrogen "absorber";
5. water treatment by SPE (solid polyelectrolytes) membrane;
6. elimination of nitrates;
7. electroinduced gas separation;
8. sensors;
9. fuel cells.

4.2.4 Controlled release of species with membranes

The principle of controlled release is the precise and long-term distribution of certain substances to the specified objects. Controlled release (dosage) is an important tool in the implementation of sustainable technologies. Controlled release is used in such fields as environmental protection, agriculture, forestry, farming, medicine, pharmacy. The first commercially successful goes back to the forties in the use of long-acting drugs and biocides, preservatives for various materials such as rubber, paper, fabric. Controlled dosage is conveniently carried out using membranes. For this purpose, the microcapsules are prepared either from microemulsions or the active material is coated with a permeable polymer layer (see Fig. 4.10).

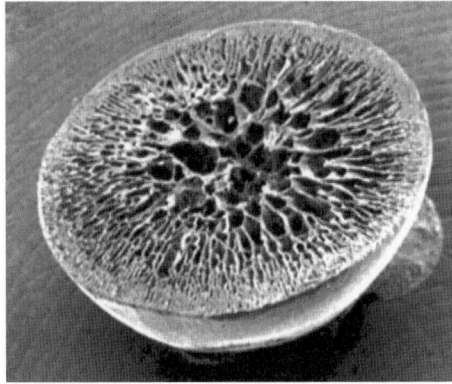

Fig. 4.10. Microcapsule cross-section.

In agriculture, gardening and households plants need certain chemicals and fertilizers only for a certain period of vegetation. To ensure that the plants receive the proper dose at this time such substances are often used excessively. Consequently the unused substances are flushed into the groundwater, which violates the ecological balance. Remember that conventional farming is one of the largest sources of pollution of the world. The controlled dosing of pheromones as repellents against troublesome insects is useful in breeding animals. In forestry pheromones are used to lure the pest-insect to the specific pitfalls. This is much better than using poisons which always go into the environment. Controlled dosage of pheromones must be very precise in order to be effective.

In pharmacy and medicine (see Section 6.1.2), controlled release is applied for drugs which require a certain constant level in the body. This prevents rapid injection and raising the concentration, which in the case of some drugs may cause dysfunction, shock or even death. In this way many psychotropic, hormonal and other important drugs are delivered, such as scopolamine, atropine, insulin, and even vitamins and microelements. Also in the case of transplants, implants and other foreign bodies introduced into the body long-term controlled dispensing of medications (such as antibiotics) may be necessary to combat immune rejection of these implants.

The controlled release society (CRS) was founded in 1978 as a not-for-profit organization dedicated to the science and technology of controlled release. The field of CRS includes scientific and technical activities to control the spatial and temporal effects in a variety of areas, including human and animal health, as well as nonpharmaceutical areas such as agriculture, cosmetics and consumer products and the environment. In all areas, controlled dosage must be based on a very precise determination of the kinetics of diffusion, which can be controlled by the appropriate membrane.

4.2.5 Membrane separation processes

In accordance with the principles of process engineering, the membrane processes can be classified into physical and chemical processes (see Fig. 4.11). Bearing in mind the type of the so-called driving force, they can be divided into pressure-driven membrane processes, diffusional, thermal and electrical processes.

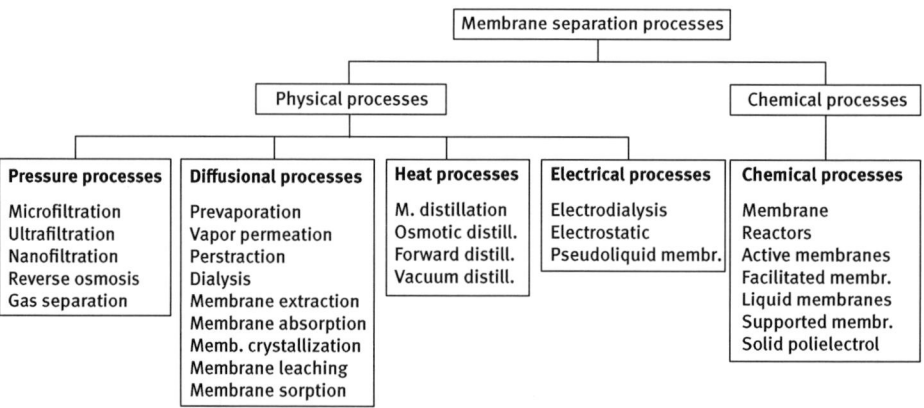

Fig. 4.11. The classification of membrane processes.

4.2.5.1 Reverse osmosis

Reverse osmosis (RO) and nanofiltration (NF) are processes for separating molecules with a minimum molecular weight (typically < 150 Daltons) from the solvent, usually water. The basic mechanism of separation is the rejection by the membrane on the basis of size and charge. RO membranes have a pore size less than 2 nm and are conventionally referred to as dense nonporous. The following materials are commonly used for the production of RO membranes of both asymmetric and composite (thin film) types: cellulose triacetate, polyamide, polysulfone and polyethersulfone. The separation ability of reverse osmosis membranes is determined in the particle size range of 1–10 Å. The driving force of this process is a pressure gradient of 15–100 bar, whereas the permeate flux of 1–30 LMH is obtained, which correspondes to specific permeability in the range of 0.05–10 LMH/bar).

The principles of separation in reverse osmosis processes (RO) are: the mechanisms of dissolution, diffusion and preferential sorption. The mass transport models used in RO are the diffusion solution model, preferential adsorption model, finely porous model, and irreversible thermodynamics models. The most commonly used membrane modules are spiral wound (74 %) and hollow fiber (26 %) modules.

RO membranes are used for desalination, recycling of hazardous waste in the metal industry (electroplating, finishing), in the electronics industry for the production of ultrapure water, purification of groundwater, purification of leachates from landfills. To estimate the costs of such processes assumed operating costs (energy consumption) are $ 0.10–$ 0.40/m^3, the cost of equipment: $ 200–800/m^3 day.

4.2.5.2 Nanofiltration
The nominal pore size of the membrane is typically about 1 nanometer. Nanofilter membranes are typically rated by molecular weight cut-off (MWCO) rather than nominal pore size. The MWCO is typically less than 1000 atomic mass units (Daltons). The required transmembrane pressure (i.e. pressure drop across the membrane) is lower (up to 3 MPa) than the one used for RO, reducing the operating cost significantly. However, NF membranes are still subject to scaling and fouling and often modifiers such as anti-scalants are required for use. NF membranes are a type of RO membranes which allow the passage of monovalent salts but retain polyvalent salts and neutral (no load) solutes ≥ 400 Daltons. Membranes for reverse osmosis (RO) have a very small pore size, and are designed to separate ions from each other. Pores smaller than 10 nm ("nanofiltration") and may be composed of nanotubes. Nanofiltration is mainly used for the removal of ions or the separation of different fluids.

4.2.5.3 Ultrafiltration
The range of pore size in ultrafiltration (UF) membranes is between 10–100 nm, but most manufacturers provide a range of molar masses of particles separated as: 1–500 kD. UF membranes are usually asymmetric, thin film, composite or porous membranes. Materials used for UF membranes are frequently the following polymers: polisulphone, polyethersulphone, polyfluorovinidylene, polyacrylonitrile, polyimide, polyamide, cellulose derivatives and also ceramic. The sieve mechanism is considered to be a mechanism of separation. Possible transport mechanisms are convection and diffusion in the pores. Adequate models for ultrafiltration are Hagen-Poiseuille and polarization concentration. The permeate fluxes in ultrafiltration are 50–1000 liters/m^2 h (LMH), whereas the driven force is transmembrane pressure (1–10 bar). The power consumption is 0.3–1.5 kWh/m^2, (10–150 W/m^3) and the cost of mem-

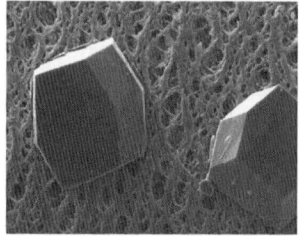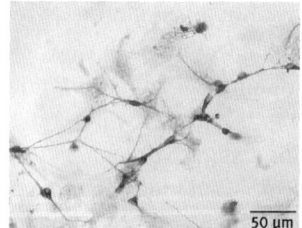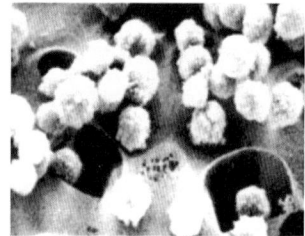

Fig. 4.12. Application of MF and UF membranes: (a) protein crystals on UF membrane surface; (b) neuron cells cultured on membrane surface; c) microorganisms on membrane surface.

branes: $ 100–200/m^2 membranes. Ultrafiltration membranes are designed for the concentration and separation of biopolymers e.g. dextrans and proteins. Therefore ultrafiltration is used most often in biotechnology, pharmaceuticals, food processing, environmental protection (separation of oil emulsions), electroplating plants, dairies, textile production (see Fig. 4.12).

4.2.5.4 Microfiltration
Microfiltration (MF) is a method of filtering the smallest particles that cause turbidity in the range of 0.05 and 10.0 microns. The sieve mechanism of separation is that the particles are retained, because they are larger than the pores in the filter. MF membranes effectively remove major pathogens and contaminants such as Giardia lamblia cysts, Cryptosporidium oocysts, and large bacteria. For this application the filter has to be rated for 0.2 µm or less.

4.2.5.5 Dialysis
The driving force of dialysis is the concentration gradient (chemical potential). The transport mechanism is diffusion in pores. The separation principle is hindered diffusion, i.e. different rates of diffusion for components and also sieving. Pore size in microporous dialysis membranes is between $1.5 < d < 10$ nm, ~ 500 MW. When dense membranes are used this is diffusion dialysis. The types of membrane modules used in dialysis are plate frame or hollow fiber. The cost of dialyzers for medical applications are lower ($ 10–35/m^2) than industrial which is around $ 100/m^2 in hollow fiber. Dialysis is a membrane process which is mainly used in medicine for renal disease therapy (artificial kidney). The global dialysis market was around $ 69 billion for 2010 with continuous growth of around 4 % per year. However, dialysis is also used as a process of industrial separation of bases, acids, salts and metals from process media (see Fig. 4.14). The free strong acids (acid dialysis) can be recovered by using anion exchange membranes and free strong bases (base dialysis) by using cation exchange membranes. During electrolytic copper refining, dialysis is used for the separation of

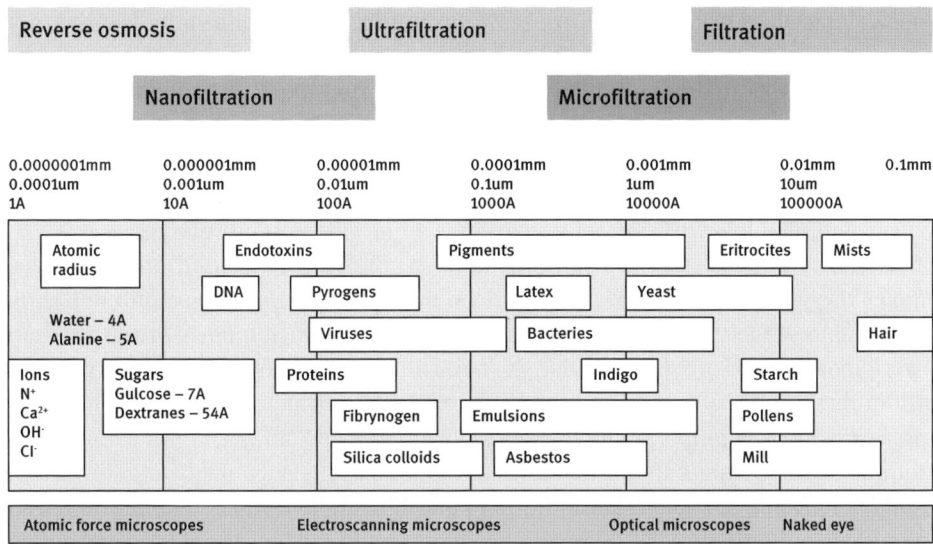

Fig. 4.13. The range of separation abilities of different pressure-driven processes.

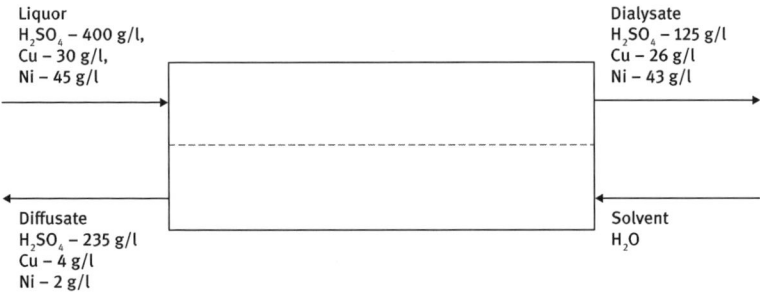

Fig. 4.14. Dialysis cell for separating metal ions and free acids from a copper refinery liquor.

nickel sulphate from sulphuric acid. The hybrid processes are used by combining diffusion dialysis and nanofiltration that have been developed for operations with a low salt concentration and high volume flows, hence reducing investment costs effectively.

4.2.5.6 Electrodialysis

Electrodialysis (ED) is used to transport salt ions from one solution through ion-exchange membranes to another solution under the influence of an applied electric potential difference (see Fig. 4.15). The driven force of electrodialysis is electrical potential gradient (electro osmosis). The transport of counter ions through the ion-exchange membrane and coupled transport is the dominant mechanism. The separa-

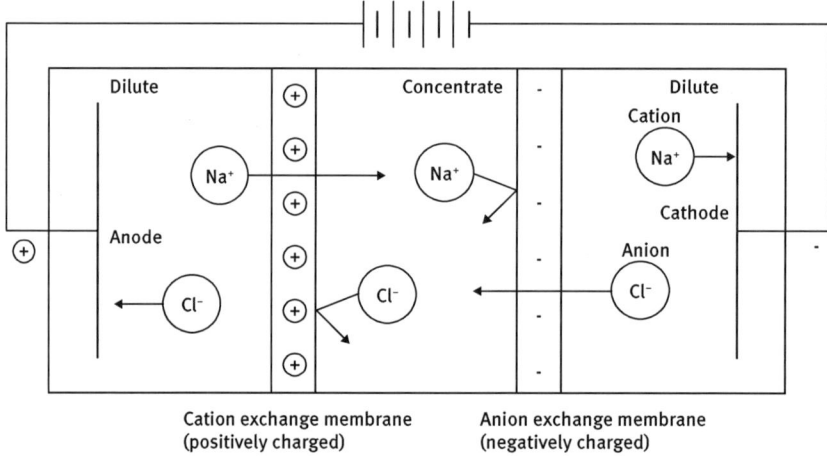

Fig. 4.15. The principle of operation of an electrodialysis process.

tion principle is Donnan exclusion mechanism and equilibrium. Membrane materials used in electrodialysis are cross-linked copolymers based on divinylbenzene (DVB), with polystyrene or polyvinyl pyridine. Due to the charge of ions transferred membranes are divided into cation-exchange and anion-exchange. Cation-exchange membranes include in its construction negatively-charged groups, such as sulfonic acid groups or carboxyl – group. Anion-exchange membranes are positively charged, e.g. quaternary ammonium groups. The most commonly used polyelectrolytes are Nafion them from Du Pond copolymers of polytetrafluoroethylene (PTFE) and polysulfonyl fluoride-vinyl ether. Membranes for electrodialysis are nonporous and membrane modules are flat. Flux is dependent on the required separation. Energy consumption depends on feed concentration, approximately ~ 0.4 kWh/m³. Operating costs may be evaluated based on feed flow rate amounts $ 0.10/m³, and the total is approximately $ 0.34/m³. Applications: brackish water, table salt, wastewater treatment, desalination of solutions in food, chemical and pharmaceutical industries, concentration of brines. In almost all practical electrodialysis processes multiple electrodialysis cells are arranged into a configuration called an electrodialysis stack, with alternating anion and cation exchange membranes forming the multiple electrodialysis cells. Electrodialysis is a very useful technology for the recovery of pure components of solutions, even very diluted.

4.2.5.7 Pervaporation
The driving force in pervaporation process is a chemical potential difference between the liquid feed/retentate and vapor permeate on either side of the membrane. This driving force can be maintained in very simple and economical ways, i.e. by mild heat-

ing of the retentate and by maintaining low pressure on the permeate side or by sweeping of the permeate by inert gas. The principle of pervaporation is solution diffusion of given components. Vapor permeation is essentially the same process as pervaporation taking the same membranes and similar conditions into account. The only difference is using gas on the feed side which reduces the fouling problem and means membranes may be used for longer.

The pervaporation process can be extremely selective. This is due to the special separation mechanism in which the component must first dissolve in the membrane and then diffuse through it. Due to the big impact on the solubility of specific components in the given polymers we can easily affect the selectivity of the membranes through proper selection and modification of polymers. Hydrophobic membranes are frequently made of polydimethylsiloxane-based polymers, whereas hydrophilic membranes are based on polyvinyl alcohol or polyimide. Ceramic membranes are also available which consist of nanoporous layers on top of a macroporous support based on tailored zeolites or silica layers of appropriate molecular selectivity. These membranes are fabricated by sol-gel chemical processes. The combination of organic-inorganic is known as hybrid membranes. Hydrophobic pervaporation is effective in cases of extremely diluted solutions to recover organic substances from contaminated waters. Hydrophilic pervaporation is frequently used for dehydration of alcohols containing small amounts of water. The main applications of pervaporation processes are:
1. solvent dehydration;
2. continuous ethanol removal from yeast bioreactors;
3. continuous reaction product removal to enhance conversion and rate;
4. concentration of hydrophobic flavors;
5. purification of refinery streams;
6. breaking of azeotrope;
7. purification of extraction media;
8. purification of organic solvents.

4.2.5.8 Membrane distillation

Membrane distillation is a process in which each unique pore is a distillation assembly, because it acts as an evaporator from the one end and a condenser on the other. To make this possible the pores must be filled with a mixture of steam and gas only, and thus the membrane must be a liquid repellent (say hydrophobic) on both sides of the pores. The second condition of operation of membrane distillation membranes is temperature difference in liquids on both sides of the membrane. Therefore, it is said that the temperature (gradient) is the driving force of membrane distillation. Where the temperature is greater, the evaporation is more intense and the partial vapor pressure is greater on that side. Since the membrane is thin, even a small difference in temperature allows creation of a large gradient. The basic mechanism of mass transport is Knudsen diffusion, and the principle of separation is the vapor-liquid equilibrium. The nature of the membranes is hydrophobic microporous, with a pore size of 0.1–0.45 microns. Materials used for the production of hydrophobic membranes are polytetrafluoroethylene, polypropylene, polyvinylidene PVDF. The most commonly used modules are spiral-wound and hollow fiber (ENKA AG). Membrane distillation is used for the production of distilled water from wastewater using waste heat so i.e. the temperature difference from the environment of a few degrees. Membrane distillation is used in the semiconductor industry, to desalinate sea water, boiler water and the concentration of aqueous solutions.

4.2.6 Mass transport in membranes

There are two main criteria to assess membrane quality. These are the transport properties and separation (see Fig. 4.16). The value of the permeate flux contains information on the kinetics of the process and the required membrane surface of the membrane. Knowing the permeate flux value, one can easily estimate the cost of membrane surface for any project in practice. Additionally, it is helpful that the scale-up is almost always linear in membrane processes.

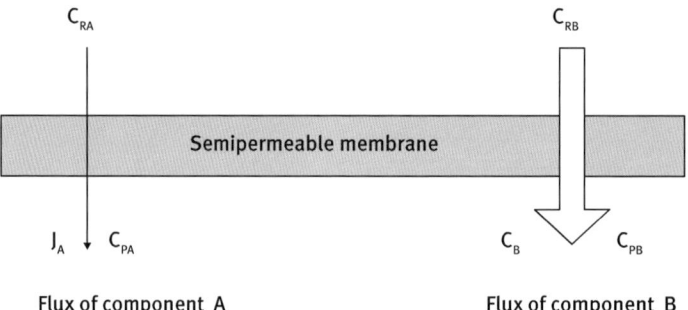

Fig. 4.16. Relationship between transport and separation of components passing through semipermeable membrane.

The volumetric flux of given component that passes through the membrane (under the action of any given driven force) can be defined as a flowrate per unit area of the membrane:

$$J_A = \frac{1}{F} \frac{dV_A}{dt}. \tag{4.5}$$

Separation depends on the ratio of the molecule to the pore size (sieve mechanism), solubility, wettability, electrical charge and chemical or other properties. The selectivity of separation for two components A and B can be defined by means of concentrations in permeates and retentate:

$$\alpha_{AB} = \frac{\dfrac{C_{PA}}{C_{PB}}}{\dfrac{C_{RA}}{C_{RB}}}. \tag{4.6}$$

The separation effect also results from properties of the membrane and feed components. In the case of liquid separation the selectivity can be expressed directly by the fluxes as follows:

$$\alpha_{AB} \cong \frac{J_A}{J_B}. \tag{4.7}$$

The separation effect is most often described by means of the retention coefficient:

$$R_A = 1 - \frac{C_{PA}}{C_{RA}}. \tag{4.8}$$

This is convenient parameter which is used to characterize membranes. Retention (or rejection) coefficients are used commercially by membrane manufacturers to provide membranes which retain standard molecules with a retention factor of 95 %. There are various standard molecules, such as salts (NaCl), dextran, proteins or others. For microfiltration membranes the molar mass of the particle is given instead of the linear dimensions.

The rate of mass transport through the membranes depends on the driving forces and properties of the given component and membrane. In dense homogenous membranes the mass transport is carried out by molecular diffusion enforced by the concentration gradient. The separation effect may be achieved by solubility of the separated components in the membrane. Within microporous membranes the mass transport is mainly realized by volumetric flow with diffusion superimposed on it. Diffusion may occur simultaneously in membrane flesh and within the liquid inside the membrane pores.

The driving forces for mass transport of the given component inside the membranes can be:
1. pressure gradient;
2. concentration gradient;
3. electrical potential gradient;
4. temperature gradient.

Depending on the dominant driving forces, the membrane processes are classified as follows: pressure processes, diffusional, electrical or thermal conductivity. With this classification, we can predict other processes which will have different driving forces in the future such as dry electrostatic, electromagnetic etc. A model of statistical mechanics is a convenient tool for the prediction of dry processes, in which there is room for yet undefined external forces. Such a model was developed by Mason and Lonsdale [65] (see Eq. (2.240), Section 2.5.5) based on the Maxwell model. The statistical-mechanical model describes the transport of molecule population with average velocity (u), as a result of different driving forces. This model describes the motion of molecules of N components in three-dimensional space, taking different interactions and driving forces into account. The general equation describes the balance between the dimensionless driving forces and the related resistances during the multicomponent transport through the membrane.

The different separation effects of the components in the membranes are expressed by dimensionless separation coefficients. The statistical-mechanical model of mass transport in membranes describes all cases and mechanisms which could play roles in membrane processes (see Section 2.5.5, p. 110). In practice the model is very useful for generating the relevant set of equations depending on the number of components (rejected and passing through the membrane) and existing mechanisms of mass transport. In the majority of cases, the membrane acts as a semi-permeable barrier for the components of the solution; it may also play an active role in the transport of mass. These are cases of active or facilitated transport.

Thus basic ways of mass transport through the membranes (Fig. 4.17) may be defined as follows:
1. passive transport – when the membrane is only an inert barrier for some components to pass through;
2. facilitated transport – when the fluid elements pass through the membrane according to the driving force, which may be a chemical potential gradient; in this case, diffusion is enhanced by the reversible reaction with a carrier which is fixed inside the membrane;
3. active transport – similar to facilitated but in the opposite direction; this transport type occurs in the cells of living organisms.

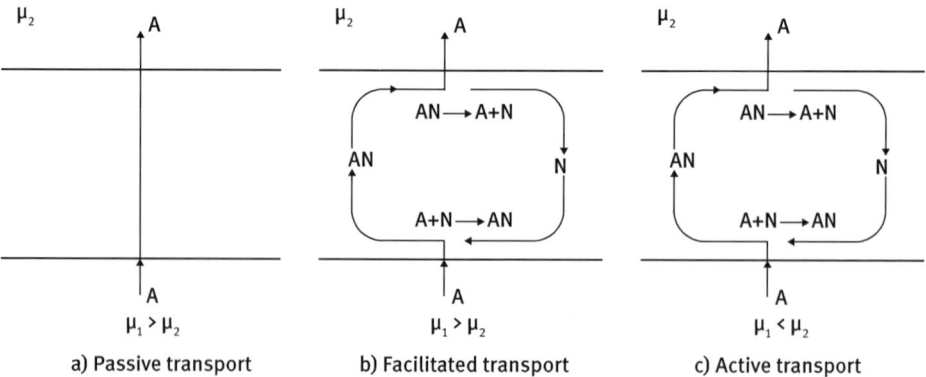

a) Passive transport   b) Facilitated transport   c) Active transport

Fig. 4.17. The overall classification of transport through the membranes.

Basically there are no continuous materials at all and therefore each membrane is porous in a certain sense. The division of membranes into dense and porous only has contractual significance. The mechanisms of particle transport in the membranes are therefore based on pore size, as presented in Tab. 4.2.

Table 4.2. Effect of pore size on mass transport in membranes

Pore radius r	Ratio λ/r	Transport mechanism
up to 5 Å	∞	Solution & diffusion
5–30 Å	> 10	Knudsen diffusion
30–30,000 Å	10–0.001	Transition area (coexistence of different mechanisms)
More than 30,000 Å	< 0.01	Convection flow

The mean free path in the gases is dependent on pressure (concentration) as follows:

$$\lambda = \frac{kT}{\sqrt{2} \cdot \pi d^2 P}. \tag{4.9}$$

The mass transport of components in membranes can be carried out in many ways simultaneously, depending on membrane structure and solute properties (see Fig. 4.18). Diffusion and volumetric flow can occur at the same time. Diffusion may occur within the membrane, and in the fluid filling the pores. Depending on pore size it can be molecular diffusion, Knudsen diffusion, selective adsorption or other mechanisms.

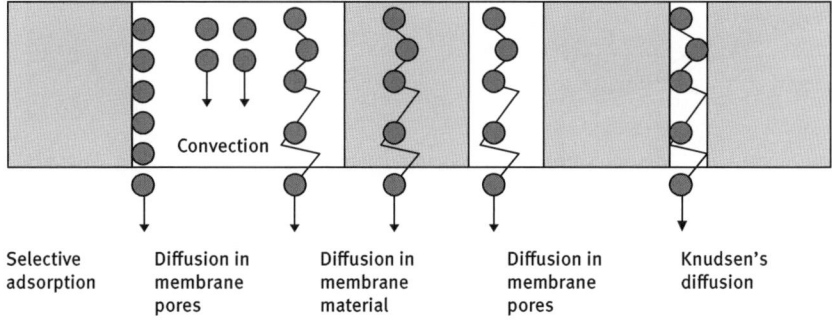

| Selective adsorption | Diffusion in membrane pores | Diffusion in membrane material | Diffusion in membrane pores | Knudsen's diffusion |

Fig. 4.18. Mass transport mechanisms through porous membranes.

4.2.6.1 Irreversible thermodynamic models

Mass transport is connected with irreversible processes and may be described by phenomenological equations. These relate mass or energy fluxes to relevant driving forces. In cases when several driving forces contribute to mass transport, the thermodynamics of irreversible processes expresses the linear relation between fluxes (J) and thermodynamical driven forces (X):

$$J_i = \sum_k L_{i,k} \cdot X_k.$$ (4.10)

The kinetics coefficient (L) reveals the symmetry properties discovered by Onsager (see Section 2.2.9) which are now one of the basic thermodynamic rules known as the 5th thermodynamic principle:

$$L_{ik} = L_{ki}.$$ (4.11)

The following driving forces are most important: pressure gradient, concentration temperature and electrical potential gradients.

4.2.6.2 Friction models of mass transport

Mass transport is a movement of molecules under the driving forces which are counterbalanced by the force of friction between other components. The forces of friction also occur between molecules in movement and membranes, and are always proportional to relative mean values of molecular velocity:

$$X_i = \sum_j f_{ij} \left(u_i - u_j \right) + f_{im} \cdot u_i.$$ (4.12)

The friction coefficients are related to various phenomena which hinder mass transport. Practically the friction coefficients within membranes are always higher than those between molecules.

4.2.6.3 Solution diffusion model

In fact, diffusional transport which results from the stochastic movement of molecules in the concentration gradient area may be expressed as

$$J_i = \frac{D_i}{l} \left(C_{Ri} - C_{Pi} \right).$$ (4.13)

Diffusional transport in membranes takes place inside the membrane material, inside membrane pores and in the membrane boundary layer (see Fig. 4.19). The kinetics of diffusion depends on the shape and size of the molecules. Temperature always enhances diffusion.

The condition for diffusion in any given space is the solubility of the component in the appropriate environment. The solubility constant may be defined as the relationship between partial pressure and concentration in the equilibrium state:

$$S = \frac{C_i}{P_i}.$$ (4.14)

It is convenient to use the term of permeability (K) for every pressure-driven process, which is defined as

$$J_i = K_i \cdot \frac{P_{Ri} - P_{Pi}}{l}.$$ (4.15)

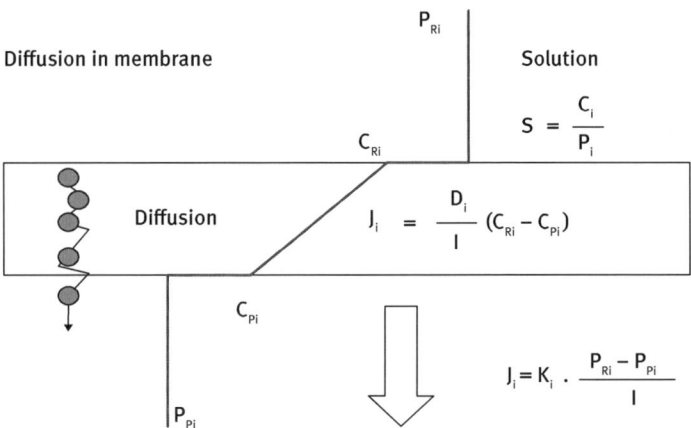

Fig. 4.19. The solution-diffusion model.

Hence the diffusion and solubility coefficients influence the permeability as a product:

$$K_i = S_i \cdot D_i. \tag{4.16}$$

Separation selectivity is always dependent on the difference between permeabilities of separated species. Therefore this selectivity is equally dependent on diffusivity and solubility. In practice it is much easier to affect selectivity of separation by the proper selection of materials with different solubility for two components, than to control the mobility of molecules during diffusion. This is the major reason for the invention of composite membranes, where selectivity is attained by covering the membrane surface with the appropriate material.

4.2.6.4 Knudsen diffusion

This type of diffusional transport occurs when the pore size is smaller than the free path of the molecules (see Fig. 4.20):

$$d_{pi} < \lambda_i, \tag{4.17}$$

where the free path is considered to be the distance between two subsequent intermolecular impacts and may be determined as follows:

$$\lambda_i = \frac{\mu}{\rho} \sqrt{\frac{\pi \cdot M_i}{2kT}}. \tag{4.18}$$

In this case the number of impacts between diffusing molecules and membrane prevail over the intermolecular impacts. Hence the diffusion coefficient may be evaluated based on pore diameter and the size of the molecule:

$$D_{Ki} = \frac{d_p}{3} \sqrt{\frac{8RT}{\pi \cdot M_i}}. \tag{4.19}$$

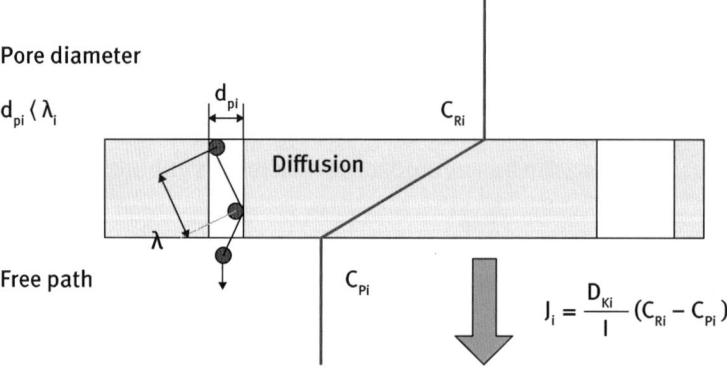

Fig. 4.20. Knudsen diffusion.

For intermediate cases between conventional and Knudsen diffusion, and especially for liquids, the Knudsen diffusion may be determined based on the self-diffusion coefficient:

$$D_{Ki} = D_{ii} \cdot \frac{d_p}{\lambda}. \tag{4.20}$$

As the pores get smaller the mechanisms of diffusion will change. The same happens when pores became bigger because molecular diffusion of mass transport is accompanied by volumetric flow. This makes distinguishing between so-called dense membranes and porous membranes difficult. Some attempt has been made by Bean, who introduced useful criterion for distinguishing porous and dense membranes. This criterion is based on comparison of permeabilities due to diffusional and convectional transport. Pressure-driven convectional permeability prevails over the diffusional transport $L_P > L_D$ in porous membranes. In contrast, the membranes may be considered dense when diffusion prevails over convection $L_P < L_D$.

4.2.6.5 Fine pores model

When membrane pores are of the same order as the molecules it is very difficult to predict the transport mechanism because the size of the pores may change very rapidly due to adsorption of solvent molecules on the pore walls. Water molecules can be bonded to membrane material by hydrogen bonds creating an additional layer inside the pores. This layer changes the separation effect on any ions to be separated. Simultaneously the ions themselves are surrounded by hydration layers of Ist or IInd orders. Ions containing a 2nd order hydration layer are of 1.5–2 nanometers dimension and are comparable with the dimensions of nanofiltration membranes. It was found that separation selectivity depends on hydration enthalpy, i.e. the heat released during creation of the hydration layer. The fine pores model presents the relationship between hydration enthalpy and separation effects for cations and anions of the salts according to the formula

$$\log(1 - R) = K_1 - K_2 \cdot \log[f(\Delta H_{hydr})],$$ (4.21)

where constants (K) are dependent on membrane properties and a function f(ΔH) can be expressed as follows:

$$f(\Delta H) = \Delta H_I \cdot \Delta H_{II}^m.$$ (4.22)

Table 4.3. Enthalpies of ion hydration

Anion	H kJ/mol	Cation	H kJ/mol
Li^+	−636	F^-	−449
Na^+	−454	Cl^-	−325
K^+	−363	Br^-	−303
Rb^+	−337	I^-	−274
Cs^+	−286	NO_3^-	−310
Ca^{++}	−1615	SO_4^-	−1074

Table 4.4. The m-constant of ion hydration

Cation	Anion	m
1	1	0.51
1	2	0.51
2	1	0.47
2	2	0.33
3	2	0.33

4.2.6.6 Selective adsorption models

In cases where pore size is comparable to molecular size the transport kinetic can be affected by adsorption effects on pore walls (see Fig. 4.21). Separation selectivity then depends on the difference between molecular mobilities along membrane walls. The flux of a given i-th component which passes near the pore walls can be described by the equation:

$$J_i = -D_{Si} \frac{dC_i}{dP} \frac{dP}{dl}$$ (4.23)

Diffusion coefficients near the walls depend on surface activation energy:

$$D_{Si} = D_{S0} \cdot e^{-\frac{E}{RT}},$$ (4.24)

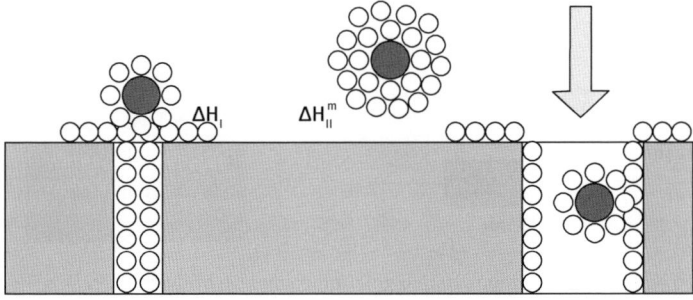

Fig. 4.21. Fine pores model.

where D_{S0} depends on free path and concentration on membrane surface and the activation energy can be expressed as

$$E = \frac{q}{m},$$
(4.25)

where q is the differential heat of adsorption and m is dependent on the type of surface bond.

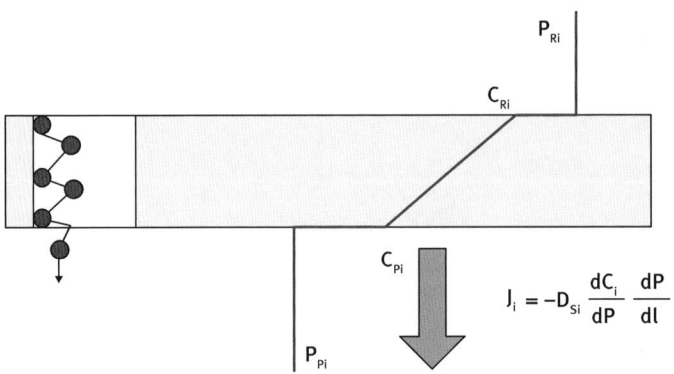

$$J_i = -D_{Si} \frac{dC_i}{dP} \frac{dP}{dl}$$

Fig. 4.22. Selective adsorption.

4.2.6.7 Porous layer model

The hydraulic permeability of the membrane is considered to be a porous layer and can be determined using the Hagen–Poiseuille equation:

$$L_P = \frac{A_p}{l_p} \frac{r_p^2}{8\mu},$$
(4.26)

where the dynamic viscosity (μ) of the liquid which flows through the capillary pores of the length is lp, the radius is rp, and the cross-section area is A_p.

Diffusional permeability of the membrane may be expressed as

$$L_D = \frac{A_p}{l_p} D (1 - q)^2 = \frac{D_m}{l_p},$$
(4.27)

where q is the ratio of molecule size to pore size, and D is the self-diffusion coefficient, which can be determined from the formula

$$D = \frac{k \cdot T}{2\pi\mu} \cdot \sqrt[3]{\frac{N}{\upsilon}},$$
(4.28)

where k – Boltzmann constant and N – Avogadro number. Based on this criterion the threshold value of the pore size 0.35 nm has been determined for water, where diffusional and convectional fluxes are equal. This means that membranes with smaller pores may be considered dense membranes. The permeability of membranes for pressure-driven processes may be determined from an equation similar to the Ohm law:

$$J = \frac{\Delta P_{TR}}{R_m}.$$ (4.29)

The membrane resistance may be determined by the Carman–Kozeny equation based on porosity (ϵ), specific surface (Sc), tortuosity (τ), and length (l):

$$R_m = K\frac{(1 - \epsilon)^2 \cdot S_m^2 \cdot l_p \cdot \tau}{\epsilon^3}.$$ (4.30)

The constant (K) depends on pore shape and K = 2 for cylindrical pores. Flow resistance through the very thin layers such as in composite membranes must involve the "flange effect":

$$R_m = \frac{3\pi\mu}{r\epsilon}.$$ (4.31)

The same concerns diffusional resistance:

$$R_D = \frac{\pi \cdot r_p}{4D\epsilon}.$$ (4.32)

In cases when components of the solution penetrate the membrane pores, the intrinsic resistance of the membrane can be added to membrane resistance:

$$R_m = \frac{C_p \cdot f \cdot \mu \cdot l_p}{M_i \cdot \epsilon}.$$ (4.33)

The separation effects in porous membranes may be evaluated based on the ratio of molecules to pore size:

$$q = \frac{r_i}{r_p}.$$ (4.34)

The Peclet number is also a criterion for transport mechanism through the porous layer:

$$Pe = \frac{J_v \cdot r_p}{\epsilon \cdot D}.$$ (4.35)

In cases when the Peclet number is smaller than 1, it is assumed that separation effects are solely dependent on diffusion and the retention coefficient may be determined as follows:

$$R = 1 \qquad \text{for} \quad q \geq 1$$ (4.36)

$$R = \left[1 - (1 - q)^2\right]^2 \quad \text{for} \quad q \leq 1.$$ (4.37)

As is well-approximated by the formula

$$R = 1 - \left\{1 - [q\,(q - 2)]^2\,e^{-0.7146q^2}\right\}.$$ (4.38)

In cases when convectional (volumetric) transport dominates, the separation effect may be determined based on the following formula:

$$R = \left[1 - (1 - q)^2\right]^{1.5} \quad \text{for} \quad q \leq 1.$$ (4.39)

These equations may only be applied when pores are of the same size. When size distribution can be described by logarithmic-normal distribution:

$$n(r_p) = n_0 \cdot e^{-\left[\frac{\log\left(\frac{r_p}{r_m}\right)}{\log(\sigma)}\right]^2},$$

(4.40)

where

n_0 – number of pores with average radius r_m;

r_m – average radius size;

σ – standard deviation.

For the distribution of same-sized pores the flux can be evaluated from the equation

$$J = \Delta P \frac{\pi \cdot n_0}{8\eta \cdot l_p} \int_0^\infty n(r_p) \cdot r_p^4 dr.$$

(4.41)

In this case the retention may be determined as follows:

$$R = \frac{\int_0^\infty n(r_p) \cdot r_p^4 \left[1 - (1-\lambda)^2\right]^2 dr_p}{\int_0^\infty n(r_p) \cdot r_p^4 dr_p}.$$

(4.42)

The problem of determining the size of molecules remains. There is a relationship between the size and molecular mass in a limited number of cases only. For albumins the shape is close to globular. This relationship may be expressed as

$$\log(r_i) = 0.47 \cdot \log(M_i) - 0.513.$$

(4.43)

For dextranes with chain and branch structure this relationship yields

$$r_i = 0.33 \cdot M_i^{0.46}.$$

(4.44)

The above-mentioned formulas concerned the sieve mechanisms of separation. Other mechanisms that should be involved are molecular mobility, adsorption, affinity effects, and molecule shape.

4.2.7 Model of constant pressure membrane filtration

Hermia [120] introduced the generalized model which describes all cases of constant pressure filtration. This model describes the variation of the yield using a differential equation with two constants, where V denotes volume of the filtrate and t is the process time:

$$\frac{d^2 t}{dV^2} = k \left(\frac{dt}{dV}\right)^n.$$

(4.45)

The constants in this equation characterize the four main cases of membrane filtration (see Fig. 4.23). Hence they may be used for identification of the mechanisms of flow

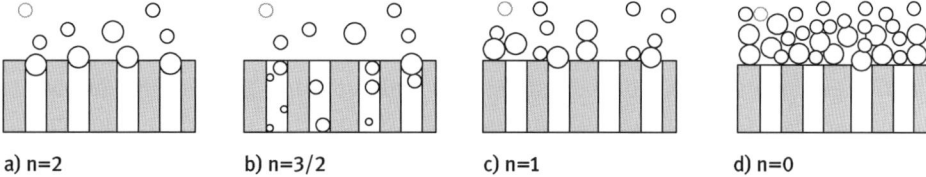

a) n=2   b) n=3/2   c) n=1   d) n=0

Fig. 4.23. The mechanisms of permeate flow resistance: a) complete pore blocking; b) standard pore blocking; c) transient pore blocking; d) cake filtration.

resistance based on experimental results. It should be mentioned that the values of the parameter n may only be 2, 1.5, and 1 i 0, whereas the constant (K) has different dimensions in each case. This model may be used based on experimental data of flux decline under constant pressure.

Although identification of the parameters n and k in Hermia's equation seems to be a simple method of determining the actual mechanism of flux decline, it was found that in practice small fluctuations in flux lead to large errors in the second derivative. An effective method for analyzing flux decline mechanisms was proposed by Koltuniewicz and Field [121], and its feasibility has been tested experimentally (see Fig. 4.23):

$$J(t) = \left[J_0^{n-2} - k \cdot A^{2-n} \cdot (n-2) \cdot n\right]^{\frac{1}{n-2}}. \tag{4.46}$$

This equation has been derived from the Hermia model, however its usage is more accurate and convenient because direct flux decline measurement is used for determination of the parameters n and k of the model.

As shown in Fig. 4.24, in actual cases, filtration mechanisms change during the microfiltration process. To begin with the membrane pores are perfectly blocked and

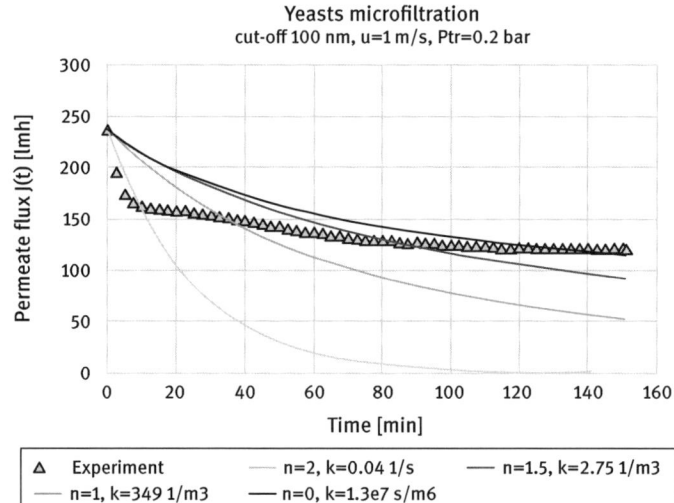

Fig. 4.24. The graphical interpretation of flux decline during yeast microfiltration, by means of different mechanisms of Hermia model according to equation proposed by Koltuniewicz and Field.

finally the "cake layer" grows on the membrane surface. However, it can be concluded that the model could be suitable for the proper selection of the membrane. Well-chosen membranes should have the smallest pore blockage and the largest share of cake mechanism. Membranes selected in this way are resistant to the accumulation of substances in the pores. However, the build-up on the surface of the material to be retained as a result of separation is normal and is called concentration polarization. It can be remedied by considering the flow conditions over the membrane. The concentration polarization phenomenon is natural and occurs whenever separation on the membrane takes place and therefore will be described in detail in the next section (Section 4.2.8).

4.2.7.1 Complete pore-blocking mechanism

In the complete pore-blocking mechanism ($n = 2$) (see Fig. 4.23a), it is assumed that all particles in suspension participate in blocking pores. Each particle entering the pore causes complete blockage. The particles do not collect on one another and do not interfere. This results in a linear relationship between filtrate volume and blocked surface. The proportionality constant σ depends on suspension properties according to the relationship:

$$\sigma = 1.5\frac{\rho_s\phi}{\rho_0 d\psi},$$ (4.47)

where

ρ_s and ρ – densities of suspended matter and the suspension;
ϕ – the mass ratio of the suspended solids;
d – equivalent particle diameter;
ψ – shape coefficient.

The relevant constants in the general equation may be expressed as

$$n = 0 \quad \text{and} \quad k = \sigma \cdot u_0,$$ (4.48)

where u_0 is the superficial velocity of the filtrate.

4.2.7.2 Standard mechanism of pore-blocking

In the standard mechanism of pore blocking (see Fig. 4.23b), it is assumed that pore blockage takes place inside the pores, and volume is reduced permanently and proportionally to the filtrate volume. All pores are of the same size and length to begin with. The parameters of the general equation can be expressed for this case as

$$n = \frac{3}{2} \quad \text{and} \quad k = 2\frac{C}{L}\sqrt{\frac{J_0}{F_0}},$$ (4.49)

Where L is the pore length, F_0 is the cross-section surface of the pores at the beginning, J_0 is the initial permeate flux, C is the ratio of the volume of retained particles to the volume of permeate. Taking nonideal separation of the membrane into account, this constant may be determined from the equation

$$C = \frac{(C_0 - C_p)}{\rho_s}.$$

(4.50)

4.2.7.3 Transient pore-blocking

In transient pore-blocking (see Fig. 4.23c) it is assumed that all particles enter the pores and can settle on one another. In this case, the probability of pore-blocking is permanently reduced with the reduction of the cross-section area of the pores. The corresponding constant in general equation gives

$$n = 1 \quad \text{and} \quad k = \frac{\sigma}{F_0}.$$

(4.51)

4.2.7.4 Cake filtration mechanism

The cake filtration meachanism (see Fig. 4.23d) assumes that flux declines due to the growth of cake resistance, which has built-up on the membrane surface during the filtration process.

The entire flow resistance is then the sum of membrane and cake resistance. It is assumed that cake resistance is proportional to the mass of particles accumulated on membrane surface, which is dependent on the filtrate volume:

$$R(t) = R_0 + \frac{\alpha \cdot \rho \cdot \phi \cdot V(t)}{(1 - m \cdot \phi) \cdot F},$$

(4.52)

where α – the specific resistance of the cake, ρ – density of the filtrate, ϕ – volumetric ratio of solid particles in the cake, and m – ratio of wet to dry mass of the cake. The values of the parameters in the general equation (Eq. (4.53)) are presented below:

$$n = 0 \quad \text{and} \quad k = \frac{\alpha \cdot \rho \cdot \phi}{F^2 \cdot R_0 \cdot J_0 \cdot (1 - m \cdot \phi)}$$

(4.53)

4.2.8 Concentration polarization

Concentration polarization is the occurrence of the concentration gradient in the vicinity of the membrane during separation of the components from solutions or suspensions. The concentration profile at the membrane boundary layer depends on the balance between inflow and outflow of the separated solute. A schematic flow sheet of the membrane surface is presented in Fig. 4.25. The mass balance of the solute involves the possibility of solute accumulation on the membrane surface during flow of the solvent through the membrane with separation of the solute on it.

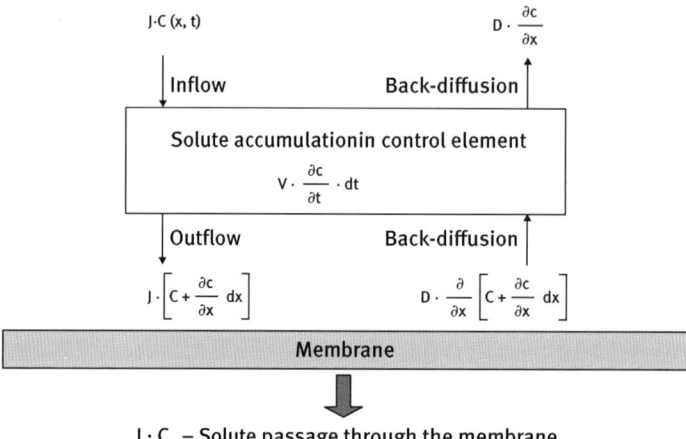

$J \cdot C_p$ – Solute passage through the membrane

Fig. 4.25. Mass balance of the solute at membrane surface.

In this case the concentration profile may be determined as a solution of the differential equation as follows:

$$\frac{\partial C}{\partial t} = D\frac{\partial^2 C}{\partial x} - J\frac{\partial C}{\partial x}.$$ (4.54)

This is a partial differential equation describing diffusion from the membrane surface to the retentate simultaneous with the convection movement of this component permeating the membrane. In order to solve this equation it must first be simplified to ordinary transport equation, as shown in Section 2.2.7 of this book. The appropriate boundary conditions must then be defined to obtain the function of concentration of time and space. In any case, solution is very troublesome. In practice the process is stationary and the balance of the component on the membrane (x = 0) can then be expressed by means of ordinary first-order differential equation:

$$J \cdot C_w = D \cdot \left.\frac{\partial C}{\partial x}\right|_{x=0} + J \cdot C_p,$$ (4.55)

which can be solved by adopting only one obvious boundary condition:

$$C = C_b \quad \text{for} \quad x = \delta.$$ (4.56)

In stationary conditions when inflow and outflow of the solute from the membrane layer are mutually counterbalanced the permeate flux may be determined after integrating

$$J = k \cdot \ln\left|\frac{C_w - C_p}{C_b - C_p}\right|,$$ (4.57)

where the mass transfer coefficient (k) may be defined as

$$k = \frac{D}{\delta}.$$ (4.58)

The value of the mass transfer coefficient may be calculated from dimensionless criteria equations derived using the heat/mass transport analogy:

$$Sh = \alpha \cdot Re^m \cdot Sc^n \cdot \left(\frac{d_h}{l}\right)^p. \tag{4.59}$$

The constants α, m, n, and p in Eq. (4.59) are determined experimentally. Eq. (4.58) enables the thickness of the concentration polarization layer to be determined. It is obvious that elevated concentration at the membrane surface leads to additional flow resistance, but the interpretation of this phenomenon is disputable. Two different explanations occur which are known in the literature as the osmotic and gel models of concentration polarization.

4.2.8.1 Osmotic model of concentration polarization

According to the **osmotic model** of concentration polarization, mass accumulation on membrane surface hinders permeate flux due to the appearance of osmotic pressure. The difference between concentrations at the retentate and permeate side of the membrane lead to an osmotic pressure difference which reduces the driving force and subsequently the permeate flow:

$$J = \frac{\Delta P_{TR} - \Delta \pi}{R_m}. \tag{4.60}$$

Van't Hoff's equation may be used for osmotic pressure determination for diluted systems only:

$$\pi = RT \cdot C. \tag{4.61}$$

In real conditions the nonlinear viral equations describe this relation:

$$\pi = \sum_{j=1}^{N} a_j \cdot C^j. \tag{4.62}$$

Permeate flux, mass transfer coefficient, the thickness of the concentration polarization layer, and the osmotic pressure may be obtained by simultaneously solving the three previous equations. In steady state conditions the flux and solute concentrations at the membrane surface (wall) are stabilized at a level dependant on hydrodynamic conditions.

4.2.8.2 Gel model of concentration polarization

According to the **gel model** explanation, the concentration at the membrane surface instantly achieves maximum value, which is known as a gel concentration:

$$C_w = C_g \quad t \geq 0. \tag{4.63}$$

The solute accumulation at the membrane surface (wall) is achieved by increasing the gel layer, which causes additional resistance to the permeate flow:

$$J = \frac{\Delta P_{TR}}{R_m + R_w}.$$

(4.64)

The permeate flux may be determined based on Eqs. (4.64) and (4.65). The resistance of the gel layer can be determined using the Carman–Kozeny equation, even for macromolecules and particulates of dimension (d_i), taking the porosity of the layer (ϵ) and thickness (δ) into account:

$$R_g = \frac{180 \cdot (1 - \epsilon)^2 \cdot \delta_g}{\epsilon^3 \cdot d_i}.$$

(4.65)

This is commonly used to interpret the concentration polarization phenomenon based on the film model.

4.2.9 Concentration polarization according to surface renewal theory

In the film model a constant film thickness of the boundary layer is assumed, in which only pure diffusion occurs (see Fig. 4.26). In the case of turbulent flow or colloidal suspensions this assumption cannot be satisfied. However, the surface renewal model or critical flux model may be used as discussed in the following sections.

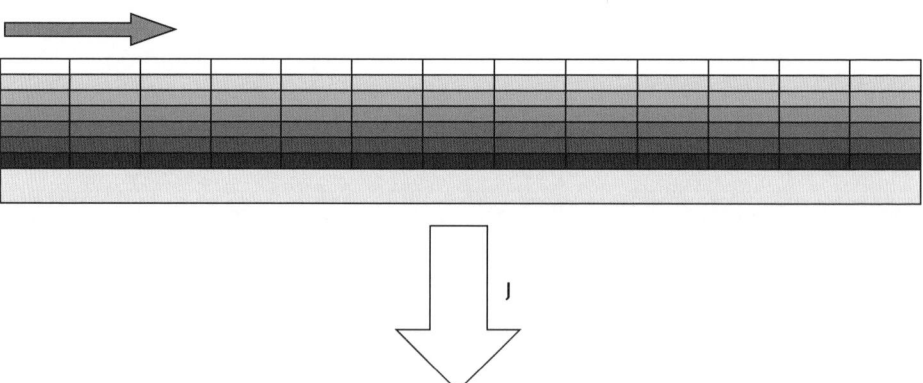

Fig. 4.26. Graphical interpretation of concentration polarization by means of film model.

According to the surface renewal model [122–126] (see Fig. 4.27), the material retained on the surface of the membrane forms a heterogeneous layer consisting of a mosaic of elements of different ages. The distinct permeate flux occurs through each element

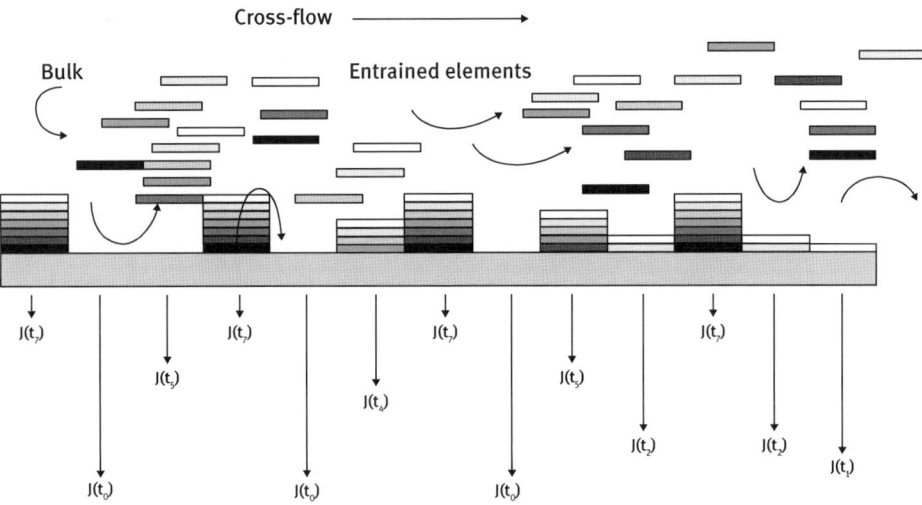

Fig. 4.27. Graphical interpretation of concentration polarization by means of a surface renewal model.

of the surface, depending on the age of that item. More substances, which have been retained by the membrane, are accumulated on older surface elements. The age of each element starts at the moment of renewal under the influence of the vortex. Each vortex sweeps the entire content of the element mass from the portion of surface by mixing it with the bulk of the retentate immediately. It is assumed that vortices are present on the entire surface stochastically and irrespective of time and place.

The average flux from the entire membrane surface may be expressed as the average of the statistical distribution (see Fig. 4.27) thus:

$$\bar{J} = \int_0^\infty J(t) \cdot f(t)dt, \tag{4.66}$$

where $J(t)$ is dependant on the permeate stream from the time and $f(t)$ is the age distribution function.

Danckwerts derived age distribution function $f(t)$, by definition on the so-called "rate of surface renewal" (s), and assuming its constant value s = const in time and on entire surface:

$$s = \frac{f(t)dt - f(t + dt)dt}{[f(t)dt]dt}. \tag{4.67}$$

Another assumption is to compare the integral of the entire distribution, to unity, i.e. 100 %, a condition which meets each distribution function:

$$\int_0^\infty f(t)dt = 1. \tag{4.68}$$

The age function defines the percentage of the surface occupied by the elements of a given age (t) and is determined by the formula, valid only for steady-state process after infinite time of the process:

$$f(t) = s \cdot e^{-st}.$$ (4.69)

4.2.9.1 Steady state processes

Danckwerts theory was originally presented for the absorption process, where instantaneous mass stream was calculated by solving the transport equation, which gave a rather complicated formula. Koltuniewicz adapted Danckwerts surface renewal model, to the mass transport on the membrane surface within "the concentration polarisation layer", as described in a series of papers [121, 123–130]. It was very important for turbulent flow, in which the homogeneous and the stationary film layer are assumed, which is far from reality. The surface renewal model also allowed description of a permeate flux decline, within the initial nonstationary (start-up) period in membrane processes [127]. This unsteady state always occurs at the beginning of all membrane processes and is associated with the formation of the polarizing film on the membrane. It should be distinguished from the "fouling" problem.

The semi-permeable membrane requires the convection element of the transport equation to be taken into account, which further complicates the problem. The convectional element of the transport equation must be eliminated by appropriate substitutions, as shown in Section 2.2.7 of this book. It should be added that the transport equation applies only to those processes in which particles are subject to diffusion (i.e. solutions). This problem is simplified considerably when we use the empirical formula of the form of exponential decay (which appears to be much simpler than solving the transport equation) (see Fig. 4.28). Instantaneous flux in a unique element can be written as follows:

$$J(t) = (J_0 - J^*) \cdot e^{-At} + J^*.$$ (4.70)

The two streams "(flux (J_0) and final (J^*)) can be determined very simply during initial testing in the system dead-end. It should be noted that in the same test the parameter (A) is defined as the rate of accumulation of A, which counteracts the surface renewal (s). The graph in Fig. 4.28 shows the relationship between the instantaneous permeate flux and the mass accumulation on the membrane.

Thus in view of surface renewal theory, permeate flux can be expressed by a simple formula which expresses the weighted average between the two streams, i.e. the initial permeate flux J_0 and final flux J^* (see also in Fig. 4.29):

$$\bar{J}_a = (J_0 - J^*)\frac{s}{A + s} + J^*.$$ (4.71)

It must be emphasized that the formula is valid only for stationary conditions, which can be reached after an infinite time.

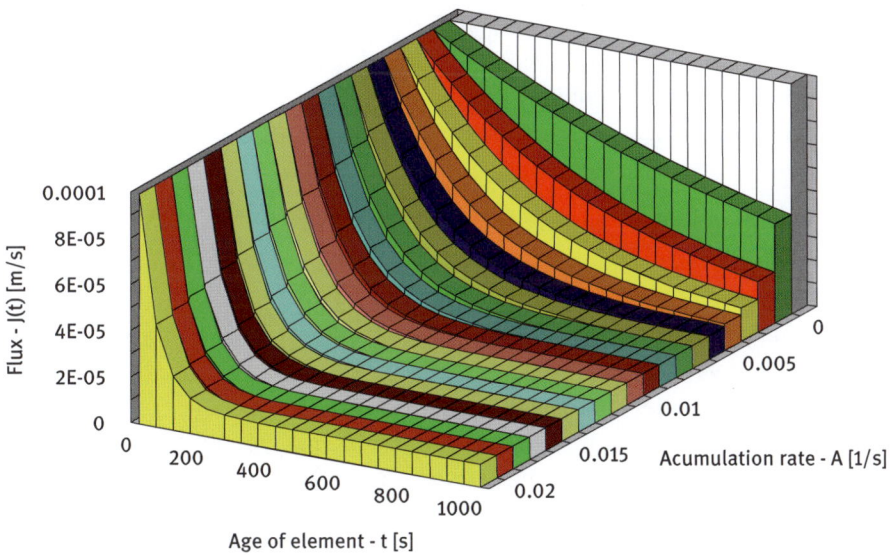

Fig. 4.28. Instantaneous flux within unique surface element as effect of the accumulation rate during yeast microfiltration. ($J_0 = 10^{-4}$ m/s; $J^* = 10^{-5}$ m/s).

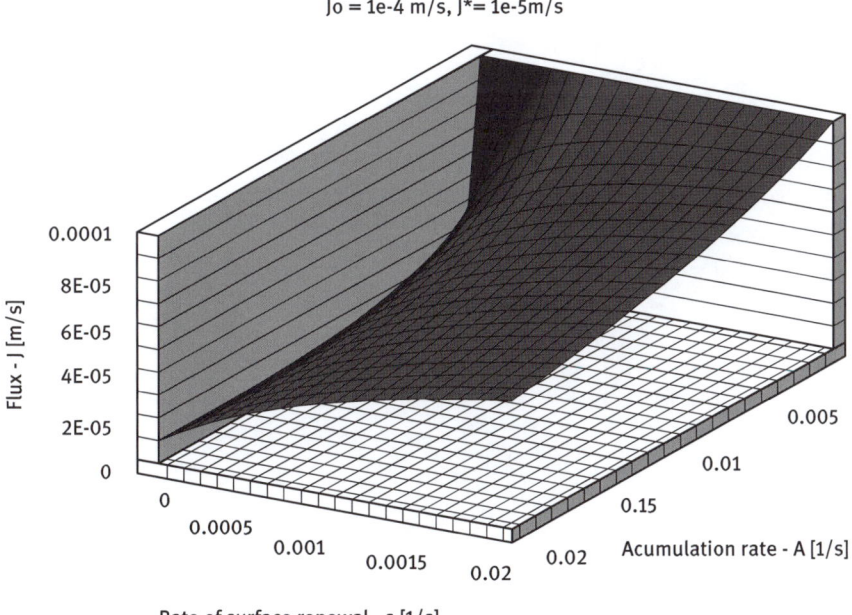

Fig. 4.29. Permeate flux as a function of A – mass accumulation rate and s – surface renewal rate.

4.2.9.2 Surface renewal model for unsteady state processes

If we assume that the age of the element may not exceed the process time, then the age distribution is contained within a closed interval $0 < t < t_p$ which leads to the time-dependent function of the age distribution:

$$f(t) = \frac{se^{-st}}{1 - e^{-st_p}}. \tag{4.72}$$

This formula allows description of the phenomena during the initial formation of the boundary layer, but is also valid for the entire process time range. After substituting an infinitely long process time (t_p = infinity), the formula becomes a primary function according to Danckwerts equation, which determines the age distribution of elements. But in contrast to the film-model, this formula can take the initial effects into account, what allows the automatic procedures washing of the membrane CIP to be applied more effectively. The change in the age distribution is shown in Fig. 4.30.

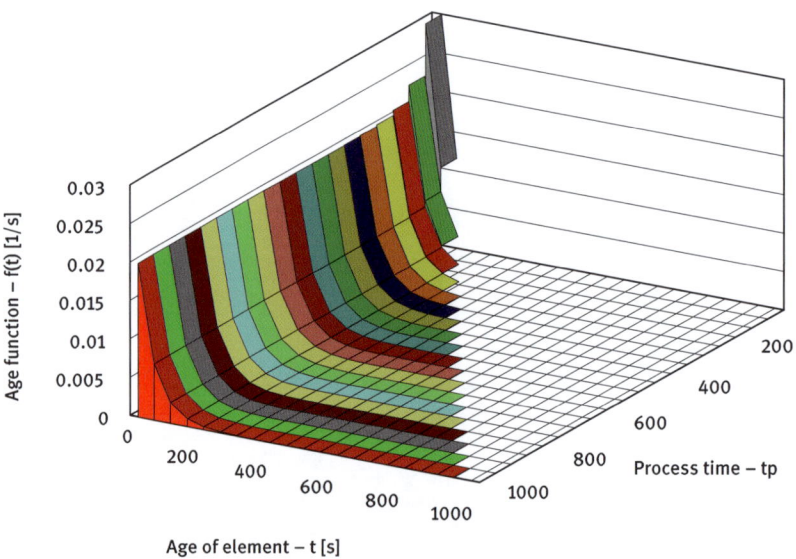

Fig. 4.30. The modified age distribution function.

The modified age distribution function (Eq. (4.72)) is used when calculating the average flux. Ultimately, after integration of the equation 4.66 (with the modified age function) we will arrive at a new equation (Eq. (4.73)) which takes the duration of the process (tp), i.e. the effect of changes to the age distribution (aging), into account. At the beginning of the process all the surface elements are young and the permeate fluxes are greatest. Over time, the age distribution of the elements becomes more and more diffuse. In this way, the modified age function takes the initial period into account when the film is formed:

$$\bar{J}\left(t_p\right) = \left(J_0 - J^*\right) \frac{s}{A+s} \frac{1 - e^{-(A+s)t_p}}{1 - e^{-st_p}} + J^*. \tag{4.73}$$

Thus the average flux can describe the dynamic properties of the membrane systems because it is time dependent. Strictly speaking, the evolution of age distribution (with process time), corresponds to the aging of the layer (see Fig. 4.31).

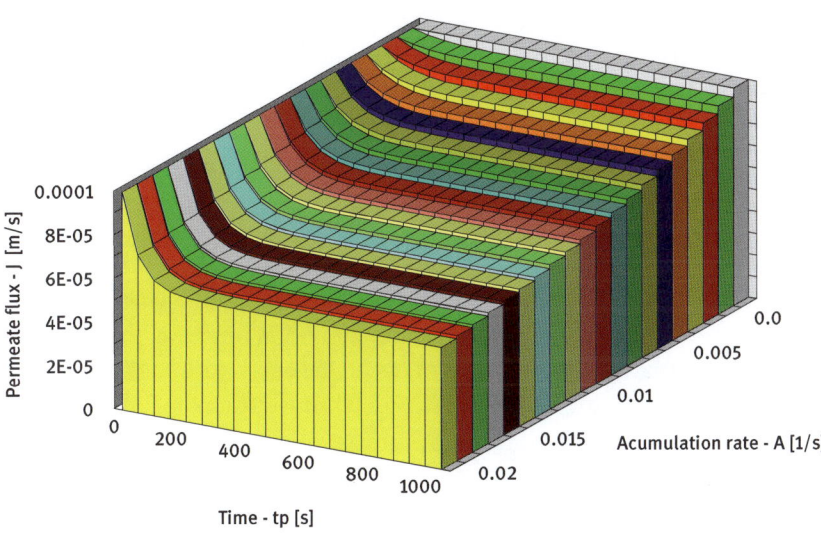

Fig. 4.31. Permeate flux from the whole membrane surface, dependent on process time – t_p and mass accumulation – A (see Eq. (4.73)).

The opportunity to take the dynamic (changes in time) of the phenomena in the vicinity of the membrane surface into consideration has helped describe the time-dependent sorbent saturation in the membrane biosorption process (see Section 4.5) and also the time-dependent rate of flushing in the leaching process (see Section 4.6). Moreover, the ability to describe dynamic properties [127] is very important for systems operating periodically with automated cleaning. In this case, it is possible to optimize the appropriate selection of the washing time (pause) and the permeation time:

$$\overline{J_{av}} = \int_0^{t_{pi}} \bar{J}(t_p)dt_p. \tag{4.74}$$

Moreover, taking changes in the age distribution function of the elements on the membrane surface into account is essential for hybrid membrane processes such as membrane biosorption [128, 129] and membrane leaching [130], which are described later in this book.

4.2.9.3 The critical flux model

Systematic investigations of the colloidal particles on membrane surface proved the existence of lifting forces which enhance permeate flux (see Fig. 4.32).

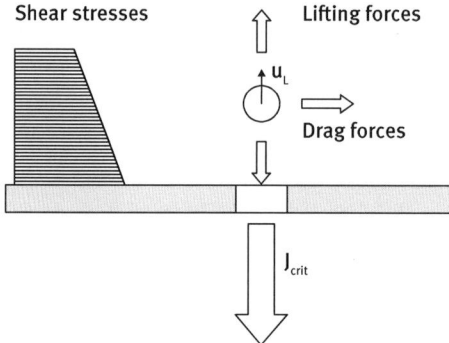

Fig. 4.32. The main forces acting on the colloidal particles at membrane vicinity.

Lifting forces appear in the field of shear stresses on the membrane surface. They are dependent on module configuration, hydrodynamic conditions and rheological properties of the fluid. Particulate behavior is dependent on the net force including inertial, shear, and drag forces. The lifting velocity can be determined from the formula

$$u_L = C \cdot \frac{\rho \cdot d^3 \cdot \gamma^2}{\mu} \cdot f(y'). \tag{4.75}$$

In cases when the lifting forces are strong enough, the particulate can be picked-up despite the drag force which pushes the particle towards the membrane surface. The flux corresponding to the equilibrium between the lifting and drag forces is known as a critical flux. When the flux is below the critical value the highest water permeability may be achieved and the process is very efficient. Exceeding the critical flux value causes cake formation and leads to excessive permeate flow resistance (see Fig. 4.33).

The lifting forces and hence the value of critical flux depend on particle diameter, shear rate, concentration of the particles in bulk and on the membrane surface and the length of the module. The main advantage of the critical flux phenomenon

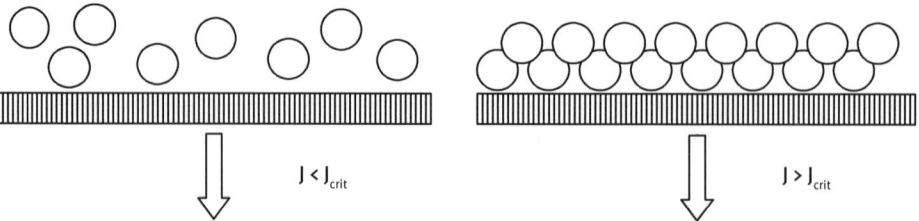

Fig. 4.33. Comparing the collection of the mass retained a) below the critical flow, b) above the critical flux.

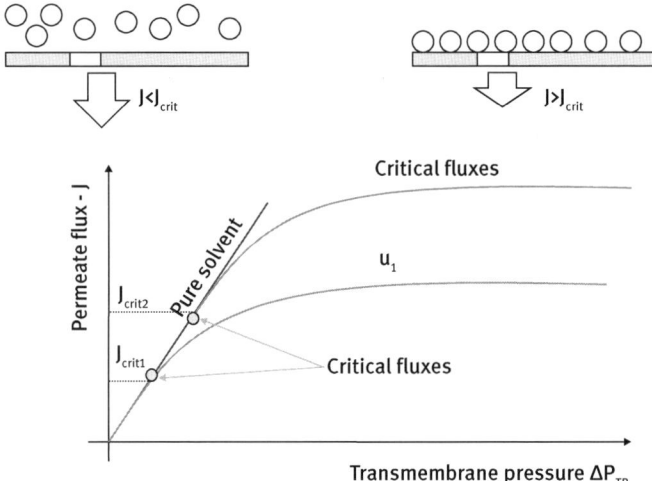

Fig. 4.34. The explanation of critical flux for various cross-flow velocities.

is the possibility of process control to ensure the highest yield with moderate energy consumption thus enabling cost reduction (see Fig. 4.34).

4.2.10 Engineering of membrane processes

Prior to designing a membrane system, the separation target must be defined; first of all what is the main product, permeate or retentate? If the permeate is the product the most important factors are the retention coefficient and recovery factor. If the retentate is to be the product the most important parameter is the concentration factor or direct concentration in the retentate. The designer must also take the fact into account that, despite their many advantages, membranes are sensitive to a variety of stresses such as concentration polarization, viscosity, pH, concentration, pressure and temperature. The effectiveness of the separation processes depends on many factors which can be divided into three main groups, i.e., membrane properties, operation conditions and system configuration.

4.2.10.1 Selection of membranes and the cleaning methods

The major problem of membrane processes is the sensitivity of the membranes and susceptibility to various stresses, such as chemical stresses (pH), and physical stresses (temperature, pressure and pH). These adverse factors and conditions are compounded by the increasing concentration of the compounds retained on the membrane. Therefore, over time a continuous decline in performance is observed, known as the flux decline (see Fig. 4.35).

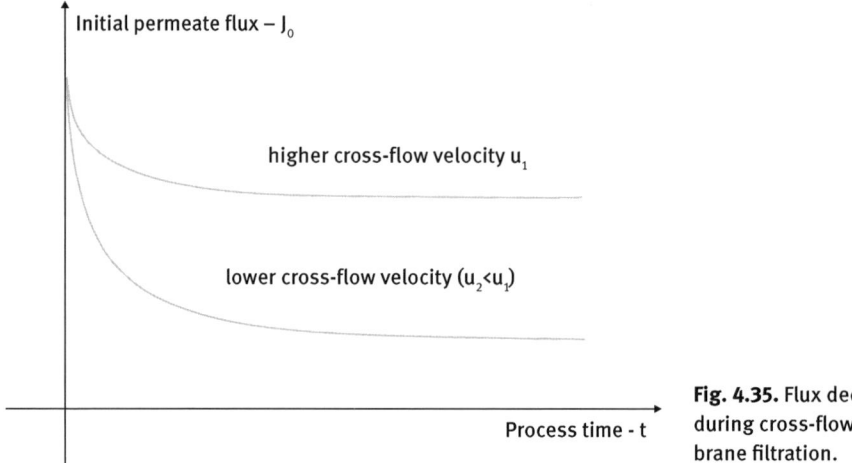

Initial permeate flux – J_0

higher cross-flow velocity u_1

lower cross-flow velocity ($u_2 < u_1$)

Process time - t

Fig. 4.35. Flux decline during cross-flow membrane filtration.

Cross-current flow is used in all types of membrane modules to reduce the effects of concentration polarization (see Fig. 4.36). It should be noted that efforts must be counterbalanced by the flux increase, and usually the circulation flow rate is about 50–100 fed flow rate.

Cross-flow modules are not the only way to combat the effects of concentration polarization. Another option is immersed modules. These configurations of membrane processes have reached minimum exploitation costs. The submerged modules use a relatively low transmembrane pressure (often a vacuum + hydrostatic pressure) and reduction of concentration polarization (or a filtration cake) with the support of a gentle movement of membranes induced by the flow of air bubbles. Even the best-

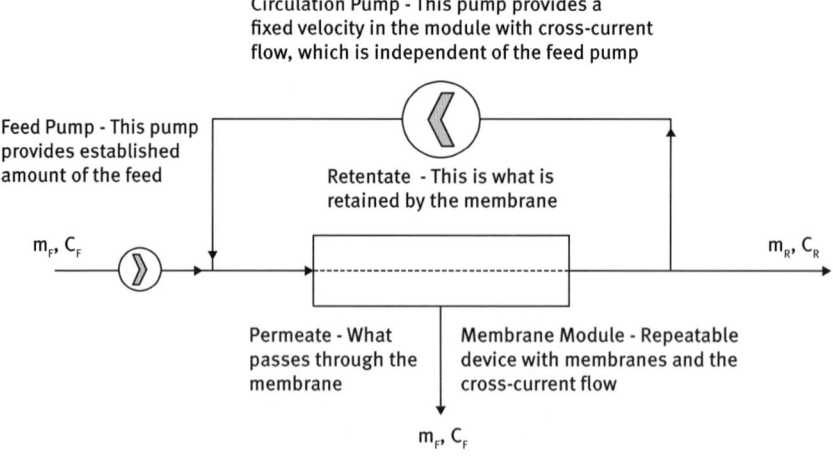

Circulation Pump - This pump provides a fixed velocity in the module with cross-current flow, which is independent of the feed pump

Feed Pump - This pump provides established amount of the feed

Retentate - This is what is retained by the membrane

m_F, C_F

m_R, C_R

Permeate - What passes through the membrane

Membrane Module - Repeatable device with membranes and the cross-current flow

m_F, C_F

Fig. 4.36. Schematic diagram of cross-flow module.

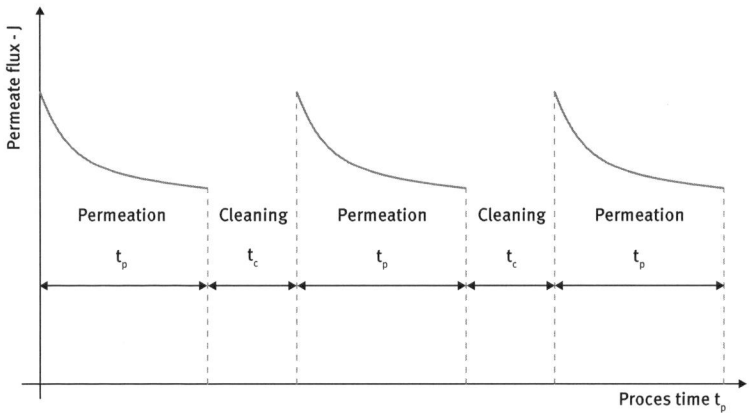

Fig. 4.37. Permeate flux during periodic cleaning and permeation.

designed membrane processes experience a fall in permeability known as membrane fouling after some time. There are many causes of fouling. The membrane cleaning procedure and chemicals should be individualized in each case. So in fact permeate flow can be represented in the form of a saw tooth (see Fig. 4.37):

$$\bar{J}_{av} = \frac{1}{n} \sum_{i=1}^{n} \frac{1}{t_{pi} + t_{ci}} \int_{0}^{t_{pi}} \bar{J}(t_{pi})dt_{pi} \ . \tag{4.76}$$

Figure 4.38 demonstrates that the cleaning time may be an optimization parameter of the process during cyclic operation, provided that the flux decline function is

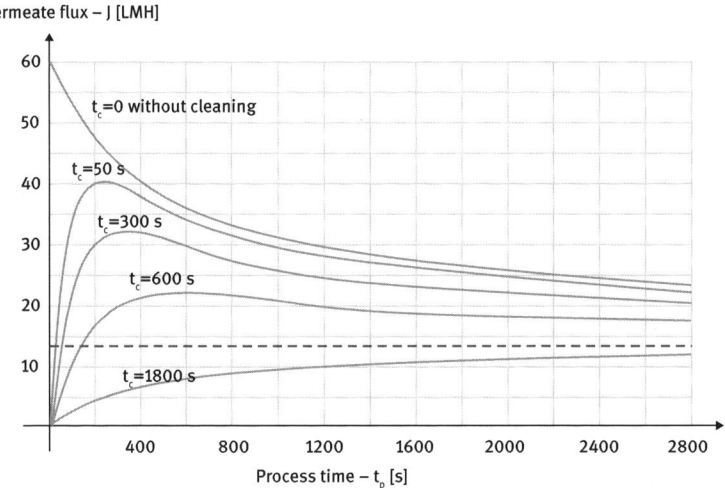

Fig. 4.38. The effect of the time of cleaning on the average flux. J0 = 60 LMH, J* = 0.60 LMH, limiting flux is denoted by the dashed line, Jlim = 13 LMH, the rate of surface renewal s = 0, 0014 s^{-1}.

known. In this case, the surface renewal model was applied, and the whole has been thoroughly described in the literature [127]. Average permeate flux has been calculated using Eq. (4.76).

4.2.10.2 Selection of operation conditions

Proper membrane selection, i.e. material and separation selectivity, permits long-term operation. However, whether or not the solution components that are to be in contact with the membrane could possibly be harmful to it must always be checked with the manufacturer. This may seem paradoxical, but in the majority of cases the most important parameters of membrane processes do not depend on the membrane itself but on the working conditions of the membrane. Regardless of the type of membrane process, certain restrictions for the driving forces should never be exceeded. Deviation from the linear relationship of permeate flux can be observed in pressure membrane processes, when the transmembrane pressure is gradually increased (see Fig. 4.39).

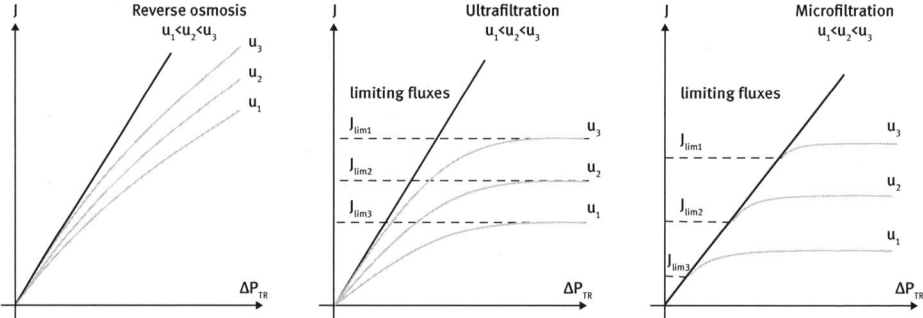

Fig. 4.39. The impact of cross-flow velocity on the permeate stream in different processes.

To determine the effect of cross-flow velocity on the polarization of concentration, one of the models describing mass transport in the boundary layer of the membrane must be assumed. If we assume the simplified (R = 1) film model (see Eq. (4.76)) with the ability to determine the mass transfer coefficient from the criterial equations, then the two characteristic diagrams may be determined (see Fig. 4.40).

The border areas of the flow, i.e., laminar and turbulent, may be determined based on the graphs (see Fig. 4.40a). Whereas the maximum achievable concentration may be characterized based on Fig. 4.40b. These issues are applied to optimize membrane processes and properly select working concentration and the cross-flow velocity of retentate. This is important because both parameters have a large impact on the permeate flux and thus membrane surface, but also on pressure loss, energy consumption and the cost of the membrane process.

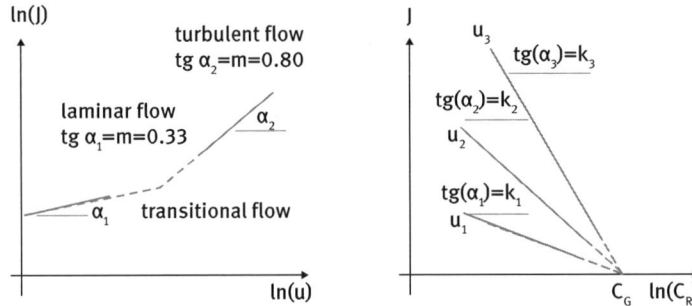

Fig. 4.40. The effects on the permeate flux of (a) the cross-flow velocity for laminar and turbulent flow, and (b) the concentration of the retentate (CR) for the different cross-flow velocities.

4.2.10.3 Selection of system configuration

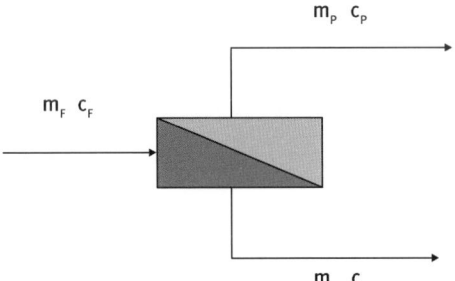

Fig. 4.41. The single stage membrane system (configuration A).

The large potential of process optimization lies in system design. The different configurations may be duly compared; keeping in mind that mass balance must be fulfilled for each of them. The overall mass balance of the separation of binary systems may be written as follows:

Balance of whole streams (A + B):

$$m_F = m_R + m_P. \tag{4.77}$$

(a) Balance of solute (A):

$$m_F \cdot C_F = m_R \cdot C_R + m_P \cdot C_P, \tag{4.78}$$

where:
m_F – mass of the feed;
m_R – mass of the retentate;
m_P – mass of the permeate;
C_F – concentration of the feed;

C_R – concentration of the retentate;
C_P – concentration of the permeate;

In order to evaluate the system behavior, the following four factors are determined:
1. CF – concentration factor for the system:

$$CF = \frac{C_R}{C_F};$$

(4.79)

2. RC – recovery of the system:

$$RC = \frac{m_P}{m_F};$$

(4.80)

3. R – retention of the system:

$$R = 1 - \frac{C_P}{C_F};$$

(4.81)

4. SF – segregation factor of the system:

$$SF = \frac{\text{mass of separated component in retentate}}{\text{mass of separated component in permeate}} = \frac{m_R \cdot C_R}{m_P \cdot C_P}.$$

(4.82)

The segregation factor (SF, the ratio of the mass of the component liable to separation) and the concentration factor are dependent on the pre-defined factors. The segregation factor can be used primarily when the product is the permeate, while the concentration factor is used when the product is the retentate.

The value of the retention coefficient depends primarily on the proper selection of the membrane. Achieving high rates of RC is valid in all cases of separation but requires additional measures to reduce the concentration polarization, which is usually very expensive. In order to achieve the goal of separation at minimum cost, the designer can also take the appropriate flow diagram of the system into account. In order to optimize the system, the various options should be considered taking the recirculation, by-pass and multi-stage systems into account. In practice, the following systems are most frequently used (see Figs. 4.41 and 4.43–4.47).

The recirculation rate is defined as the ratio of the recycle stream (m_C) to the retentate stream (m_R from module):

$$n = \frac{m_C}{m_R}$$

(4.83)

Assuming a recovery rate (RC) and the recirculation rate (n), we can obtain new values of the all streams as follows:

$$m_P = m_F \cdot \frac{RC}{1 - n \cdot (1 - RC)},$$

(4.84)

$$m_R' = m_F \cdot \frac{(1 - RC) \cdot (1 - n)}{1 - n \cdot (1 - RC)},$$

(4.85)

$$C_P = C_F \cdot \frac{(1 - R) \cdot (1 - n \cdot (1 - RC))}{n \cdot (RC \cdot (1 - R) - 1) + 1},$$

(4.86)

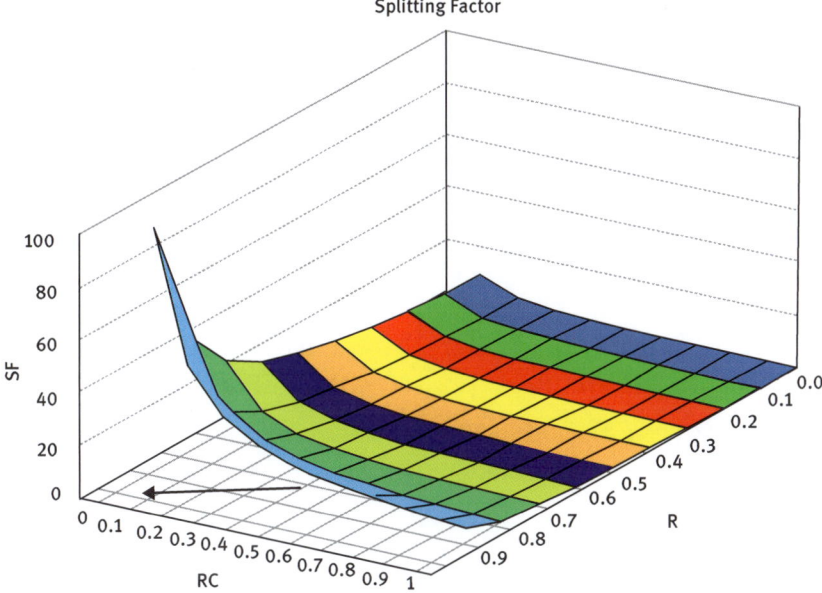

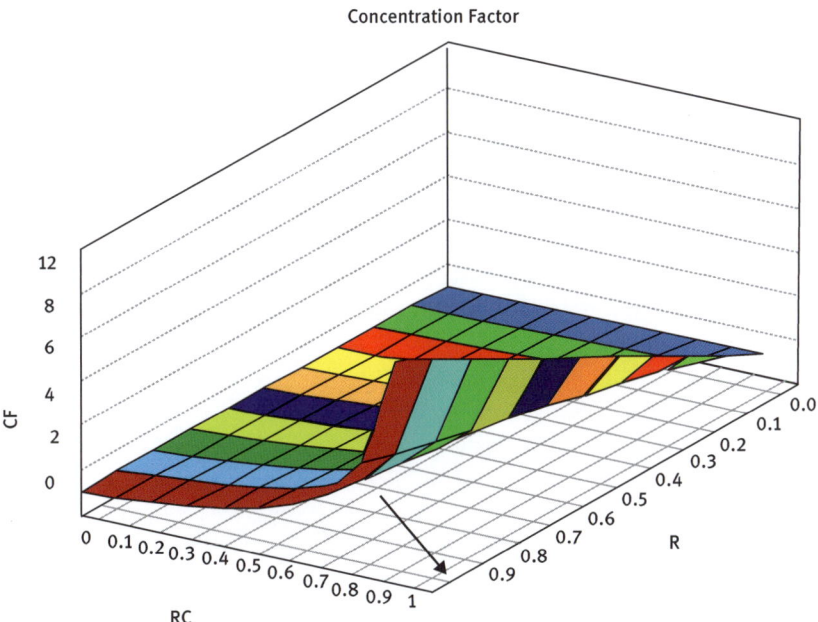

Fig. 4.42. The influence of retention factor (R) and recovery factor (RC) on splitting factor (SF) and concentration factor (CF).

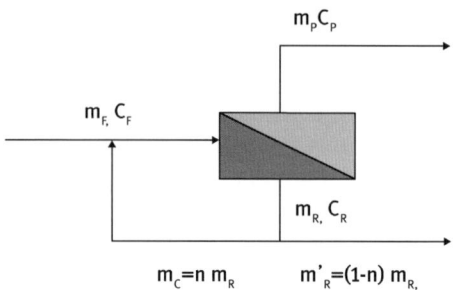

Fig. 4.43. Single-stage membrane system with recirculation (configuration B).

$$C_R = C_F \cdot \frac{(1 - n \cdot (1 - RC)) \cdot (1 - RC \cdot (1 - R))}{(n \cdot (RC \cdot (1 - R) - 1) + 1) \cdot (1 - RC)}. \tag{4.87}$$

Single stage system with recirculation (SSR) is a very simple and economical solution because it exhibits good concentration ability and good recovery simultaneously.

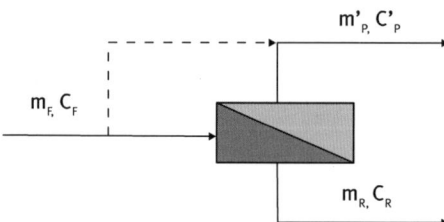

Fig. 4.44. Single-stage membrane system with by-pass (configuration C).

The by-pass rate (j) is defined as the ratio of the by-pass stream (m_B) to the feed stream (m_F):

$$j = \frac{m_B}{m_F} \tag{4.88}$$

Assuming a recovery rate (RC) and the by-pass rate (j), we can obtain new values of all new streams and concentrations as follows:

$$m'_P = m_F \cdot (RC \cdot (1 - j) + j), \tag{4.89}$$

$$C'_P = C_F \frac{RC \cdot (1 - j) \cdot (1 - R) + j}{RC(1 - j) + j}, \tag{4.90}$$

$$m_R = m_F \cdot (1 - RC \cdot (1 - j) - j), \tag{4.91}$$

$$C_R = C_F \cdot \frac{1 - RC \cdot (1 - j) \cdot (1 - R) - j}{1 - RC \cdot (1 - j) - j}. \tag{4.92}$$

This solution may be applied to increase the recovery factor.

In this two-stage system, the permeate is received separately from each step and mixed, and the retentate is twice concentrated. Assuming such a two stage separation system, providing further data on retention (R) and the recovery (RC), and the initial data on the stream and the concentration of streams and the final operating parameters can be calculated with the formulas

$$m_P = m_F \cdot RC \cdot (2 - RC), \tag{4.93}$$

$$C_P = C_F \cdot \frac{2 - RC \cdot (1 - R)}{2 - RC} \cdot (1 - R), \tag{4.94}$$

$$m_R = m_{R2} = m_F \cdot (1 - RC)^2, \tag{4.95}$$

$$C_R = C_{R2} = C_F \cdot \left[\frac{1 - RC \cdot (1 - R)}{1 - RC} \right]^2. \tag{4.96}$$

This arrangement provides the greatest recovery of all systems.

In this system, the permeate from the first stage is redirected as the feed stream in the second stage while the retentates of the two steps are collected. Fluxes and concentrations for the entire system can be calculated using the formula

$$m_P = m_{P2} = m_F \cdot RC^2, \tag{4.97}$$

$$C_P = C_F \cdot (1 - R)^2, \tag{4.98}$$

$$m_R = m_F \cdot (1 - RC^2), \tag{4.99}$$

$$C_R = C_F \cdot \frac{1 - RC^2 \cdot (1 - R)^2}{1 - RC^2}. \tag{4.100}$$

This system ensures the greatest separation of all systems.

The system can be characterized by the passage of the permeate through the cascade consisting of two stages of membrane processes, while the retentate from the second stage is recycled completely and mixed with the feed stream. The following formulas help to calculate the flows and concentrations in the permeate and the retentate of the system. In order to compare all these systems the calculation examples below were prepared. These calculations are performed with the same initial data in all six cases, i.e.:

1. feed flowrate $m_F = 50$ [m³/h];
2. feed concentration $CF = 10$ [g/l];

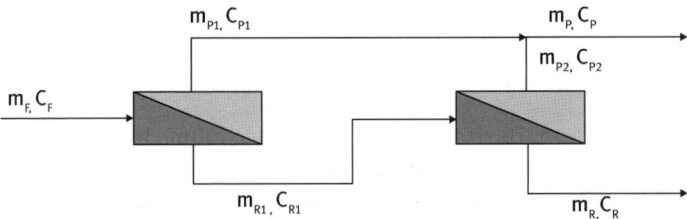

Fig. 4.45. Two-stage system with permeate collecting (configuration D)

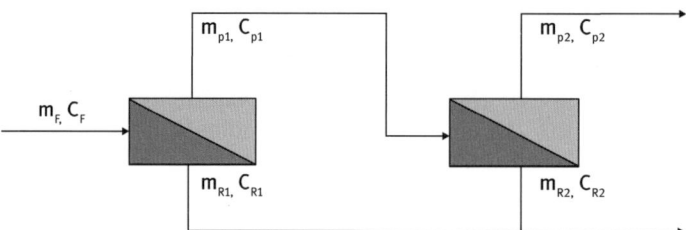

Fig. 4.46. Two-pass membrane system with retentate collecting, (configuration E).

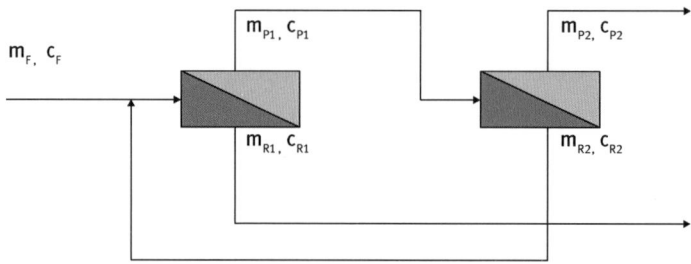

Fig. 4.47. The two-pass membrane system with retentate recirculation (configuration F).

3. retention coefficient R = 0.9;
4. recovery of the unit RC = 0.7;
5. by-pass ratio was assumed to be j = 0.5;
6. recirculation ratio n = 0.5.

The results of the calculations are collected in Tab. 4.5.

Table 4.5. Comparison of different membrane systems

	A	B	C	D	E	F
m_P	35.00	42.5	41.18	45.5	24.50	31.01
m_R	15.00	7.50	8.82	4.50	25.50	18.99
C_P	1.00	6.3	1.59	1.48	0.10	0.08
C_R	31.00	31.00	49.25	96.10	19.51	26.20
CF	3.10	3.10	4.93	9.61	1.95	2.62
SF	13.29	0.87	6.64	6.40	203.10	189.80
RC	0.70	0.85	0.82	0.91	0.49	0.62
R	0.90	0.37	0.84	0.85	0.99	0.99

The following conclusions can be derived from the results:
1. the system D has highest recovery, highest concentration factor, highest permeate and highest concentration of retentate;

2. E and F are similar, but E is better for concentrating F for purification;
3. E has the highest retention, highest segregation and highest retentate stream;
4. F has the highest retention, good segregation, smallest C_p and higher permeate stream than E;
5. C has good permeate recovery and retention;
6. B has good recovery but poor segregation and the worst C_p.

4.2.11 Sustainable applications of membrane processes

These days it is hard to name a single sector which could operate without membranes and membrane processes Efforts were made to show this graphically in Fig. 4.48 where colors correspond to the major areas of human activity. Red refers to industrial activities, yellow to agricultural activity, green is associated with health protection, and blue corresponds to environmental protection. Each of these areas is associated in some way with membranes, while all of them are interlinked. Numerous applications the membrane processes and the hybrid processes have been carefully discussed in the literature [131, 132].

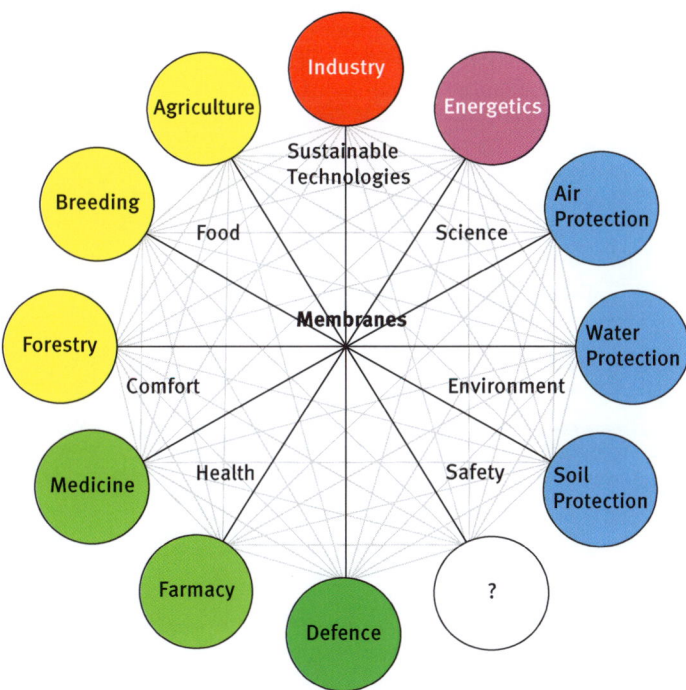

Fig. 4.48. Areas and sectors of membrane applications.

The industrial applications of membranes are numerous and they play an increasingly important role in sustainable development. These processes participate in the creation of clean technologies in industry. Clean technologies rely on solving the problems of industrial pollution at the source, through the recycling of process fluids for re-use. The classical solution is the recovery of paints during electrophoretic coating. Paint diluted by flushing is concentrated on membranes and then recycled for re-use. At the same time, the water recovered by the membrane, is also recycled for re-use in flushing excess paint (see Fig. 4.49).

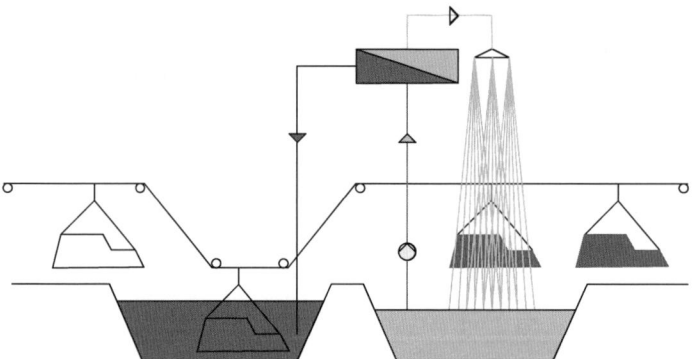

Fig. 4.49. Electrophoretic painting.

Recovery of acids, bases and salts in galvanic baths should be carried out similarly (see Fig. 4.50). In this way, not only is the problem of waste solved, but this technology also becomes increasingly the viable and profitable. Otherwise it could be very dangerous for the environment. Multiple examples of similar applications are possible, for example in the textile industry to recover dyes, cleaning and finishing, in the leather industry, where the process media are extremely dangerous (e.g. chrome), or in the chemical industry (solvents, catalysts, unreacted substrates) and in many others.

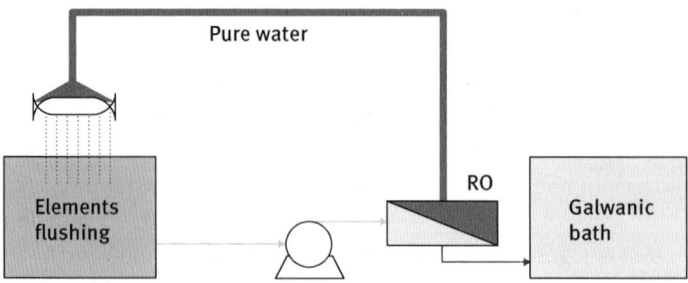

Fig. 4.50. RO for recovery the galvanic baths.

In the food industry the extremely high quality and simultaneously high consumption water is important and exerts large impact on the profitability of the production. The highest water quality can be obtained from reverse osmosis, which is commonly used also in the pharmaceutical and the electronic industries. Different types of boosters are used in order to reduce the cost of these high-pressure processes for the retrieval of energy (see Fig. 4.51).

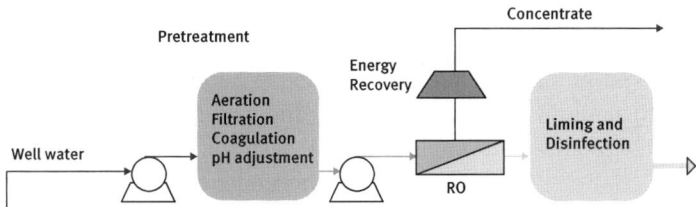

Fig. 4.51. Energy recovery during water treatment.

Membrane processes are widely used in the food industry including the dairy industry, where microfiltration is used to sterilize milk from the beginning (i.e. already in the barn). Figure 4.52 shows all the possible applications of membrane processes in the dairy industry. This scheme is purely virtual, but each single separation process has been described in the subject literature.

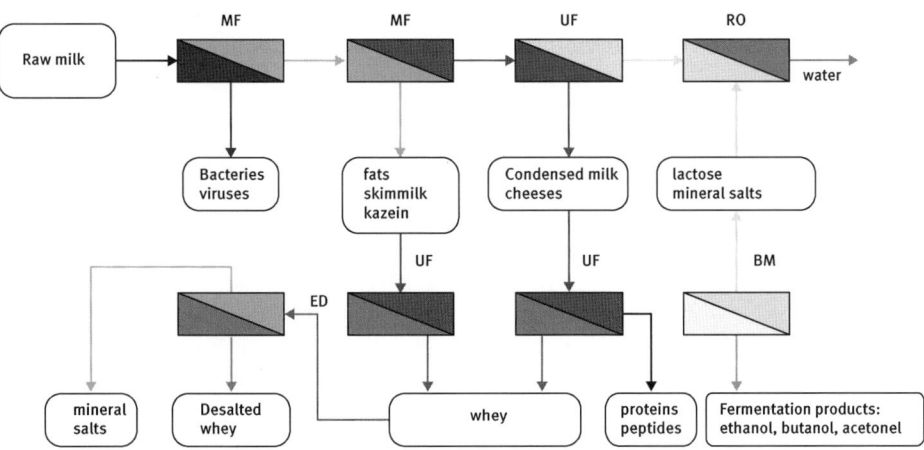

Fig. 4.52. Milk processing factory.

Hydroponics is a method of growing plants using mineral nutrient solutions in water, without soil. Terrestrial plants can grow in an inert medium, such as perlite, gravel, mineral wool, or coconut husks. Almost any terrestrial plant will grow on the basis of

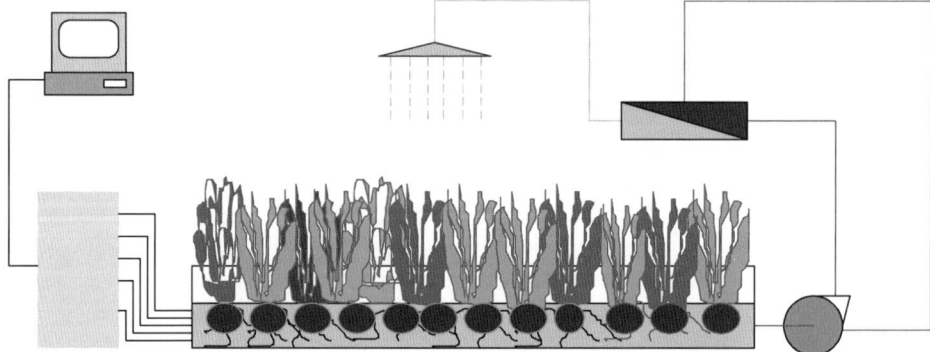

Fig. 4.53. Hydroponic culture.

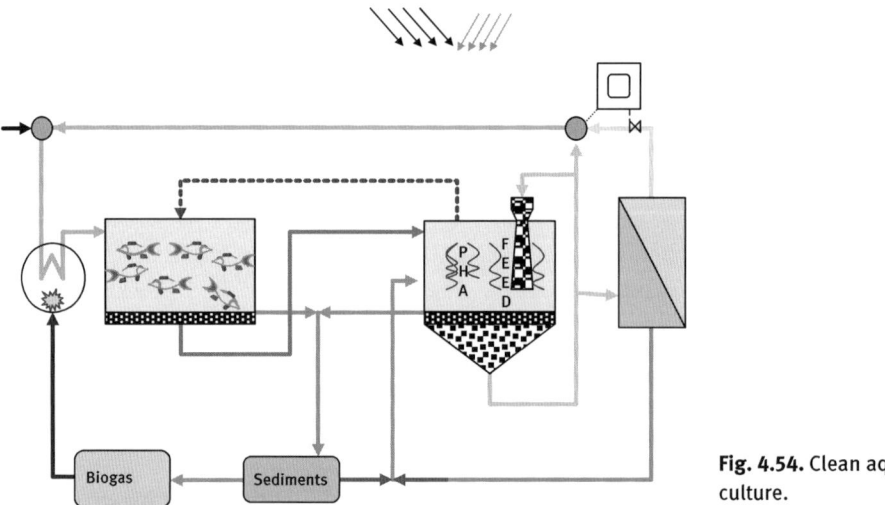

Fig. 4.54. Clean aqua-culture.

hydroponics. Hydroponic agriculture is entirely sustainable because no soil is needed and no soil contamination is expected. The water can be entirely in closed circuit so that its content may be controlled, giving a chance for healthy food. The particular nutrients may be delivered exactly when they are needed during growth. The environment as a whole can reap the benefits of hydroponics because it does not release fertilizers or other plant protection products in to the environment. Plants are healthier and easier to harvest. The cost of such crops is lower because they use fewer chemicals. Water consumption is negligible because it can be circulated in a closed circuit and may be content-controlled.

Thus aquaculture can be completely balanced, especially when a combination of hydroponics and algae is applied.

4.3 Hybrid processes

Hybrid processes are defined in literature [133] as a process package consisting of different unit processes and operations which are interlinked and optimized to achieve a predefined task. There are two types of hybrid processes:

1. **Type 1:** Hybrid processes consisting of processes which are essentially performing the same function. In this case all processes in the package would be separation processes.
2. **Type 2:** Hybrid processes which are the offspring of two different processes. In the case of membrane-based hybrid processes this group includes the combination of membranes and a reactor.

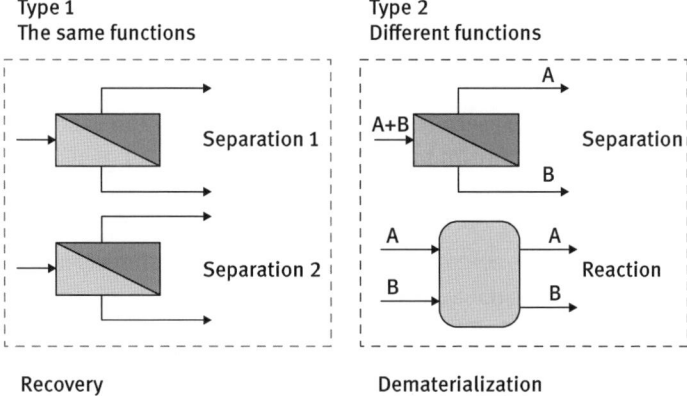

Fig. 4.55. Two types of hybrid systems involving membranes.

Generally, a hybrid process can be called a combination of several processes which aim towards the same functional objectives, but in physically different ways. Thus even the combination of the reactor with the process of separation, can be considered to be the same goal if we are considering the elimination of unwanted components.

4.3.1 Hybrid processes with low integration degree

Hybrid systems of this type are formed when unit processes are combined directly by coupling the equipment. Further development of hybrid processes can be expected as shown in Figure 4.56.

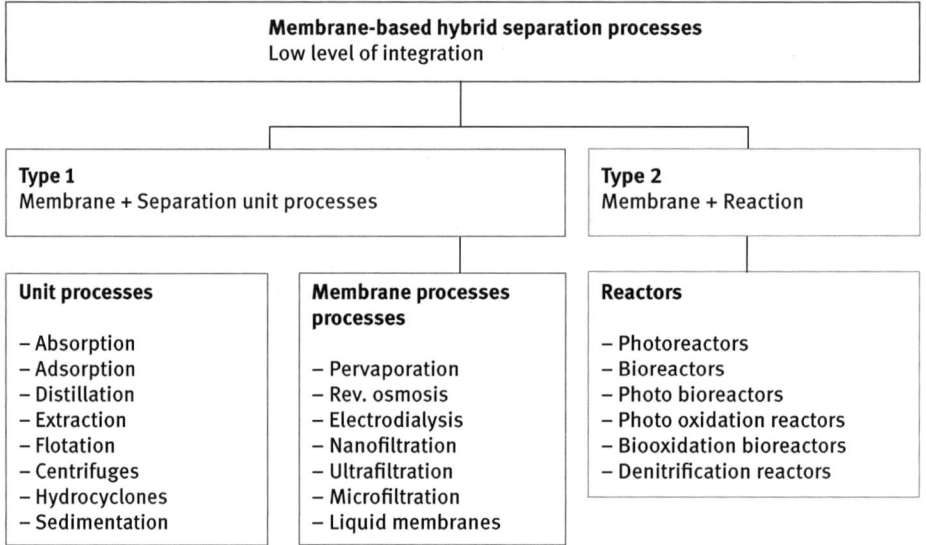

Fig. 4.56. Possible systematic hybrid processes.

The first and most successful hybrid system was a combination of distillation and pervaporation (see Fig. 4.57). This system allowed the separation of azeotropes in an economical way.

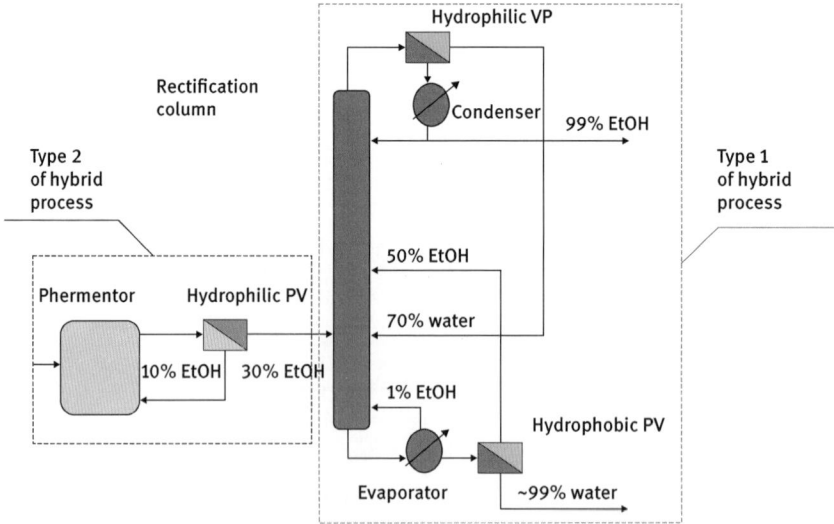

Fig. 4.57. Hybrid system for the production of ethanol consisting of type 1 and type 2 hybrid subsystems.

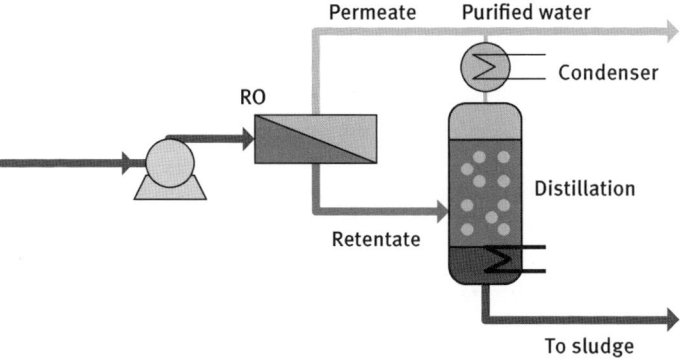

Fig. 4.58. Hybrid system for distillation and reverse osmosis.

The combination of reverse osmosis distillation (see Fig. 4.58) allows cheaper desalination of water. Reverse osmosis water should focus only on the lower ranges of concentrations, when negative effects such as osmotic backpressure or even crystallization have no significant impact. In such cases evaporation is preferable, which yields much less. It is best to take advantage of the optimization algorithm to simultaneously minimize the surface of the membrane, the height of the column and also energy expenditure.

In a similar manner, membrane processes are used for the initial dewatering prior to drying in the food industry technology, such as powdered milk or other instant products like tea, coffee, cocoa, soup or juice. Hybrid processes bring great benefits in each of these cases.

The best way, however, is the closed water cycle in which water is recovered from the process water which is subsequently treated and then returned to production. In some cases, by using appropriate hybrid systems valuable chemical reagents, detergents, dyes, organic solvents etc. can be recovered.

Combining two membrane processes, such as reverse osmosis and pervaporation (see Fig. 4.59), enables the separation of two types of impurities (mineral and organic)

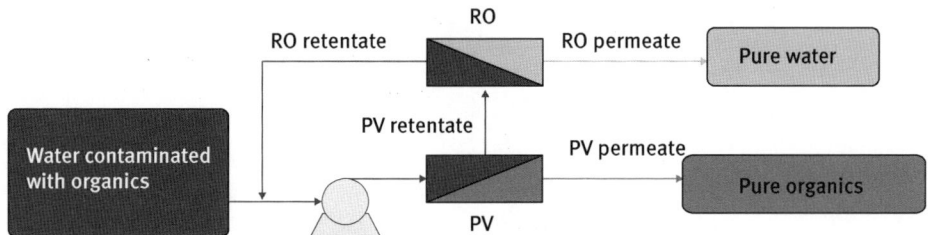

Fig. 4.59. The hybrid system comprising of pervaporation and reverse osmosis, for the reclamation of water and organic components.

from the water intended for recirculation. Moreover, by carefully selecting the membrane material and the appropriate membrane processes, separation of various other mixtures, such as bigger molecules from smaller, polar from nonpolar, soluble or not soluble in the materials of both diaphragms may be achieved in this way. Hybrid systems should assist in the recovery of valuable substances such as oil emulsions with water. Here it is important that the separation was made at source and mixing of different types of wastes such as oil emulsions, solvents and detergents is avoided. The proper installation of plants for the recovery of valuable substances (see Fig. 4.60) is a prerequisite for the effective implementation of clean technologies.

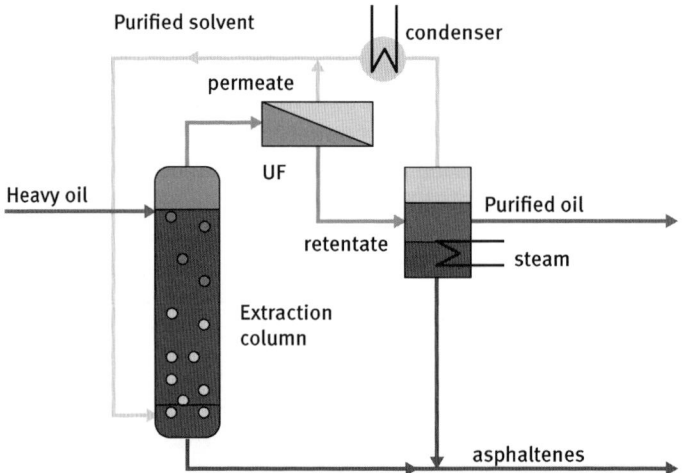

Fig. 4.60. Plant for solvent recovery during extraction in chemical industry.

In this scheme (see Fig. 4.61), flotation was used for the effective separation of small droplets of the emulsion. This is consistent with the principles of hybrid processes that the functions of the processes that are cooperated as components of the hybrid

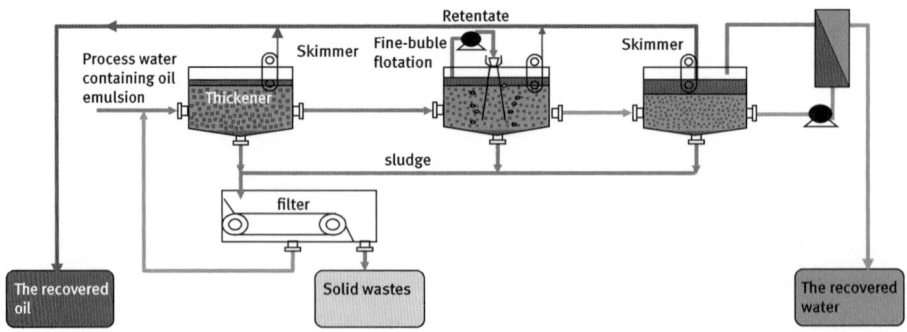

Fig. 4.61. Hybrid system for oil recovery from wastewater.

Mechanical separation   Membrane separation   Biological separation

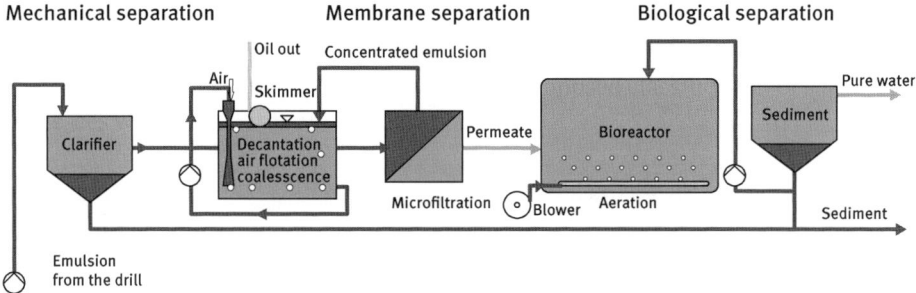

Fig. 4.62. Hybrid system ultrafiltration/bioconversion for groundwater purification.

should be duly spread out. The hybrid system for the decontamination of groundwater contaminated by fuel has been constructed in a similar way [134]. In this project, particular attention was paid to the long-term operation of the hybrid system, which mainly involved biological conversion combined with ultrafiltration (see Fig. 4.62). In order to relieve the membranes an innovative solution was used involving flotation with fine bubbles and labyrinth baffles to increase the water residence time. This solution was very effective. The ultrafiltration membrane (50 kD) rejected only very small drops of emulsion while maintaining stable permeate flux for a long time. Bioconver-

Fig. 4.63. Ultrafiltration of the recycled wastewater.

sion of organic residues was complete because only organic components which were dissolved in the permeate were used.

A similar solution, i.e. a hybrid system which included ultrafiltration and bioconversion, was used to recycle wash water from railway wagons (see Fig. 4.63). The problem in this case was the undefined composition of the water, which was contaminated by a variety of loads. Due to the high cost of large quantities of water and the disposal of sewage, it was decided to use water in a closed circuit. Long-term operation in a closed circuit was possible thanks to the use of ultrafiltration in combination with an open storage tank containing algae culture. The solid residue is removed in the conventional manner, i.e. with grit chambers and the removal of sludge from the bottom of the tank. In this way, the water composition was stable in the closed circuit over a long period of operation.

4.3.2 Hybrid processes with high integration degree

Technological developments have made it possible to achieve an even greater integration of unit processes by combining several functions in one unit (see Fig. 4.64). These functions are physical phenomena such as absorption, adsorption, extraction, distillation combined with the membrane. In this way, new types of hybrids are developed, such as membrane contactors and reactive membranes. The same group may

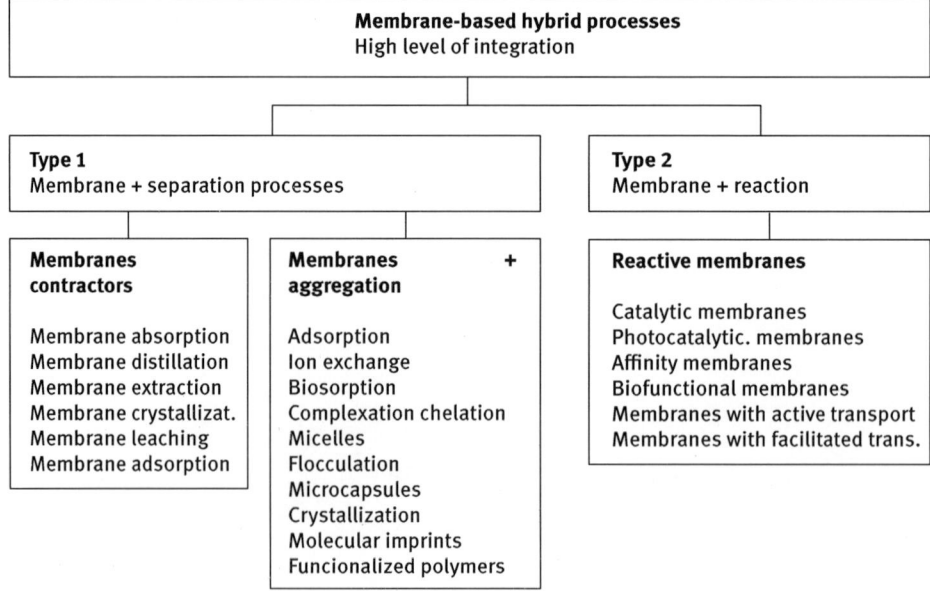

Fig. 4.64. Hybrid processes with a high level of integration.

also include a new type of hybrid consisting of the first stage, in which the particle size is increased in various ways, followed by the efficient separation of such aggregates in the second stage. Figure 4.64 shows the proposed classification of such hybrid processes with a high degree of integration.

4.3.2.1 Membrane contactors

The essential characteristic of membrane contactors is close integration of membranes with the given separation process such as distillation, extraction, absorption. The principle of operation of known membrane contactors have been discussed in the general Section (4.2.2) on membrane processes. The book by Drioli et al. [135] entitled "The Membrane contactors: Fundamentals, Applications and Potentialities" is the most comprehensive, systematic and extensive book on this subject. It was demonstrated that these processes are economically justified when used in applications for environmental protection. The use of simple systems with submerged membranes in wastewater and water treatment systems is particularly favored [136, 137] (see Fig. 4.65).

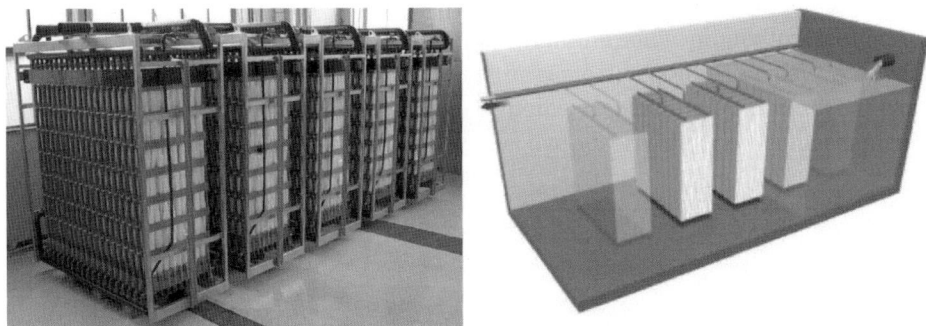

Fig. 4.65. Submerged modules of membrane reactors (photo courtesy of Professor Drioli).

Membrane contactors can be combined in multi-stage systems. Connecting the modules simultaneously increases the capacity of the system, and in order to increase the separation efficiency they are combined in cascades of counter-current flow. The "Membrane adsorption" of VOCs in the membrane cross-counterflow system was discussed in the paper [138]. The paper proposes a new type of membrane contactors which work in solid-liquid systems consisting of the countercurrent cascade, with the cross-flow membrane modules. Numerous examples of membrane extraction, membrane absorption [139], membrane adsorption [129], membrane biosorption [128, 140], ion exchange [141] and micellar enhanced membrane processes [142–144] for water and air purification from such contaminants as heavy metals and organic components can be found in the literature. In the following chapters new processes such as membrane biosorption and membrane leaching will be discussed.

Membrane contactors are preferred to gas absorption in the case of medium and large scale installations which are guaranteed to require a high degree of purification (for instance $H_2S < 4$ ppm) (see Fig. 4.66). They have all the advantages of the absorption columns in less space, less burden on the environment, lower consumption of sorbents and lower operating costs.

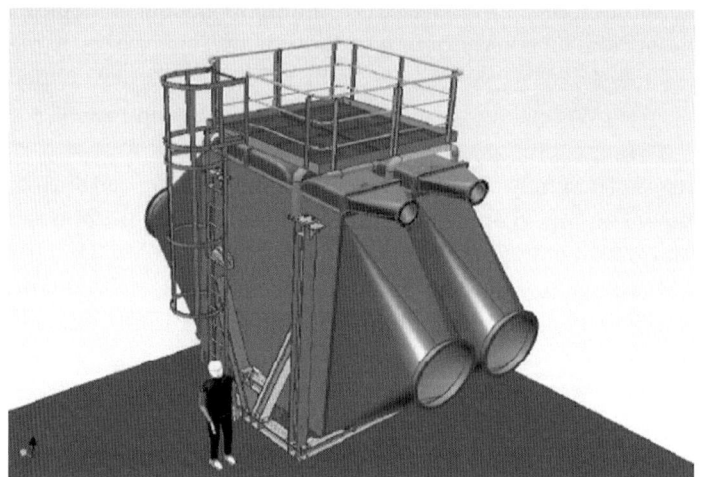

Fig. 4.66. Membrane contactor for absorption of gases CO_2 and H_2S.

4.3.2.2 Catalytic membranes

The catalytic membrane acts as reactor and separator simultaneously (see Fig. 4.67). This equipment can therefore be classified as a hybrid device with a high degree of integration. Most of the catalytic membranes are inorganic membranes due to their excellent resistance to elevated reaction temperature. The membranes used can have a catalyst immobilized on the surface or in the pores, and can also be filled with a bed of catalysts in the form of granules. Catalytic membranes offer not only greater conversion and rate of reaction, but also simpler reactor construction. Ceramic materials such as alpha or gamma alumina inorganic oxide or silicon are commonly used for the preparation of catalytic membranes. However, the allotropic variation of gamma alumina, for example, superior for the separation of hydrogen. The membrane pores, in which Knudsen diffusion occurs, are smaller than 5 nm. The membranes can be coated with the metals listed in Table 4.6, namely: platinum, palladium, vanadium, silver, ruthenium, chromium, and various alloys of these metals.

Catalytic membranes play a particularly important role in many aspects of sustainable development such as production of new generation safe fuels, decomposi-

Fig. 4.67. Catalytic membranes (courtesy of Kamal Surface Coating Accessories Pvt. Ltd.).

Table 4.6. Catalytic membranes

Membrane	Catalyzer	T	
SiO$_2$/Vycor	Cr$_2$O$_3$/Al$_2$O$_3$	450–550	[554]
Zeolite	Pt/Sn/g-Al$_2$O$_3$	450	[555]
CrO$_3$/alumina	Cr$_2$O$_3$/K$_2$CO$_3$/α-Al$_2$O$_3$	450–500	[556]
Pd/alumina	Cr$_2$O$_3$/Al$_2$O$_3$, Pt/Al$_2$O$_3$	450–550	[557]
Pd/alumina	Pt/In/SiO$_2$	400–500	[558]
Pd	Pt/g-Al$_2$O$_3$, Pt or Ru monolith	400–700	[559]
Ru/Pd	Pt/Al$_2$O$_3$	400	[560]

tion of dangerous substances, and also have great importance for clean technologies and green chemistry in general [145]. Moreover, catalytic membranes can improve process performance by up to 50 % more than in conventional reactors. The ideal catalytic membrane should be permselective to one component, for example hydrogen during dehydrogenation. The applications of catalytic membranes usually concern the following chemical reactions: oxidation, dehydrogenation and hydrogenation, oligomerisation and dimerization of isobutene fuel additives, epoxidation of ethylene, and reforming (steam and dry).

Dehydrogenation reactions concern oxidative dehydrogenation of alkenes, methanol, ethanol, andpropane, decomposition of hydrogen iodide. Examples of the applications of catalytic membranes are: the combustion of volatile organic compounds, oxidation of methane, oxidation of propene, oxidation of benzene to phenol, oxidation of sulphides in the liquid phase, catalytic oxidation of n-butane to maleic anhydride, selective oxidation of light paraffins and decomposition of NO$_x$. Dehydrogenation reactions are used to produce valuable olefins from paraffins more cheaply

[150–152]. Catalytic membranes are used for selective oxidation to convert natural gas components into liquid products, which is very cost effective, since resources of natural gas are generally not found in the areas of use. However, paraffins are poorly reactive and therefore drastic reaction conditions must be used in conventional reactions [153, 154]. The oxidative dehydrogenation of gaseous alkanes is preferably carried out on platinum (Pt), copper (Cu) and zinc (Zn) alloys deposited on a ceramic monolith [155], while magnesium oxide, vanadium (V/MgO) are used for the oxidative dehydrogenation of butane to butene and butadiene. The hydrogenation reaction chosen is the catalytic hydrogenation of benzene to cyclohexane for styrene production [156]:

$$C_6H_{12} \rightarrow C_6H_6 + 3H_2. \tag{4.101}$$

Other examples are dehydrogenation of butane to butene with a Pt catalyst [157], or dehydrogenation of isobutane to isobutene [158], a reaction which plays a significant role in the production of methyl t-butyl ether (MTBE), which is a gasoline octane enhancer. The general formula for the dehydrogenation reaction is as follows:

$$C_nH_{2n+2} \Leftrightarrow C_nH_{2n} + H_2. \tag{4.102}$$

The equilibrium of the hydrogenation reaction should be shifted in the direction of the product. In order to improve the efficiency of carbon and thermal efficiency and to reduce investment costs reformer to produce synthesis gas catalytic partial oxidation of methane to synthesis gas has been proposed as an alternative to the widely used steam cracking process [159]:

$$CH_4 + \frac{1}{2}O_2 \rightarrow CO + 2H_2 \quad \Delta H_{1000\,K} = -21.8 \text{ kJ/mol}. \tag{4.103}$$

Catalytic membranes may be used for steam and dry reforming of methane to produce hydrogen-rich synthesis gas (hydrogen and carbon monoxide). Steam reforming can be written in the form of several reactions as follows [160, 161]:

$$CH_4 + H_2O \rightarrow 3H_2 + CO \quad \Delta H_{298} = 206.0 \text{ kJ/mol} \tag{4.104}$$

$$CO + H_2O \rightarrow H_2 + CO_2 \quad \Delta H_{298} = -41.0 \text{ kJ/mol} \tag{4.105}$$

$$CH_4 + 2H_2O \rightarrow 4H_2 + CO_2 \quad \Delta H_{298} = 164.9 \text{ kJ/mol}. \tag{4.106}$$

The dry process of reforming is of great importance for the protection of the environment, due to the possibility of removing CO_2 from the simultaneous alteration of CO_2 into valuable synthesis gas (syngas). This can be done by using a catalyst Ni/La_2O_3:

$$CH_4 + CO_2 \rightarrow 2H_2 + 2CO \quad \Delta H_{298} = 247.3 \text{ kJ/mol}. \tag{4.107}$$

Catalytic membranes fulfill another role for the environment because they allow decomposition of harmful gases. They are used for denitrification on ceramic membrane coated with Pd moderate reactions conditions (eg, $T = 278$–298 K, $p \leq 1$ bar):

$$NH_3 \rightarrow \frac{1}{2}N_2 + \frac{3}{2}H_2 \quad [-\Delta H = 54.6 \text{ J/kmol}] \tag{4.108}$$

Catalytic combustion is the preferred method for removing volatile organic compounds from the air, especially when complete destruction of a highly toxic compound present in small concentrations [146–149] is required. The limits for volatile organic compounds in the atmosphere are generally still being reduced in order to protect human health. Selective oxidation is also attractive for economic reasons. Propane can be used instead of propylene for the production of acrylic acid, which is an important intermediate. Other examples of the use of catalytic membranes for environmental protection are: oxidation of H_2S [162] on stainless steel membrane with pores > 1 microns, oxidation of ammonia [163] catalyst in form of pallets inside the lumen of tubular membrane. MTBA decomposition [164] was even performed on the polymeric membranes PPO polyphenylene with mesoporous silica composite catalysts with controllable loadings $H_3PW_{12}O_{40}$ (PW) [165]. Methanol is an inhibitor of the degradation reaction and must be removed in this case by the catalytic membrane.

4.3.3 Photocatalyst

Photocatalysis is a reaction which uses light to activate a substance which enhances the rate of a chemical reaction. Chlorophyll of plants is a typical natural photocatalyst. Photocatalytic processes have emerged as the technology for destroying refractory materials which are harmful to the environment. This enables conversion of the pollutants into harmless substances directly within the contaminated source.

The photocatalysis reaction occurs on the catalyst's surface. In solid-state physics a band gap or an energy gap is an energy range where no electron states can exist. When light falls on the photocatalyst with greater energy than the band-gap (E.g. $\geq hv$), the photocatalyst generates free electrons and electron holes called H^+ which then take part in the reaction. The "electron hole" describes the lack of an electron at a position where one could exist in an atom or atomic lattice. The electron hole pair is the fundamental unit of generation and recombination, corresponding to an electron transitioning between the valence band and the conduction band. In photocatalysis the pairs of electron holes diffuse to the surface of the photocatalyst, thus initiating a series of chemical reactions based on free radicals.

Free radicals are usually created by the breakdown of larger molecules into the primary molecule with the use enough high energy, such as ionizing radiation, heat, electrical discharges, electrolysis and chemical reactions. Radicals are often the intermediate stages in many chemical reactions. The concept of free radicals was introduced by Moses Gomberg (1866–1947), who was professor of chemistry at Michigan University. A free radical is an atom, molecule or ion that has unpaired valence electrons or an open electron shell. These bindings make free radicals highly chemically reactive among other substances, or even towards themselves. Free radicals can spontaneously dimerize or polymerize. Examples of free radicals are the hydroxyl radical

(HO•) or the carbene molecule (CH_2), Photocatalytic activity (PCA) depends on the ability of the catalyst to form electron-hole pairs, which generate free radicals.

The practical application of photocatalysis was made possible by the discovery of water dissociation by Fujishima in 1960, who found that titanium oxide irradiated by light could break down a water molecule into oxygen and hydrogen gas. This is called advanced oxidation process (AOP) and need not necessarily be linked to TiO_2 and even to UV. A basic condition for AOP is the production and use of hydroxyl radicals.

Advanced oxidation processes, in the broadest sense, refer to a set of procedures for water and wastewater treatment. Frequently, however, the term AOP refers to chemical processes which use ozone (O_3), hydrogen peroxide (H_2O_2) and/or UV light.

Photocatalysts are typically heterogeneous oxides and transition metal sulfides. The best known is titanium dioxide, which is one of the most effective photocatalysts and is used for H_2 photocatalytic production from water or ethanol. Some additional metals such as platinum Pt/TiO_2 enhance the splitting of water. TiO_2 absorbs only ultraviolet light due to its large band gap (> 3.0 eV), but it is better than most of the photocatalysts that work with visible light. The splitting of water can also be accomplished by the use of cobalt oxide [166, 167]. This catalyst was discovered in 2009, by the US Department of Energy in experiments on electrolysis of a solution containing dissolved salts of cobalt. Also the Bismuth titanate pyrochlore ($Bi_2Ti_2O_7$ – BTO) is a cheap and efficient photocatalyst with a direct band gap of 2.6 eV, which corresponds to a red light of 70 nm in absorption activity compared to titanium dioxide (TiO_2). Several impurity elements are used to enhance the visible light absorption. Production of hydrogen by photocatalytic water splitting on the catalyst $NaTaO_3$: La [168] had the highest efficiency at a wavelength of 270 nm. The photocatalyst $K_3Ta_3B_2O_{12}$ was used for the same purpose with the co-catalyst Ta: La, using a stoichiometric ratio of 1:2. In this case, the quantum efficiency and kinetics of the separation of water is considerably less than that without the aid of a co-catalyst. The rate of water decomposition was 0.4 mmol/h, which determines the efficiency of the photocatalyst ($Ga_{82}Zn_{18}$) ($N_{82}O_{18}$) [169].

Photocatalysis can also be homogeneous when the reagents and photocatalysts are present in the same phase. The most commonly used photocatalyst is homogeneous ozone. The mechanism of hydroxyl radical production by ozone can follow two paths [170]. O (1D) means the first excited state of atomic oxygen, designated as the 1D ("single D") state [171]:

$$O_3 + h\nu \rightarrow O_2 + O\,(1D) \tag{4.109}$$

$$O\,(1D) + H_2O \rightarrow •OH + •OH \tag{4.110}$$

$$O\,(1D) + H_2O \rightarrow H_2O_2 \tag{4.111}$$

$$H_2O_2 + h\nu \rightarrow •OH + •OH. \tag{4.112}$$

Henry John Horstman Fenton invented reagents which may be used for the destruction of organic compounds such as trichlorethylene (TCE) and perchlorethylene (PCE)

[172]. Iron (II) is oxidized by hydrogen peroxide to ferric iron (III), resulting in creation of the hydroxyl radical and the hydroxyl anion. Iron (III) is then reduced back to iron (II), the superoxide radical and the proton. The net effect is a disproportionation of hydrogen peroxide to create two different oxygen-radical species, with water ($H^+ + OH^-$) as a byproduct. Disproportionation or dismutation is a kind of chemical change in which one element (or ion compound) undergoes a chemical change at the same time for two different products [173]:

$$(1) \quad Fe^{2+} + H_2O_2 \rightarrow Fe^{3+} + HO \bullet + OH^- \tag{4.113}$$

$$(2) \quad Fe^{3+} + H_2O_2 \rightarrow Fe^{2+} + HOO \bullet + H^+ \tag{4.114}$$

$$(3) \quad Fe^{2+} + HO\bullet \rightarrow Fe^{3+} + OH^-. \tag{4.115}$$

In photo-Fenton type processes, additional sources of OH radicals should be considered: through photolysis of H_2O_2, and through reduction of Fe^{3+} ions under UV light:

$$(4) \quad H_2O_2 + h\nu \rightarrow HO \cdot + HO\cdot \tag{4.116}$$

$$(5) \quad Fe^{3+} + H_2O + h\nu \rightarrow Fe^{2+} + HO \bullet + H^+. \tag{4.117}$$

The oxidation of organic compounds by Fenton's reagent is fast and exothermic, and thus leads to the oxidation of organic pollutants in the end, i.e. to carbon dioxide and water [174, 175]. The efficiency of the Fenton process depends on the concentration of hydrogen peroxide, pH and UV radiation. The main advantage of this method is the possibility of using natural sunlight (to 450 nm), which greatly reduces the cost compared with the use of UV light and electricity. The disadvantage of the Fenton method is the necessity of removing iron and acid reaction medium.

Heterogeneous photocatalysis includes a lot of different reactions from mild to complete oxidation, dehydrogenation, isotope exchange, removal of contaminants from water and gases [176]. In the oxidation reaction the positive hole reacts with moisture in the surface to form hydroxyl radicals. Oxidative reactions due to photocatalytic effect may also be presented as

$$UV + MO \rightarrow MO \left(h\nu + e^- \right), \tag{4.118}$$

where MO is the metal oxide, h^+ represents the hole and e^- represents an electron:

$$h^+ + H_2O \rightarrow H^+ + \bullet OH \tag{4.119}$$

$$2h^+ + 2H_2O \rightarrow 2H^+ + 2H_2O_2 \tag{4.120}$$

$$H_2O_2 \rightarrow HO \bullet + \bullet OH. \tag{4.121}$$

The reductive reactions due to photocatalytic effect are

$$e^- + O_2 \rightarrow \bullet O_2 \tag{4.122}$$

$$\bullet O_2^- + HOO \bullet + H^+ \rightarrow H_2O_2 + O_2 \tag{4.123}$$

$$H_2O_2 \rightarrow HO \bullet + \bullet OH. \tag{4.124}$$

Photolysis is a chemical decomposition reaction under the influence of light energy. Each photon has sufficient energy to affect the chemical bonding compound. Since photon energy is inversely proportional to the wavelength of electromagnetic waves, shorter visible light such as ultraviolet light has sufficient energy. For example, nitrogen oxide can be decomposed in accordance to the following reaction:

$$NO_2 + h\nu \rightarrow NO + O \bullet.$$
(4.125)

Photosynthesis is the reaction which takes place in green plants and certain other organisms by which carbohydrates are synthesized from carbon dioxide and water using light as an energy source. Most forms of photosynthesis release oxygen as a byproduct. The general reaction of photosynthesis photolysis can be given as

$$H_2A + 2h\nu \rightarrow 2e^- + 2H^+ + A.$$
(4.126)

The chemical compound "A" depends on the type of organism. Purple sulfur bacteria oxidize hydrogen sulfide (H_2S) to sulfur (S). In the process of photosynthesis, oxygen, water (H_2O) is used as a substrate for photolysis by the generation of diatomic oxygen (O_2). This is the process that draws oxygen into earth's atmosphere.

Photodegradation of water pollutants by advanced oxidation processes is a successful method of treatment of industrial and agricultural wastewater. These harmful compounds which can be degraded by photocatalysis include alkanes, haloalkanes, aliphatic alcohols, carboxylic acids, alkenes, aromatics, halo aromatics, polymers, surfactants, herbicides, pesticides and dyes. They are commonly used as solvents, propellants, refrigerants and intermediates in industrial production. Many of them are toxic and have mutagenic and carcinogenic effects [177–180].

Hazardous environmental contaminants are also found in military waste sites, i.e. underground storage tanks and landfills which leak into our water and soil. The treatment is very awkward because it is expensive and time consuming. Already more than 1,800 military installations in the United States will have been cleaned for $ 30 billion, and the time required was estimated to be more than 10 years [181]. Removal of such impurities can be solved by the use of photocatalysts.

Photocatalysts can also be used to purify marine waters polluted by oil, using the free energy of oxygen and water. Photocatalyst particles can be placed on the floating surface, making it easier to recover and to catalyze the reaction. The use of biosorption in addition, to enable the adsorption of waste materials such as bark and wood chips, peat, chitin shells from wastes, and other biodegradable materials, all of which may act together with the photocatalysts in hybrid systems.

Photocatalysts can still have a wide variety of other practical applications. Free radicals can be used to disinfect water, surgical instruments and optical and electronic products [182]. Photocatalysis can be used for the conversion of carbon dioxide to hydrocarbon gas in the presence of water [183]. The inclusion of carbon-based nanostructures such as carbon nanotubes [184], and metallic nanoparticles [185] has shown that it is possible to increase the effectiveness of photocatalysts. Photocatalysts are often used as an ingredient in paints because they are a less-toxic alternative to tin and copper-based antifouling marine and electrical paints, which generate hydrogen peroxide by photocatalysis. TiO_2 suspension and UV light can photocatalytically degrade the PAHs.

Photocatalytic water splitting is a general term used for the dissociation of water into hydrogen (H_2) and oxygen (O_2), using either artificial or natural light. The production of hydrogen fuel is receiving more and more attention as an alternative to fossil fuels. Hydrogen is derived from water splitting and photosynthesis, and carries solar energy, which will remain available as a basic resource for a long time yet.

The photocatalytic splitting of water is simple and safe for the environment. The easiest way is to use natural photosynthesis, which is a natural method of converting solar energy into chemical energy. When H_2O is divided into O_2 and H_2, the stoichiometric ratio is a 2 : 1 product:

$$2H_2O \xrightarrow[\text{1.83 eV}]{\text{photon energy}} 2H_2 + O_2. \tag{4.127}$$

Artificial photocatalysis is much worse than natural, because artificial catalysts must meet a number of stringent requirements, which have already been met by natural catalysts as a result of continuous optimization over the years of evolution. The potential may be less than 3.0 eV for the efficient use of energy present in the full spectrum of sunlight since the minimum energy gap for successful cleavage of water at pH = 0 is 1.23 eV, which corresponds to the light wavelength of 1008 nm. These values are met only for completely reversible reactions at standard temperature and pressure (1 bar and 25 °C).

TiO_2 is the appropriate semiconductor with the proper band structure, however, due to the conduction band of H_2, the production is very slow. Therefore, TiO_2 is typically used with a co-catalyst such as Pt to increase the rate of H_2 production. Most semiconductors with band structures suitable for splitting water absorb mostly UV light; in order to absorb visible light, the band gap must be reduced. As the conduction band is fairly close to the reference potential for H_2 formation, it is preferable to alter the valence band to move it closer to the potential for O_2 formation. Photocatalysts can suffer from catalyst decay and recombination under operating conditions. Catalyst decay becomes a problem when using a sulfide-based photocatalyst such as CdS, as the sulfide in the catalyst is oxidized to elemental sulfur at the same potentials used to split water. Sulfide-based photocatalysts are therefore not viable.

Dye-sensitizing solar cells (DSSC) produce energy directly from sunlight. DSSC have a number of attractive features, such as ease of implementation using conventional printing techniques. In practice, the complete elimination of expensive materials, such as platinum, ruthenium and a liquid electrolyte is impossible. However, photovoltaic cells are an important source of energy [186].

4.3.4 Electroprocesses

In search of new, more efficient processes an electric current is increasingly used to assist conventional processes. The flow of electric current produces a number of phenomena improving mass transport. The most common hybrid processes are outlined below,

During electrolysis an electrical power source is connected to two electrodes or two plates (typically made from some inert metal such as platinum, stainless steel or iridium) which are placed in water. Hydrogen will appear at the cathode (the negatively charged electrode, where electrons enter the water), and oxygen will appear at the anode (the positively charged electrode). Due to the low degree of dissociation of pure water, as compared with sea water, the electrical conductivity is one million times smaller. Therefore, pure water electrolysis requires a large amount of energy [561]. The efficiency of electrolysis is increased through the addition of an electrolyte (such as a salt, an acid or a base) and the use of electrocatalysts. The electrolytic process is rarely used in industrial applications

Electrocoagulation is the electrically assisted hybrid process of separation, which includes: coagulation, electrolysis and flotation. Electrocoagulation does not require the use of chemicals because the dissolving aluminum or iron electrode plays role of coagulant. The primary role of the electric field is to stimulate the coagulation of the sludge particles under the influence of electrokinetic phenomena and their motion in the direction of the respective electrodes. Therein, the fine emulsions can be separated for wastewater treatment and water purification without use of additional chemicals. In addition, the fine gas bubbles released during electrocoagulation help to enhance the flotation. By means of electrocoagulation, even the stable emulsions can be separated into two separate phases, i.e. oil and water. Electrocoagulation can be competitive in relation to processes such as microfiltration, hydrocyclones and centrifuges.

Electrophoresis is the electrokinetic separation process in which charged particles move under the influence of an electric potential. The rate of movement is affected by the properties of the resort, mainly viscosity. The phenomenon of electrophoresis is widely used as the analytical technique used, for example, to identify proteins [540].

The electrofiltration process is a hybrid membrane separation process which comprises applying electrically charged membrane or other barrier filter matched to the particles to be separated. [541]. The particles are charged in the same way as the filtering septum, and are therefore repelled from it and do not form a thick filter cake. By reducing the filtration resistance, and by increasing the permeate flow filtration electro improves the separation efficiency. Electrofiltration is most commonly used for the separation of biopolymers, the concentration or fractionation in the aquatic environment.

Electroosmosis is the flow of ions through the capillary channel under the influence of electric field potential. The ions are usually covered with a double layer of oak trees. If the capillary dimensions are smaller than the dimensions of the film Debye

ions is conducive to their net charge movement, under the influence of Coulomb force. Electro-osmosis is carried out in processes such as capillary electrophoresis [542], the flow through the porous membrane layer [543] and microreactors [544], which is often used as a type of pumping device.

The process of electrodialysis [545] has already been fully discussed in Section 4.2.5.6 concerning the membrane processes. The essence of this process is to arrange alternatively the anionic and cationic selective membranes in a cascade. Under the influence of an electrical potential, all the ions migrate in the opposite direction, i.e. anions towards the anode, and the cations to the cathode. In this way, the zones of high and low ionic concentration are formed between the membranes. Electrodialysis leads to the demineralization of different solutions including desalination of water and whey and the recovery of acids, bases, and salts of various industrial waste water.

4.3.5 Particle aggregation with membrane separation

These hybrids are based on improvement of membrane separation processes through a variety of particle aggregation methods. Membrane separation processes, which are reinforced by various types of particle aggregation, have long been used under various names, such as polymer-supported ultrafiltration, membrane separation assisted by micelles. These examples can be extended to other cases of membrane separation processes which are enhanced by biosorption on microorganisms, adsorption on powdered adsorbent or bonding with powdered polymers, functionalized ion exchange resins and molecularly imprinted materials. Support of membrane separation is achieved by complexing, chelation, coagulation, flocculation, precipitation and crystallization. Despite the physical differences, in all these cases of membrane separation intentional creation of larger aggregates of substances separated in order to increase efficiency and selectivity of separation is common. Thus it is possible to create a new group of membrane contactors operating in the system solid-liquid.

The idea of a new type of membrane contactor processes for solid-fluid systems such as membrane sorption and membrane leaching is similar to other contactors which perform unit processes on membranes, such as membrane distillation, membrane absorption, membrane chromatography and membrane extraction. "Contactor processes", such as membrane distillation, membrane extraction, membrane absorption and so on do not differ physically from their conventional archetypes, such as distillation, extraction and absorption. The only difference is that in membrane contactors the two phases are separated by a membrane. Although this introduces additional resistance, the phase can be more dispersed (thus revealing interface) as opposed to the conventional devices such as distillation columns, extraction or absorption or adsorption bed or bed catalysts. Excessive dispersion is avoided in the columns or beds due to the hydraulic resistance and the number of various interferences of these devices, such as entrainment of the dispersed phase, weeping, occurrence of recycling, dead zones and longitudinal mixing. In membrane contactor, the flow of the phases is in order, they do not affect each other and the contact surface can be freely expanded by the use of capillary membranes made of hollow fibers. The same positive effect can be used in the "new type of membrane contactors for solid-fluid systems" with the use of very finely dispersed solid phase in suspension. The kinetics of membrane sorption and membrane leaching will be described in detail in later sections of this book.

In the conventional unit processes, the decrease in particle size leads to an increase of interfacial area (Eq. (4.128)) and simultaneously to the rapid increase in flow resistance, energy consumption and thus operating costs (Eq. (4.129)). On the membrane, the material layer may be spread over a large area and usually reaches a value of several microns. In such cases the layer can even be completely eliminated by adjusting the permeate stream to below the critical flux value. Whereas in columns the layers can typically be up to a dozen meters thick. Thus the use of a solid phase in powder form is possible only on the membranes and totally impossible in columns.

Some sorbents such as microbials can be used immediately without any special preparation. Further cost reductions can be achieved by appropriate grinding; however, we must consider that the process of crushing and grinding incurs additional costs. Grinding has two benefits, however. First, by increasing the total surface adsorption and there is a substantial increase of the sorption uptake, secondly accelerating the adsorption process by reducing the diffusion path for adsorbed molecules and results in a further efficiency increase.

Working with fine particulates of the sorbent allows better access to active sites and greater uptake. In the case of extracted material yield can be higher, particularly where it is present in the form of closed cells or isolated crystals in caverns.

This section presents an analysis of a new type of "solid–liquid process" which would be performed on the membranes, and are now performed on the columns. Discussion about the advantages of performance enhancement of various mass transport processes, on membranes instead of packed columns is supported by many examples drawn from experience with sorption processes. However there are strong supporting arguments that other processes can also benefit from this.

These may concern the following processes:
1. adsorption on pulverized adsorbent;
2. biosorption on microorganisms;
3. complexation;
4. chelation;
5. binding on functionalized polymers;
6. binding on ion exchange resins;
7. binding on molecularly imprinted materials;
8. coagulation, flocculation;
9. precipitation;

10. micelle solubilization;
11. heterogeneous catalysis.

The mass transport of a selected component takes place from a solution to a solid phase in all of these processes. Therefore they can be called sorption processes. Leaching processes may be considered those where mass transport occurs from a solid to a solution. Needless to say, the appropriate membranes must be applied in the respective processes. However, in these systems some contradiction typically exists between the positive impact of fragmentation and pulverization of particles, and the negative effect of excessive flow resistance in the bed formed by the fine particles. Dispersion and the small size of these active particles improve the efficiency of the above-mentioned processes. This is because when the size of the particles is reduced the specific surface area and number of active sites increases, which reduces the path of diffusion for molecules. In the case of spherical particles the specific surface can be expressed by the formula

$$a = \frac{\pi d^2}{\dfrac{\pi d^3}{6}} = \frac{6}{d}.$$

(4.128)

For other shapes the respective formulas may vary in constants but inverse proportionality with respect to the size of the particle always exists. These factors all increase the efficiency of these processes with decreasing particle size. At the same time if the costs are limited to the cost of pumping through the bed they are inversely proportional to the particle size to the second power:

$$\text{costs} \sim \frac{1}{d^2}.$$

(4.129)

In MF/UF processes when the flux is less than critical flux, flow resistance is independent of the particle size,thus increasing the efficiency of the processes with decreasing particle size. Moreover, by concentrating suspensions of active substances on the membrane surface, the overall amount used and their consumption can be reduced in the process.

An inherent phenomenon accompanying virtually all membrane processes is the concentration polarization. Concentration polarization is considered to be an inevitable but rather negative phenomenon, because it reduces the permeate flux and thereby the efficiency of membrane processes. Concentration polarization is the result of the mass accumulation substance retained on the membrane in separation processes.

However, in the new processes discussed here concentration-polarization could also play a favorable role. In these cases it is possible to use the active substances in the form of very fine particles, as opposed to the packed columns which would require tremendous energy consumption in order to maintain the flow through the porous layers of fine particles.

The behavior of very fine particles retained during microfiltration can be fully controlled. Various ways of combating the concentration polarization layer are described in the section on membrane processes. They may form the fixed layer (cake), which can periodically be subjected to backflushing during dead-end or cross-flow microfiltration. The backflushing period can be synchronized according to the kinetics of the process to achieve the desired residence time. The control of crossflow velocity within the membrane module to generate adequate shear stress or turbulence enables reduction of the polarization layer over the membrane surface.

In addition the following methods can also be used:
1. transmembrane pressure pulsations;
2. ultrasound;
3. the electric field;
4. changing the direction of permeate flow;
5. abrasive elements (betonies, foam balls, diatomaceous earth);
6. injection of gas into the retentate stream;
7. rotation of membranes.

An analysis of the kinetics of such processes is presented later, using the example of membrane sorption, although most of the conclusions of this study may also apply to other aggregation processes. Let us consider the following examples of sorption:
1. sorption in the tank for the batch mode of operation;
2. sorption in the tank during continuous mode of operation;
3. sorption at the membrane surface for the batch mode of operation;
4. sorption at the membrane surface during the continuous mode of operation.

The hybrids with a high level of integration operate on the principle of magnifying small particles into larger aggregates and then easier separation. If the materials used for such aggregation are cheap waste materials such as fly ash to adsorb, straw, bark, sawdust, for biosorption microorganisms, such separation is particularly suitable to treat large flows (which are present in environmental protection). A very useful and cost-effective technology solution is the use of submerged membranes for the protection of the environment. Advanced materials such as expensive sorbents, ion exchange resins, functionalized polymers, molecularly imprinted polymers, and microcapsules can be regenerated continuously. The technology solution in the schematic diagram (see Fig. 4.68) shows how the valuable materials and the separated contaminants can be recovered in their pure form, resulting in large savings in the cost of separation.

According to this scheme (Fig. 4.68), two new processes are described in more detail in the following sections, i.e. membrane sorption (Section 4.5) and membrane leaching (Section 4.6).

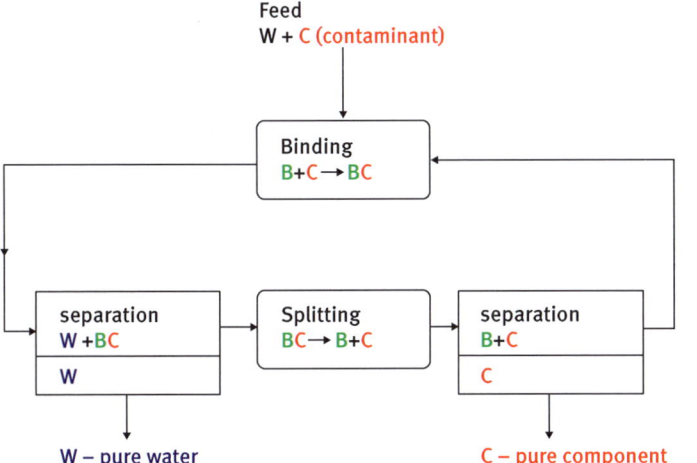

Feed
W + C (contaminant)

Binding
B+C → BC

separation
W +BC

W

Splitting
BC → B+C

separation
B+C

C

W − pure water

C − pure component

Fig. 4.68. Schematic diagram for recovery of process media in membrane hybrid systems with particle aggregation.

4.4 Sorption processes

4.4.1 Adsorption processes

The word adsorption was first introduced by the German physicist Heinrich Kayser in 1881. This phenomenon is present in nature but is also a well-known unit process in process engineering, and widely applied in various industrial technologies. Adsorption is a consequence of various bonds such as ionic, covalent, metallic, and others. Generally the surface energy of solids is similar to the surface tension on the interface between immiscible liquids, where the atoms on the surface of the adsorbent are not wholly surrounded by other atoms and therefore can attract adsorbates [187]. Adsorption, ion exchange and chromatography are called sorption processes. Adsorbates are materials which are capable of being adsorbed, whereas adsorbents are substances that can adsorb adsorbates onto their surfaces. Several porous materials are known as good adsorbents because they have high surface area in a given volume, and therefore high adsorption uptake. Many adsorbents are used in industry (see Tab. 4.7), and they have typical active area between 300–1500 m^2/g which is prepared during the activation process by oxidation of carbon with water vapor or CO_2. A good adsorbent can be characterized by high values of maximum uptake as determined by equilibrium under given conditions and the rapid kinetics of the sorption process. Adsorption uptake is incorporation of the adsorbate by adsorbent and is denoted as q:

$$q = \frac{m_{adsorbate}}{m_{adsorbent}}.$$

(4.130)

Silica gel is used for the drying of gases (such as process air, oxygen and natural gas),

Table 4.7. Sorption of chromium in various sorbents

Sorbents	Ion	q [mg/g]
Crafted activated carbon	Cr^{3+}	7,5
Activated carbon	Cr^{3+}	5,7
Ion exchanger - carboxyl resin (Purolite C106)	Cr^{3+}	0,17
Calcined clay	Cr^{3+}	4,2
Silica	Cr^{3+}	0,036
Granular sorbent polyethyleneimine	Cr^{3+}	39,4
Silica and alumina	Cr^{3+}	0,18
Australian activated carbon	Cr^{3+}	2,45
Calcium alginate	Cr^{3+}	80
Xanthate	Cr^{3+}	19,7
Activated Carbon	Cu^{2+}	31,05
Diatomaceous earth	Cu^{2+}	6,47
Polymer 8% HP + EGDMA	Cu^{2+}	68,02

and adsorption of heavy hydrocarbons from natural gas. Zeolites are manufactured by hydrothermal synthesis of sodium aluminosilicate or another silica source in an autoclave, followed by ion exchange with certain cations (Na^+, Li^+, Ca^{2+}, K^+, NH^{4+}). The channel diameter of zeolite cages usually ranges from 2 to 9 angstrom (Å). Zeolites are applied in the drying of process air, CO_2 removal from natural gas, CO removal from reforming gas, air separation, catalytic cracking and catalytic synthesis and reforming. Activated carbon can be manufactured from carbonaceous material, including coal, peat, wood or nutshells (including coconut). Activated carbon is used for adsorption of organic substances and nonpolar adsorbates, and is also usually used for waste gas and wastewater treatment. Its usefulness derives mainly from its large micropore and mesopore volumes and the resultant high surface area. Solid catalyzers such as a platinum and palladium serve as adsorbents which are used in the refining of crude oil, reforming, producing high-octane gasoline and aromatic compounds for the petrochemical industry. The chemical industry uses platinum or a platinum-rhodium alloy to catalyze the partial oxidation of ammonia to yield nitric oxide – the raw material for fertilizers, explosives and nitric acid. Viruses (virions) can also be adsorbed on cell surfaces. Viral genetic material (nucleic acid) penetrates the cell, and can be integrated into the cell genome, then directs the synthesis of viral nucleic acids and proteins and eventually release of the virus.

4.4.2 Biosorption processes

The adsorption process is generally understood as the immobilization of a selected component of the solution using various binder materials that are named sorbents. Sorption is most often the result of many physical and chemical mechanisms, dry as elementary van der Waals forces, electrostatic forces due to the presence of electric charges and the result of various chemical bonds, as discussed earlier in this manual (in the kinetics of chemical reactions). The cells of various organisms are used for this purpose in biosorption, which was extensively studied by Volesky [188]. In contrast to bioaccumulation it is a spontaneous process which may even occur in dead cells. This ability has most likely been developed through evolution, as a form of defense and protection of cells against the penetration of harmful substances. The biosorption of heavy metals has been observed in experiments in hair, feathers, in the claws and combs of hens, in fingernails and bones; i.e. in "organs of secondary importance" which are not essential for the most vital functions such as procreation. This is the case in cells of all living organisms, animal, vegetable or even microorganisms such as bacteria, fungi and yeasts. Examples of the adsorption capacity for heavy metals by natural sorbents are presented in Tabs. 4.8–4.12.

Table 4.8. Biosorption of heavy metals on bacteria

Biological absorbents – bacteria	Ion	q [mg/g]
Striptomyces noursei	Ag	38
Streptomyces noursei	Cr^{3+}	1.80
Bacillus subtillis	Cr^{3+}	118.04
Bacillus subtillis	Au	79
Bacillus licheniformis	Fe	45
Citrobacter	U	800

Table 4.9. Biosorption of heavy metals on yeasts

Biological absorbents – yeasts	Ion	q [mg/g]
Saccharomyces cerevisiae	Cu^{2+}	20.02
Saccharomyces cerevisiae	Ag^+	5
Saccharomyces cerevisiae	Cd^{2+}	20–40
Candida tropicalis	Cu^{2+}	80.06
Saccharomyces cerevisiae	Cd^{2+}	20–86
Saccharomyces cerevisiae	Ag^{2+}	5
Flocculating brewer's yeast (1)	Cr^{3+}	1.64
Flocculating yeast (2)	Cr^{3+}	13.8
Candida tropicalis	Cd^{2+}	60.13
Candida tropicalis	Cr^{3+}	4.59

Table 4.10. Biosorption of heavy metals on fungi

Biological absorbents – fungi	Ion	q [mg/g]
Rhizopus arrhizus	Au	164
Rhizopus arrhizus	Ag	54
Rhizopus arrhizus	Cd^{2+}	30.01
Rhizopus nigricans	Cd^{2+}	19.00
Rhizopus arrhizus	Cu^{2+}	16.01
Rhizopus arrhizus	Cr^{3+}	31.03
Penicillium chrysogenum	Cr^{3+}	0.33
Penicilium chrysogenum	Cd^{2+}	55.98
Penicillium spinulosum	Cd^{2+}	0.39
Cladosporium resinae	Cu^{2+}	18.01
Aureobasidium pullulans	Pb	220-360

Table 4.11. Biosorption of heavy metals on algae

Biological absorbents – algae	Ion	q [mg/g]
Spirulina sp (autotrophic)	Cr^{3+}	106
Ascophyllum nodosum (brown algae)	Cd^{2+}	215.25
Sargassum natans (brown algae)	Cd^{2+}	134.88
Fucus vesiculous (brown algae)	Cd^{2+}	73.06
Chlorella vulgaris	Au^+	80
Chondrus chrispus	Au^+	76
Sargassum natans (brown algae)	Au^+	400

Table 4.12. Biosorption of heavy metals on various biological materials and wastes

Biological absorbents	Ion	q [mg/g]
Chitosan	Cr^{3+}	92
Wool	Cr^{3+}	17
Bark	Cr^{3+}	19.5
Grass	Cr^{3+}	49.6
Straw	Cr^{3+}	36.5
Tree leaves	Cr^{3+}	46.4
Natural moss	Cr^{3+}	4.3
Sawdust	Cu^{2+}	18.94
Sunflower seed husks	Cu^{2+}	14.14
Orange peel	Cu^{2+}	21.41
Peanut shells	Cu^{2+}	42.02
Black tea dregs	Cu^{2+}	20.32
Green tea dregs	Cu^{2+}	19.84
Coffee grounds	Cu^{2+}	21.41
Starch	Cu^{2+}	5.13

Natural bio-sorbents can also be cheap plant and animal waste materials requiring minimal preparation. In the era of sustainable development it can therefore be expected that biosorption will be widely used as an important separation process to protect the environment. Different mechanisms (mainly physical and chemical) are leading to the binding of metals with the biosorbent. Possible physical mechanisms are: electrostatic attractive forces with negatively charged sites, inorganic precipitation, adsorption, gel diffusion. Chemical mechanisms may include: binding with anionic ligands (phosphoryl, carboxyl, sulphydryl and hydroxyl groups), ion exchange, complexation, coordination and chelation. Biosorption is in some cases (heavy metals) very selective for certain components of the environment, and is therefore the subject of many studies on potentially useful separation processes.

4.4.3 The properties and microstructures of sorbents

The microstructures of selected sorbents are shown in photographs (see Figs. 4.69a–e and 4.70a–f) of scanning electron microscopy. On the graphs in Fig. 4.69, the lighter line is the density of the probability distribution, and the darker curve represents the cumulative distribution function. The diagrams (Fig. 4.70a, b, c, etc.) shows the effect of pH and temperature on the biosorption equilibrium. An important property is the particle size distribution. It is worth mentioning that the sorption capacity for heavy metals increased at higher temperatures in the case of biosorbents, which means that it was not a simple physical adsorption [189].

Certain natural materials have been for a used long time as adsorbents, for example charcoal and diatomaceous earth. Activated carbon from bones or coconut shells develops most absorbent surfaces of known sorbents. Pine bark has long been used as an excellent absorbent material to remove oil from the ocean's surface after tanker spills and other disasters. Chitin and its derivatives produced from shellfish waste have recently been found to be very versatile material, but they must be subjected to pre-treatment (deacetylation) before biosorption. We must remember that the main environmental problem is the cost of the treatment of large streams. Therefore we must take advantage of all known methods of reducing costs. Biosorbents are in fact low-cost materials, moreover such reuse of waste can be highly profitable.

4.4.4 Sorption equilibrium

According to Langmuir, adsorption takes place at the surface of the adsorbent until all active sites are saturated. In 1916, Irving Langmuir published a model of isotherm for gases adsorbed on solids [190]. It is based on the following assumptions [191]:
1. the surface of the adsorbent is uniform, that is, all the adsorption sites are equivalent;

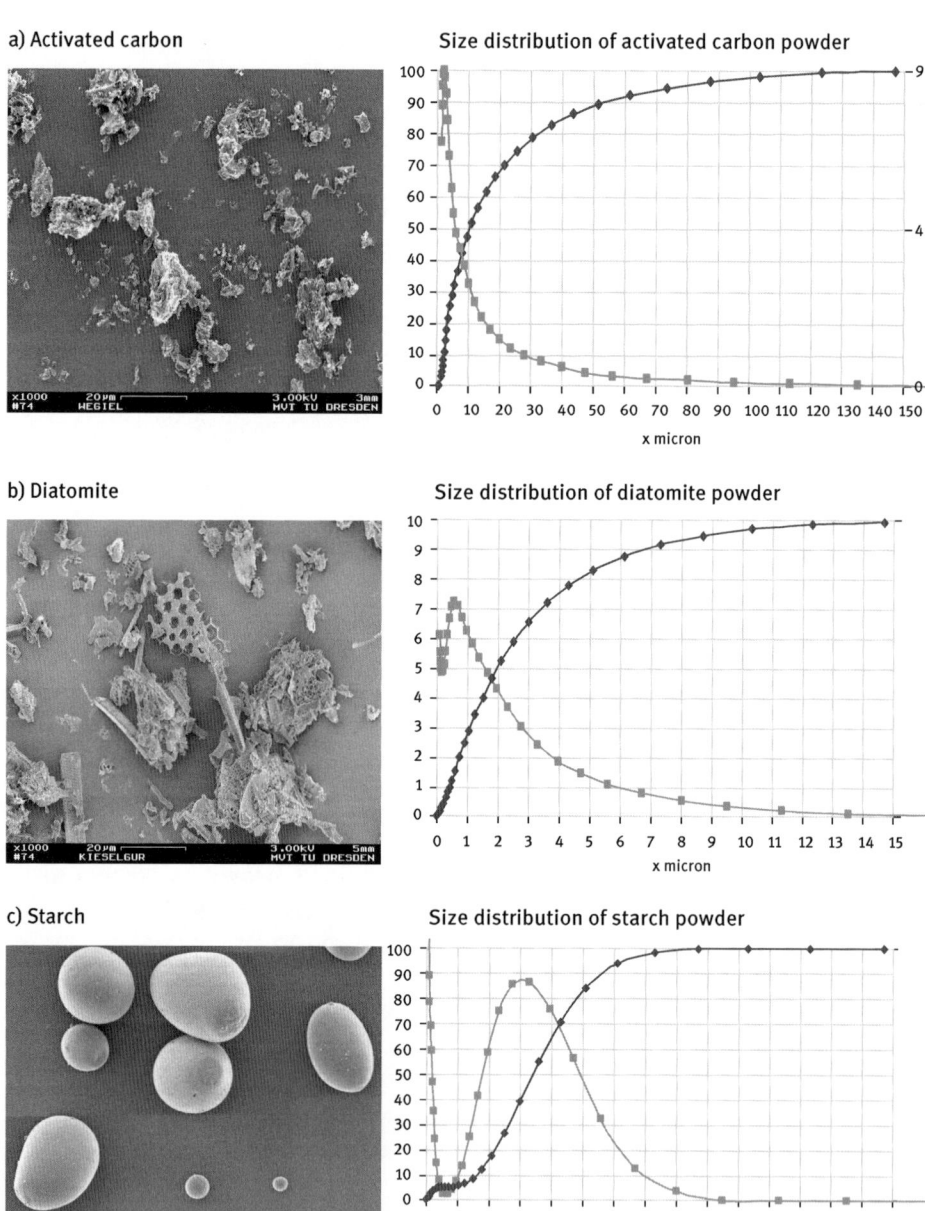

Fig. 4.69. Structures of different biosorbents.

d) Coffee

Size distribution of coffee powder

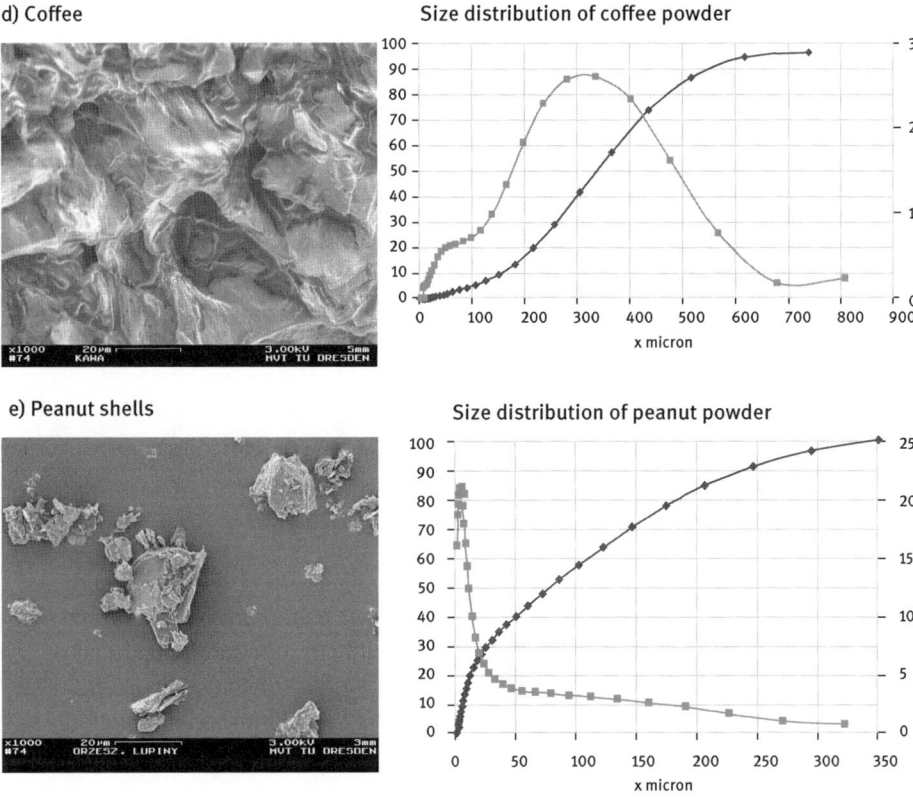

Fig. 4.69. (cont.) Structures of different biosorbents.

a) Leafs of black tea microstructure

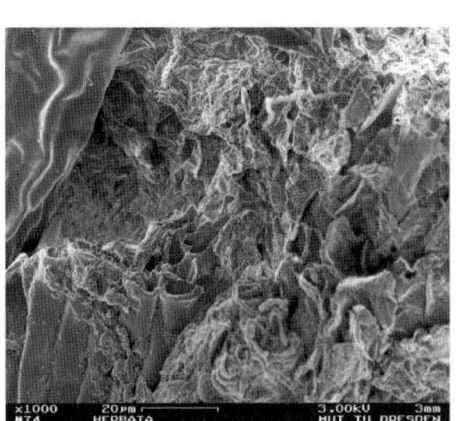

Equilibrium of chromium biosorption on tea leafs

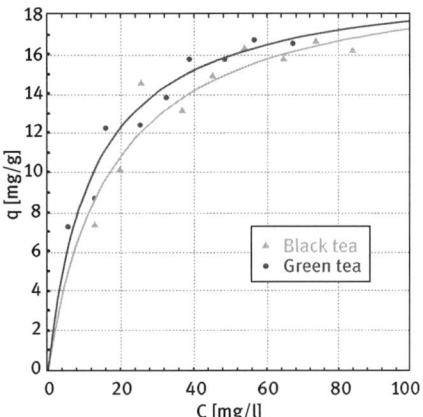

Fig. 4.70. Biosorption properties of various biosorbents.

b) Shells of sunflower seeds

Biosorption equilibrium of chromium on coffe, sunflower, diatomite and starch

c) Lemon skin microstructure

Biosorption of chromium on lemon and orange skin

Orange peel

Chromium biosorption temperature effect on uptake

Chromium biosorption pH effect on uptake

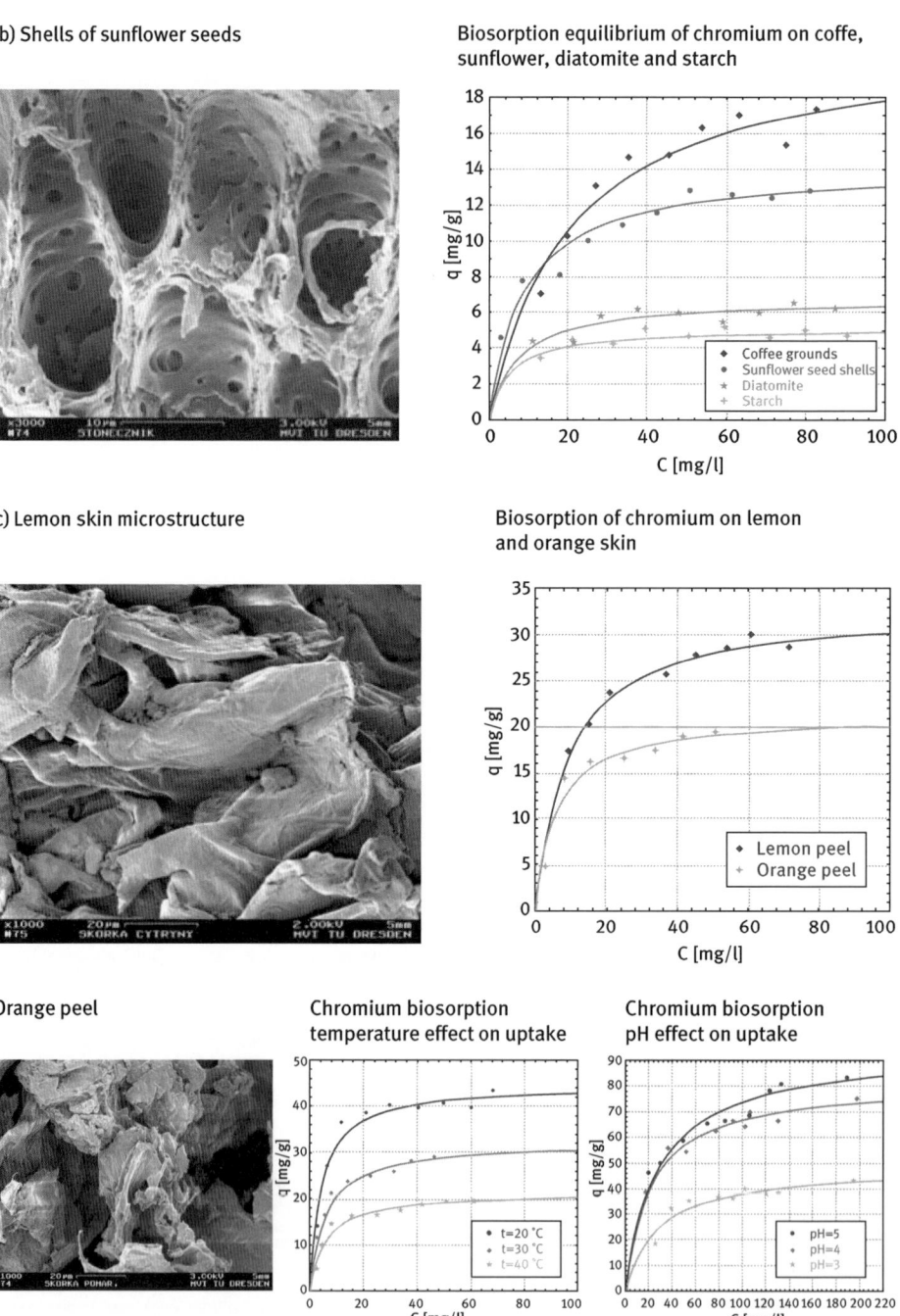

Fig. 4.70. (cont.) Biosorption properties of various biosorbents.

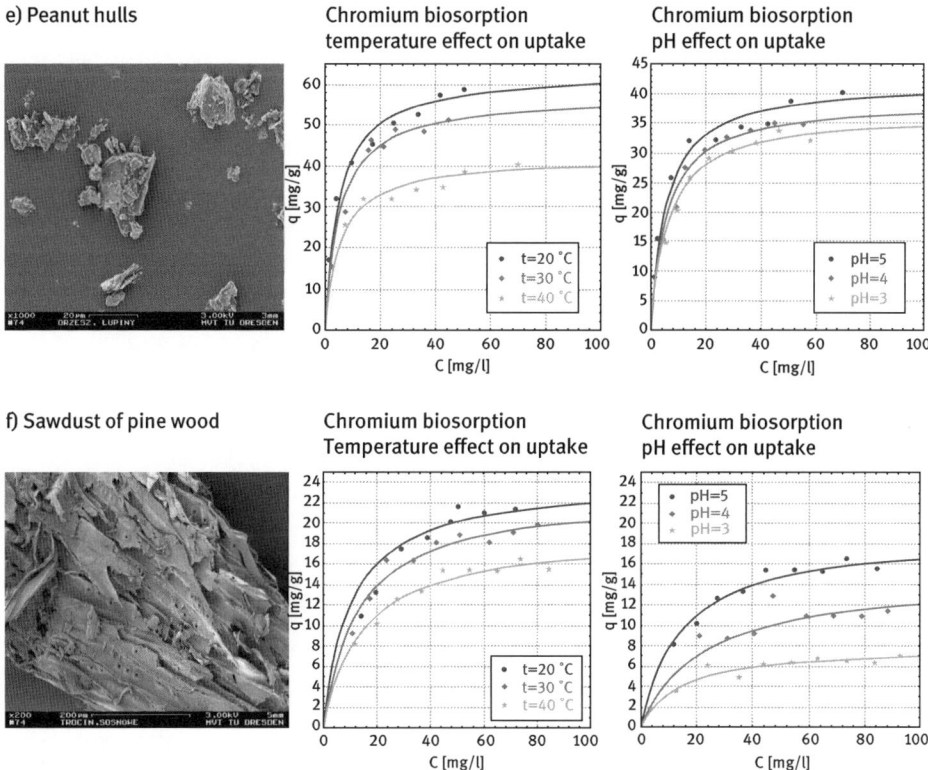

Fig. 4.70. (cont.) Biosorption properties of various biosorbents.

2. adsorbed molecules do not interact with one another;
3. all adsorption occurs by the same mechanism.

At maximum adsorption only a monolayer is formed, i.e. molecules of adsorbate do not deposit on others which have already been adsorbed. The Langmuir isotherm, Langmuir adsorption equation or Hill–Langmuir equation is a consequence of the assumption that the number of active sites is finite and may be completely saturated at equilibrium. It is often said that adsorption proceeds until the monomolecular layer of molecules of adsorbate has been formed. The assumption of the monolayer is equivalent to finite active sites during sorption. In fact, adsorption can also be expressed as a result of a dynamic balance between two opposing processes such as adsorption and desorption. No matter what the type of linkage to the active site sorbate sorbent. A balance will be attained if the two opposing processes are equally fast. No matter what type of bond exists between the active sites and the sorbate molecules, equilibrium will be attained between the two opposing processes.

The rate of sorption is proportional to the vacancies among active sites and the number of the sorbate particles, i.e. concentration near the active sites (C) (see Fig. 4.71):

$$r_{ads} = \frac{dq}{dt} = k_0 \cdot (1 - a) \cdot C \tag{4.131}$$

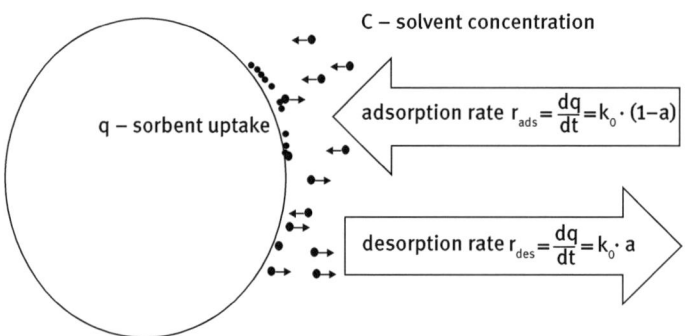

Fig. 4.71. Dynamic equilibrium of the sorption process.

However, the desorption rate is proportional to the active sites which are already occupied:

$$r_{des} = \frac{dq}{dt} = k_1 \cdot a \tag{4.132}$$

It should be noted that the active sites are distributed evenly over the entire surface of the sorbent. Let us assume that the area already occupied by sorbate molecule is proportional to the current value of q (sorption uptake), and the total area available is proportional to q_{max} (maximum uptake):

$$a = \frac{q}{q_{max}} \tag{4.133}$$

Thus we get

$$r_{sorption} = k_0 \cdot \left(1 - \frac{q}{q_{max}}\right) \cdot C \tag{4.134}$$

$$r_{desorption} = k_1 \cdot \frac{q}{q_{max}}, \tag{4.135}$$

where k_1 and k_0 are the reaction rate coefficients for both the reactions, according to the Arrhenius formula. On equalizing the rate of the reaction a state of equilibrium is reached:

$$r_{sorbtion} = r_{desorption}. \tag{4.136}$$

Isotherm	Equation
Langmuir	$q = q_{max} \cdot \dfrac{K \cdot C}{1 + K \cdot C}$
Freundlich	$q = k \cdot C^{\frac{1}{n}}$
Langmuir–Freundlich	$q = q_{max} \cdot \dfrac{K \cdot C^{\frac{1}{n}}}{1 + K \cdot C^{\frac{1}{n}}}$
Radke–Prausnitz	$\dfrac{1}{q} = \dfrac{1}{a \cdot C} + \dfrac{1}{b \cdot C^{n}}$
Reddlich–Peterson	$q = q_{max} \cdot \dfrac{a \cdot C}{1 + b \cdot C^{n}}$

Table 4.13. Different adsorption isotherms

Comparison of the two (right-hand sides) of both kinetic equations, gives the equilibrium equation, which correlates the concentration of the sorbate in the sorbent with its concentration in the solution

$$q = q_{max} \cdot \frac{k_0 \cdot C}{k_1 + k_0 \cdot C}, \tag{4.137}$$

where k_0 is the adsorption rate constant and k_1 is the desorption rate constant and their quotient is the equilibrium constant K:

$$q = q_{max} \cdot \frac{b \cdot C}{1 + b \cdot C}. \tag{4.138}$$

Since adsorption is an exothermic process, where the energy liberated is known as heat of adsorption (ΔH), enthalpy is therefore always negative. Adsorption constants are equilibrium constants; therefore they obey van't Hoff's equation:

$$\frac{\partial \ln (b)}{\partial \left(\frac{1}{T}\right)} = -\frac{H}{R}, \tag{4.139}$$

which gives the more practical relation

$$\ln \left(\frac{b_1}{b_2}\right) = -\frac{\Delta H}{R} \cdot \left[\frac{1}{T_2} - \frac{1}{T_1}\right]. \tag{4.140}$$

The type of adsorption is necessary can be demonstrated empirically on the basis of these equations, i.e. physical or chemical adsorption. While physical adsorption results in heat separation of (25–40 kJ/mol or less, in chemical adsorption more heat (80–240 kJ/mol) is separated and the process is irreversible.

4.4.5 Kinetics of the sorption process

The sorption process takes place only when the rate of adsorption is much faster than that of desorption. Sorption kinetics may be obtained from the difference in the rates of adsorption or desorption of the reaction (net result) (see Fig. 4.72).

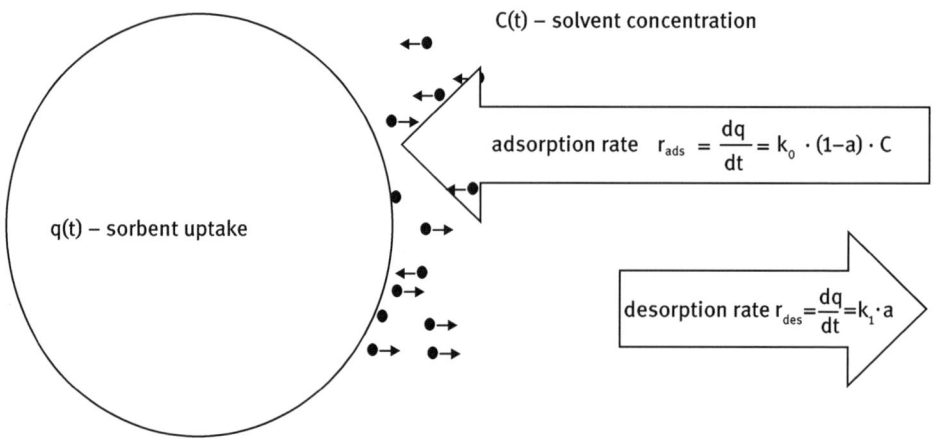

Fig. 4.72. The kinetics of the sorption process.

On this basis, we determine the sorption reaction rate which can be expressed as

$$\Delta r = r_{sor} - r_{des} = \frac{dq}{dt} = k_1 \cdot (1 - a) \cdot C - k_0 \cdot a. \tag{4.141}$$

Assuming a uniform distribution of active sites, and their constant number on any surface:

$$a = \frac{q}{q_{max}}. \tag{4.142}$$

The amount of adsorbate on the surface of the sorbent will permanently change during the adsorption, reaching equilibrium value asymptotically. The instantaneous value of q (t) at any given time can be estimated using the following formula:

$$q\,(t) = q_{max} - (q_{max} - q_0) \cdot e^{-\frac{k_0 \cdot C + k_1}{q_{max}} \cdot t}. \tag{4.143}$$

It is simply a solution of the differential equation (first row) with the initial condition

$$q\,(0) = q_0.$$

This relationship can be represented by the example of biosorption of lead on the surface of yeast (a) whole cells (bigger particles) and (b) broken cells (smaller particles) (see Fig. 4.73).

Other kinetic equations for sorption kinetics have been presented in the literature [194]

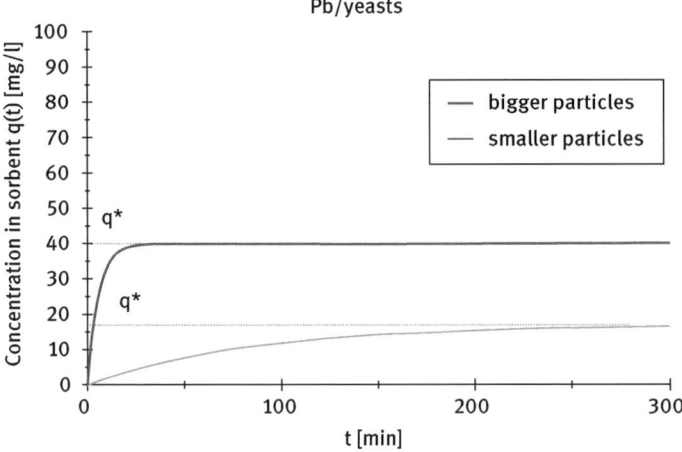

Fig. 4.73. An example of biosorption kinetics of lead on yeasts.

Table 4.14. Different kinetics of adsorption [194]

Reaction type	Differential form	Integral form
First order reaction	$\dfrac{dq}{dt} = k \cdot (q^* - q(t))$	$q\,(t) = q^* \cdot (1 - e^{-kt})$
Pseudo-second order reaction	$\dfrac{dq}{dt} = k \cdot (q^* - q(t))$	$q\,(t) = \dfrac{t}{\frac{1}{k \cdot q^*} + \frac{t}{q^*}}$

4.4.6 Experimental verification of membrane biosorption

Biosorption can be a very inexpensive tool for cleaning polluted water [192, 193]. This is mainly due to the applicability of waste materials as sorbents as described above. An additional opportunity to reduce the cost of water treatment makes it possible to use naturally occurring sorbents at no cost. Remember that biosorption may be used by different life forms of indigenous microflora and microfauna in all kinds of more or less contaminated water. Regardless of the species they are always able to reproduce, which means the biomass is permanently available for biosorption. The biosorption potential can be constantly renewed naturally (and spontaneously) by cell multiplication in water. Of course they ought to be removed from the water together with the contamination afterwards (by using microfiltration, for instance). This may be a typical way of solving the dilemma of cleaning large amounts of water containing relatively small amounts (but dangerous) of contaminants. All we need to do is to support these natural processes on the basis of our current knowledge. Biosorption has been verified as a tool for environmental protection for different contaminated waters, the parameters of which are presented in Tab. 4.15.

Table 4.15. Data for the experimental verification of the biosorption process of industrial waters (see graphs in Figs. 4.75 and 4.76)

Contaminated waters	COD mgO$_2$/l	Dissolved O$_2$ mgO$_2$/l	pH	Conduct. mS/cm	A$_{254}$	Flux LMH
Effluents from municipal solid waste landfills	37.5	5.28	7.9	4.12	0.70	435
Effluents from power plant ash dumps	11.9	5.42	7.25	3.00	0.21	367
Water after flotation of copper ore	6.6	5.28	8.6	29.00	0.04	482
Leachate from the ash heaps in ferrochromium smelter	21.3	4.71	8.86	2.25	0.41	284
Wastewater from the tractor factory	69.3	1.56	8.00	11.19	1.28	226
Circulating water in washing wagon plant	15.8	5.46	8.00	26.60	0.22	400

In all cases, the same method of purifying water without additional sorbents but with the use of existing microorganisms has been applied. Water has only been filtered through a microfiltration membrane with the possibility of separation of 0.1 microns. It was therefore not possible for the membrane to reject metal ions. Nevertheless, despite this fact there was a significant retention of metallic elements during these experiments, which can only be attributed to the biosorption by the microbial cells present in contaminated water (see Figs. 4.74 and 4.75). It should be noted that the experiment

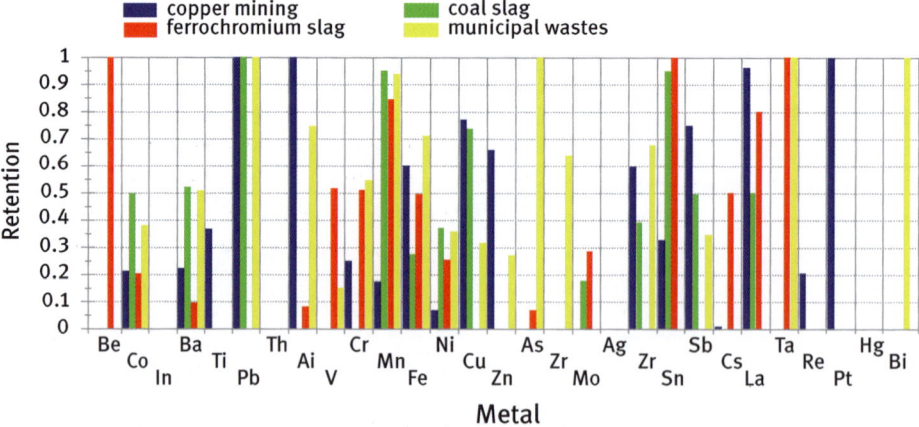

Fig. 4.74. Comparison of effects of membrane biosorption in four contaminated industrial waters, where indigenous microorganisms were used as biosorbents. Analyses of metal concentrations were determined by ICP and the membrane used was Amicon 0.1 micrometer cut-off.

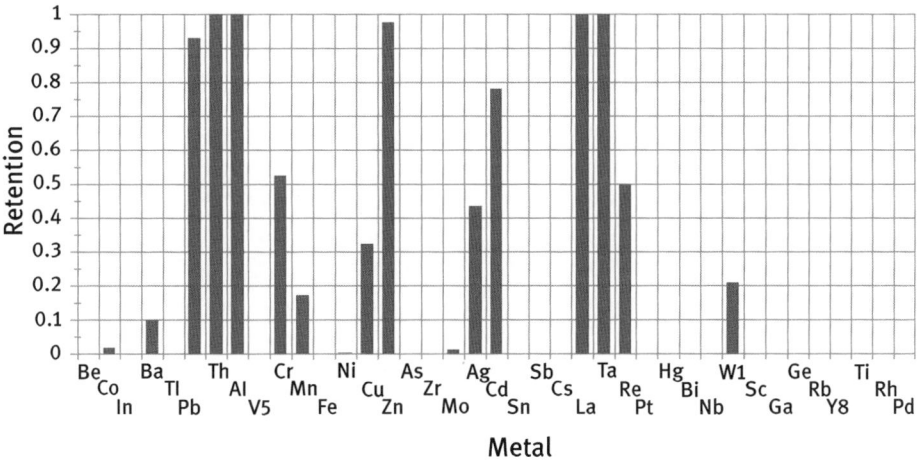

Fig. 4.75. Effects of purifying water from the wash freight wagons, using "membrane biosorption" by using native microorganisms which live in open aerated tank.

used a very accurate measurement of the concentration of metals by ICP. It should also be noted that although the biomass concentration is relatively low, (being limited to the naturally occurring biomass in the polluted water), it was highly concentrated at the place of biosroption, i.e. on the surface of the membranes. Cell multiplication made it possible to maintain sorption uptake in polluted waters that could not reach the equilibrium state while maintaining the driving force.

It must be said that as a result of reducing the size of the sorbent particulates, the sorption equilibrium (i.e. maximum sorption uptake), will be changed as well as the kinetics of biosorption. However, biosorbent milling brings not only advantages, but requires the use of other methods for further processing. Thus, separation of fine biosorbent particles from a suspension cannot be achieved by simple sedimentation, but by the use of centrifuges, hydrocyclones or membranes appear to be the most reasonable separation processes. Generally, in order to achieve maximum benefit from the natural biosorbents, a basic knowledge of process engineering is required and the whole biosorption process must to be appropriately calculated and optimized. However, it seems that the potential profits are significant.

4.5 Membrane sorption – the new sorption process in the membrane contactor

4.5.1 The mass balance in the batch mode of adsorption

Adsorption will take place in an unsteady or batch mode in a closed vessel of volume (V). The adsorbate is the solute or some kind of xenobiotic which pollutes the water. During the sorption process, the solute is removed from solution and becomes immobilized on the sorbent. During the batch mode of adsorption, instantaneous concentration of adsorbent $q(t)$ grows in the interval $[0, q*]$. At the same time, the concentration of the purified solution $C(t)$ reduces in the range $[C_0, C*]$ (see Fig. 4.76). The mass balance during the sorption in a batch mode may be expressed as

$$C_0 + X \cdot q_0 = C(t) + X \cdot q(t), \tag{4.144}$$

where
C – is the concentration of the adsorbate (solute) in solution [mg/l i.e. ppm];
q – is theconcentration of the adsorbate (solute) in the adsorbent [mg/g];
X – is the concentration of the adsorbent in the solution.

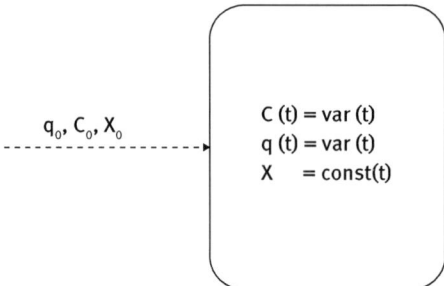

Fig. 4.76. Dead-end sorption.

After the time (t) adsorption, the concentration within the adsorbent will be $q(t)$, whereas concentration in the purified solution is $C(t)$. Both concentrations can be determined from the following formulas [129, 194]. Assuming the first order reaction type, the changes in the concentration of adsorbent could be expressed as

$$q(t) = q * - (q * - q_0) \cdot e^{-k \cdot t}. \tag{4.145}$$

Changes in the concentration in the solution can then be expressed by the formula

$$C(t) = C_0 - [q^*(C_0) - q_0] \cdot X \cdot (1 - e^{-k \cdot t}). \tag{4.146}$$

4.5.2 Mass balance of the adsorbed substance in continuous mode of operation

Adsorption can be carried out continuously, in the flow tank (see Fig. 4.77), provided that a constant supply of substrate and product is ensured. The residence time in the flow tank of volume (V), whose fluid flow rate is constant $v[m^3/s]$ is τ. Adsorption is carried out at a constant rate under stationary conditions:

$$q - q_0 = \left.\frac{dq}{dt}\right|_{t=\tau} \cdot \tau.$$

(4.147)

This gives the equation for calculating the value for final sorbent concentration as

$$q = q_0 + k \cdot (q * -q_0) \cdot e^{-k \cdot \tau}.$$

(4.148)

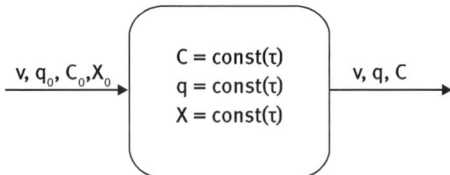

$$
\begin{array}{c}
C = \text{const}(\tau) \\
q = \text{const}(\tau) \\
X = \text{const}(\tau)
\end{array}
$$

$v, q_0, C_0, X_0 \longrightarrow \qquad \longrightarrow v, q, C$

Fig. 4.77. Continuous sorption in the flow tank.

Taking the mass balance of the sorbate at the inlet and outlet of the tank into account, one can calculate the concentration in the water leaving the flow adsorber, after the residence time τ, as

$$C = C_0 - X \cdot \tau \cdot k \cdot (q^* - q_0) \cdot c^{k \cdot \tau},$$

(4.149)

where the residence time of the fluid flowing at a flow rate v [m^3/s] in a tank having a volume V [m^3] is calculated using the formula

$$\tau = \frac{V}{v} \ [s].$$

(4.150)

4.5.3 Sorption at the membrane surface for the batch mode with backflushing

It is assumed that the sorbent is completely retained on the membrane surface F [195], and achieves constant concentration at the membrane surface X_m = const. It is assumed that the water to be treated flows through the sorbent layer, which is situated on the membrane surface and afterwards outflows from membrane as permeate (see Fig. 4.78). The function of the membrane is thereby to retain sorbent and to concentrate it on the surface. Under such conditions, the collision probability of molecules with active sites is higher than in the dilute slurry of the tank (bulk). However, the sorption capacity of the adsorbent which is immobilized in a thin layer on the membrane is very limited, and therefore the sorbent quickly becomes saturated. But it is

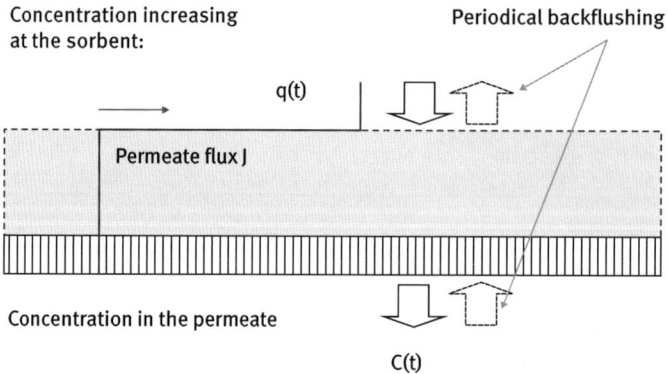

Fig. 4.78. The change in concentration of a sorbent on the membrane.

possible to control the instantaneous concentration of the permeate and the time after which it will have a minimum value. For this purpose, you can use backflushing then the sorbent is to be renewed and the adsorption process starts all over again. The time at which it is best to turn on backwash should be determined based on the instantaneous concentration of the filtrate [129].

The mass of the substance removed from the solution during the flow through membrane layer (with sorbent) in the infinitesimal time results from the difference in concentration between retentate and permeate:

$$dm = J \cdot F \cdot (C_R - C_P) \cdot dt. \tag{4.151}$$

At the same time (dt), the same mass (dm) of sorbate accumulated within the area of sorbent and solution filling control element on the surface of the membrane:

$$dm = (dC + X_{max} \cdot dq) \cdot V, \tag{4.152}$$

where q(t) – sorbate concentration in adsorbent, which depends on the rate of mass transport in the boundary layer only.

C(t) – sorbate concentration in the solution, which depends on permeate flow as well as adsorption rate in the adsorbent, whereas the volume (V) of the control element is V = Fδ. The mass balance leads to

$$\frac{dC}{dt} + X_m \cdot \frac{dq}{dt} = \frac{J}{\delta} (C_R - C). \tag{4.153}$$

Assuming the first order kinetics of sorbent saturation q(t) during the sorption,

$$q(t) = q^* - (q^* - q_0) \cdot e^{-kt}. \tag{4.154}$$

The first order ordinary equation can be obtained thus:

$$-\frac{dC}{dt} = (q^* - q_R) \cdot X_m \cdot k \cdot e^{-kt} - \frac{J}{\delta} \cdot (C_R - C)\,\theta. \tag{4.155}$$

Therefore, assuming the obvious initial condition

$$C(0) = C_R \quad \text{for} \quad t = 0, \tag{4.156}$$

one can obtain the following solution of the differential equation:

$$C(t) = C_R + \frac{K \cdot X \cdot \delta \cdot (q^* - q_R)}{J - K\delta} \left[e^{-\frac{J}{\delta} \cdot t} - e^{-K \cdot t} \right]. \tag{4.157}$$

Now the equilibrium type must be assumed, but a Langmuir equation seems to be adequate for sorption in liquid phase. This equation is very important for practical applications because it demonstrates the use of membrane sorption for water purification. Although the sorbent forms only a very thin layer with low uptake, it can be replaced by a new layer after an optimal time during backflushing [196] or pulsatile flow [197]. This time corresponds to the minimum value of contaminant concentration in water. The optimum t_{opt} is to ensure that the minimum of the function C(t), is being obtained by differentiation and comparison of the derivative to zero, which gave

$$t(C_{min}) = \frac{\ln\left(\dfrac{K\delta}{J}\right)}{K - \dfrac{J}{\delta}}. \tag{4.158}$$

4.5.4 Sorption at the membrane surface for the continuous mode in crossflow

Sorbent renewal from the membrane surface may be done by means of hydrodynamic arrangements using a number of means for reduction of concentration polarization [198], such as critical flux phenomenon [199, 200, 202, 203, 203], shear stresses, pulsations, stirring, gas sparking [204–210] and turbulences [197, 211–214] (see Fig. 4.79). The best-known model describing "concentration polarization" is the film model. It assumes a uniform and homogeneous film of constant thickness that spreads on the membrane surface. However, this is in contrast to the observations presented in numerous publications, which describe methods of removal of the polarizing layer in order to reduce hydraulic resistance in the permeate flow path. The use of Danckwerts [122] surface renewal theory, is much more realistic, assuming the occurrence of stochastic effects on removing the layer from the membrane surface (see Fig. 4.79). This model has been used for membranes, in a series of papers [121, 123–130]. The modified Danckwerts' surface renewal theory seems to be feasible for describing the rate of sorbent renewal on membrane surface, with the following assumptions:

1. The sorbent layer adjacent to the membrane surface is composed of the mosaic of infinitesimal elements.
2. Within the each individual element, the process takes place in the open flow system, which is nonstationary because of mass accumulation.

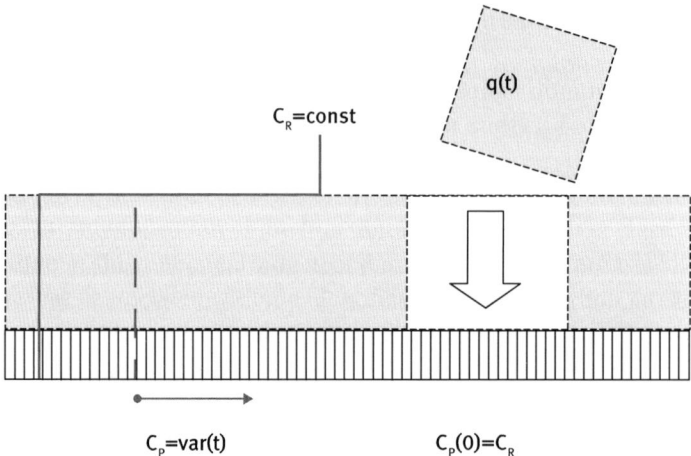

Fig. 4.79. Sorbent renewal from the membrane during crossflow.

3. These elements have different ages, therefore different concentrations of the permeate outflowing from them are present.
4. Elements are randomly entrained from the membrane, thus its surface is completely renewed and the process starts all over again. The age of any given element is the time which has passed since the last renewal.
5. The element which detaches from the membrane is immediately blended with the bulk of the retentate.

The fluxes of permeate leaving the infinitely small elements of the membrane surface have different content of the adsorbed substance (q(t) – adsorbate), depending on the age of the surface elements (t). The total concentration of the permeate that leaves the membrane (C_p(t)) is therefore the mean concentration of the age distribution of all surface elements,

$$C_P^m = \int_0^\infty C_P\,(t) \cdot f\,(t) \cdot dt, \tag{4.159}$$

where C_P(t) is an instantaneous concentration in permeate which varies in time according to the batch mode of operation (see Eq. (4.157)), and f(t) is the age function by which Danckwerts has been expressed as

$$f\,(t) = s \cdot e^{-s \cdot t}. \tag{4.160}$$

Here it should be mentioned that this is the average integral value from the age distribution (the so-called age function by Danckwerts), when time is approaching infinity. If the process takes a long time (infinite process time), the statistical distribution of the age of the surface elements is stabilized, which in turn gives a constant concentration in the permeate as follows:

$$C_P^m = C_R - X_{max} \cdot (q^* - q_R) \cdot \frac{k \cdot \delta \cdot s}{(J + \delta \cdot s) \cdot (k + s)}. \tag{4.161}$$

However, inclusion of the effects associated with the formation of the polarization layer during the process is then virtually impossible. To take the dynamic effects associated with the formation of the concentration polarization layer into account, Danckwerts' "age distribution function" should be modified to some extent. Statistical functions of probability density of each distribution, including the age distribution function of surface elements on the membrane, must fulfill one fundamental prerequisite as follows:

$$\int_0^\infty f(t) \cdot dt = 1. \tag{4.162}$$

When the assumption of an infinitely long time process ($t \to \infty$) cannot be met, the age distribution of the elements must be completely within the closed set ($0 < t < t_p$):

$$\int_0^{t_p} f(t) \cdot dt = 1. \tag{4.163}$$

Then, the age distribution is not stable, since it varies within the process duration (t_p):

$$f(t) = \frac{s \cdot e^{-s \cdot t}}{1 - e^{-s \cdot t_p}}. \tag{4.164}$$

The mean concentration (see Fig. 4.80) of the permeate based on modified age distri-

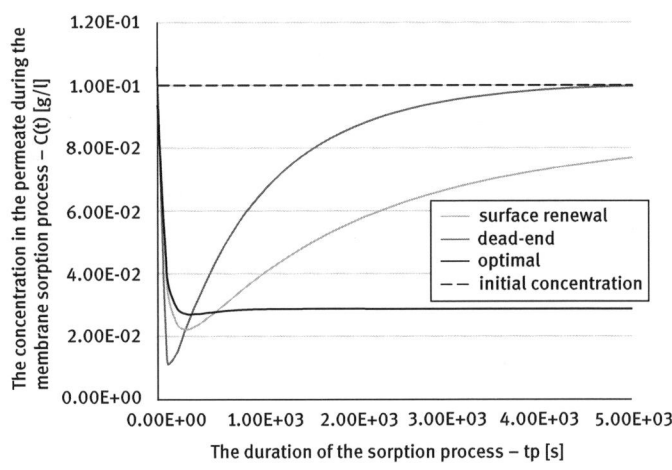

Fig. 4.80. Comparison of permeate concentration profiles during membrane biosorption in dead-end and cross-flow flow.

bution function and instantaneous concentration C(t) can then be denoted as

$$C_P^m\left(t_p\right) = C_R - \left(\frac{1 - e^{-(k+s)\cdot t_p}}{k + s} - \frac{1 - e^{-\left(\frac{J}{\delta}+s\right)\cdot t_p}}{\frac{J}{s} + s}\right) \tag{4.165}$$

$$\cdot \frac{s \cdot k \cdot \delta}{\left(1 - e^{-s\cdot t_p}\right)\cdot(J - k \cdot \delta)} \cdot X_{max} \cdot (q^* - q_R). \tag{4.166}$$

4.5.4.1 The optimal rate of surface renewal for the continuous mode of membrane biosorption

For the best sorption results, the fragments of the sorbent on the surface must be removed from it after a given time, to enable effective use of the sorption capacity of the adsorbent. Retention of the sorbent for too long is not correct, since the sorption rate decreases over time. The rate of surface renewal (s) of the sorbent layer on the membrane must be synchronized with the sorption kinetics (k) and permeate flux (J). In order to obtain optimum conditions, i.e. those at which the concentration of xenobiotics in the permeate will be minimal, the formula for a derivative of concentration in the permeate with respect to rate of surface renewal must be compared to zero as follows:

$$\text{when} \quad \frac{dC_P^m}{ds} = 0 \quad \text{then} \quad s_{opt} = \sqrt{\frac{k \cdot J}{\delta}} \quad \text{and} \quad C_P^m(s_{opt}) = \min \langle C_P^m \rangle|_{X, C_R, q^*, q_R}. \tag{4.167}$$

When the optimum value of the rate of renewal of surfaces is reached, from the values less than optimal value $s < s_{opt}$, then we obtain a series of curves, all of which eventually converge at a single site, at the initial concentration of C_R, as shown in Figure 4.81:

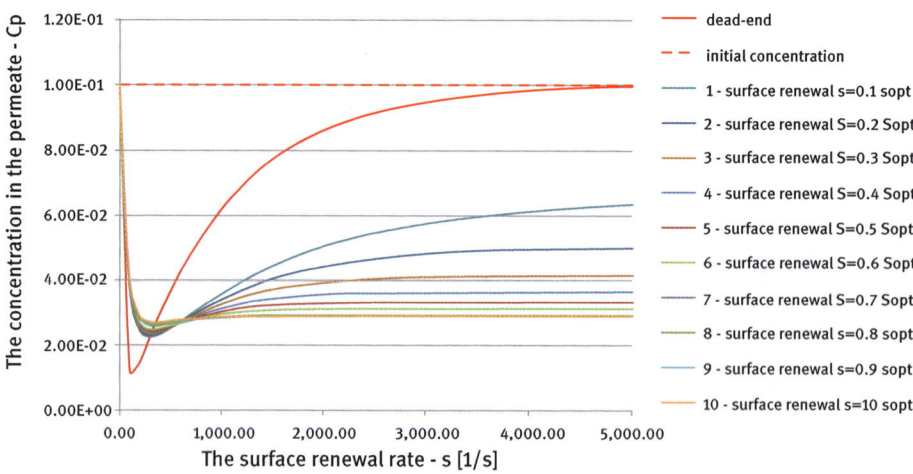

Fig. 4.81. Approaching optimal value (s_{opt}) starting from a smaller value ($s < s_{opt}$).

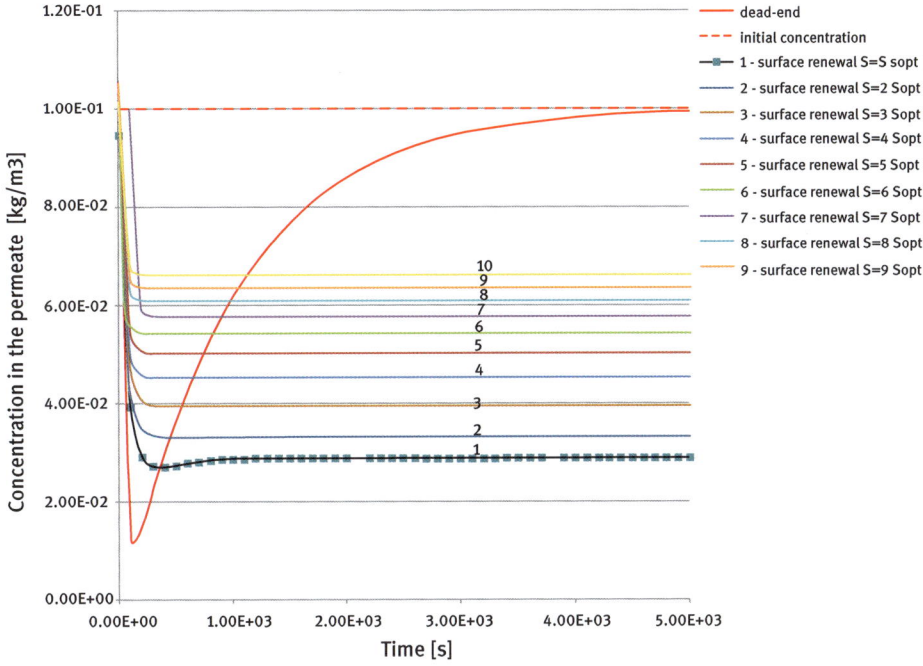

Fig. 4.82. Approaching optimal value (s_{opt}) starting from a larger values ($s > s_{opt}$).

This means that the process is not stationary if the rate of surface renewal is smaller than the optimal value. When starting from values higher than s_{opt} quite different graphs result (Fig. 4.82). We then get the characteristic stabilization of each line, i.e. "leveling up" of the different concentrations which are dependent on the surface renewal rate.

In order to know the biosorption mechanism however, one must first understand its dynamics. It is related to the dynamics of forming a mosaic of elements of sorption of different ages on the membrane surface because the formation of a stable age distribution takes time as shown in Fig. 4.84. These changes are most pronounced, especially during the first minutes of the process.

The concentration in the permeate flux, and thus the effective retention coefficient ($R = 1 - C_P/C_R$), is dependent on the concentration in the retentate stream (see Fig. 4.83).

The most important effect on sorption is exerted by the "driving force", which is the difference between the instantaneous sorbent concentration and its equilibrium value.

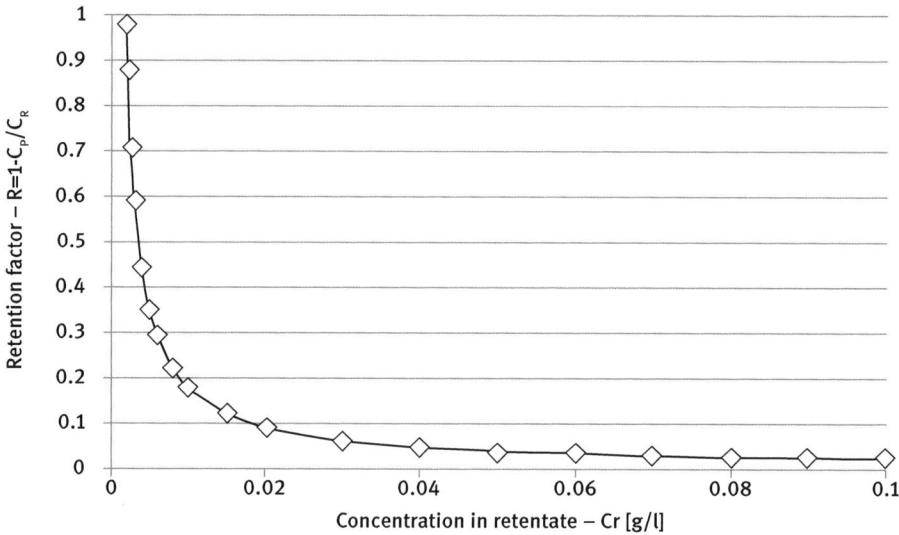

Fig. 4.83. Retention coefficient (R) of the biosorption process as a function of the concentration in the retentate (CR). Here: biomass concentration Xmax = 70 kg/m3, the sorbent uptake ▪ q = g/kg, permeate flux J = 2.8 10^{-5} m/s, ▪ = 1 mm, kinetic coefficient of the biosorption k = 10^{-3} 1/s, the optimum value of the biosorption s_{opt} = 5.29 10^{-3} 1/s (see Eq. (4.166) and Eq. (4.202)).

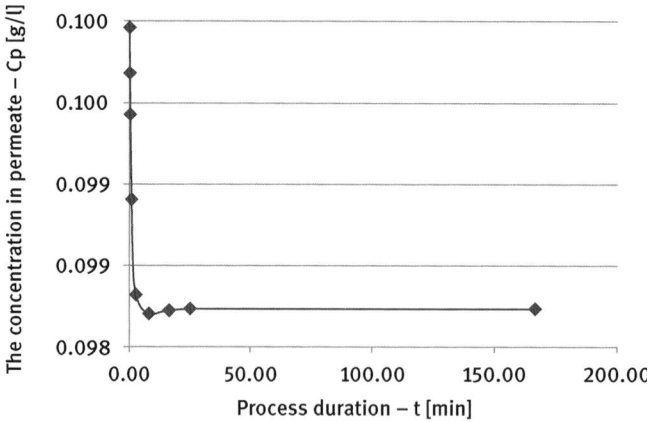

Fig. 4.84. The start-up period of membrane biosorption process.

4.5.4.2 The key parameters of membrane biosorption

The three parameters (k, J, and δ) which affect the minimal value of permeate concentration are in a mutual relationship, which the following charts attempt to demonstrate.

From Figure 4.87 it can be seen that the bigger the permeate flux, the higher the optimal value of s_{opt} is required, that is, the surface must be renewed more intensively using larger cross-flow velocity.

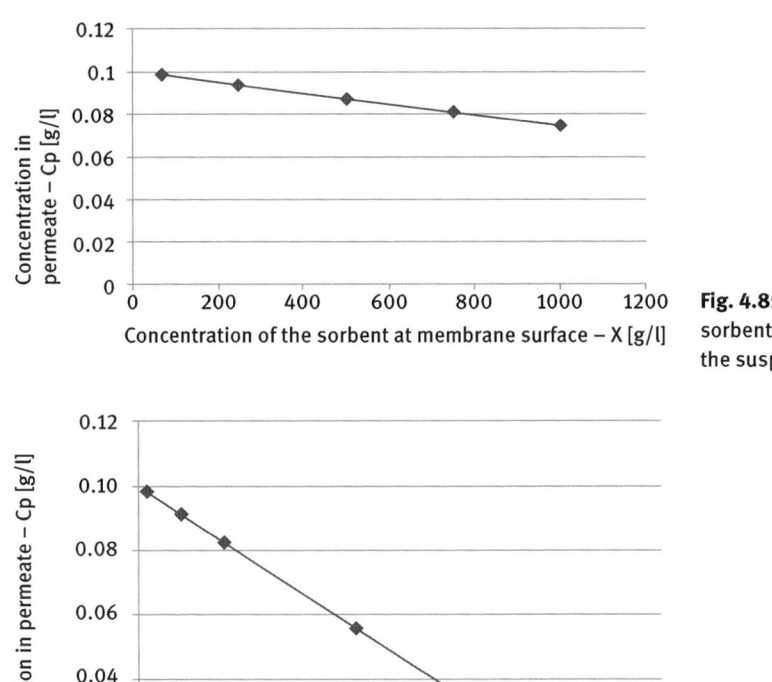

Fig. 4.85. The effect of sorbent concentration in the suspension.

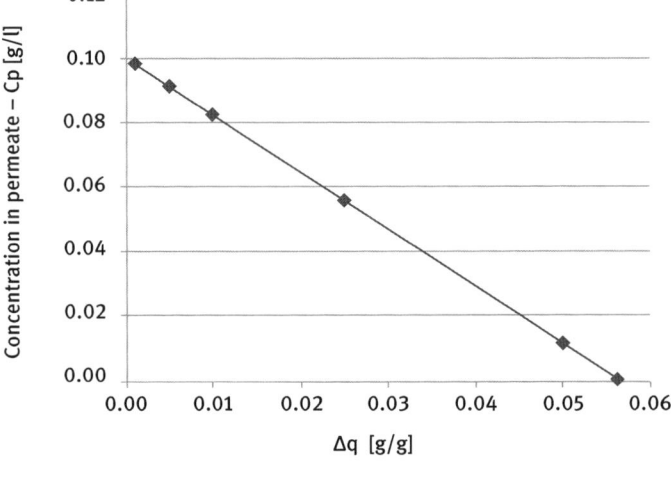

Fig. 4.86. The relationship between the driving force in the sorbent and the concentration in the permeate.

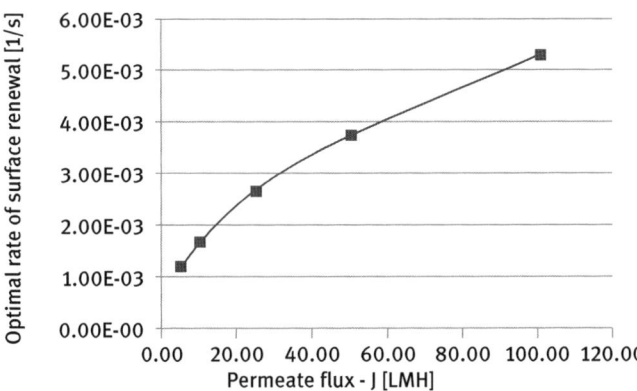

Fig. 4.87. Effect of permeate flux on the optimum surface renewal rate.

Figure 4.88 shows that a thinner layer of sorbent should be renewed faster than a thicker layer. This results from the increased sorption potential of the thicker layer, and hence the longer time needed for saturation.

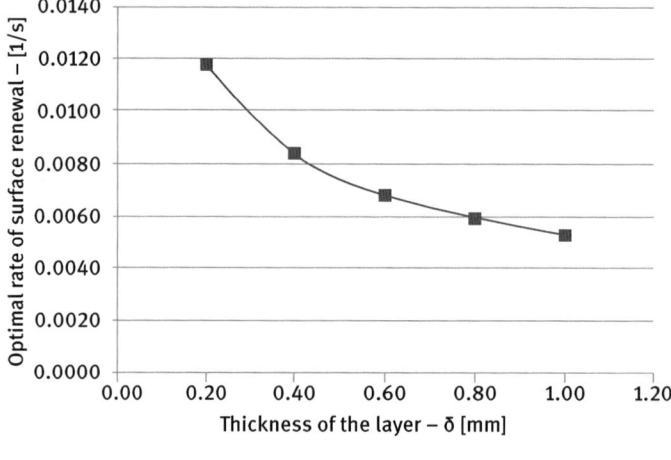

Fig. 4.88. Effect of sorbent layer thickness on the optimum surface renewal rate.

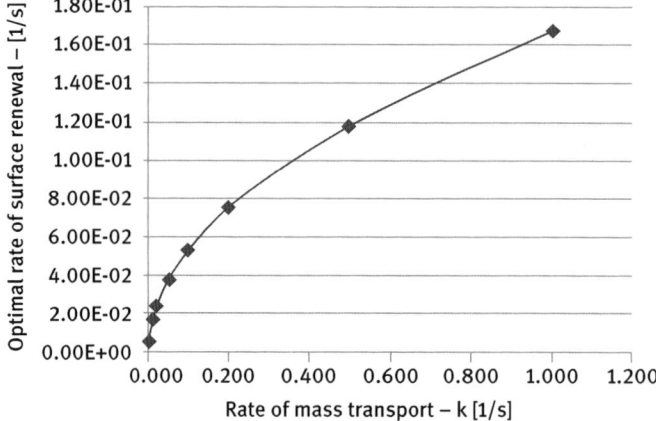

Fig. 4.89. Effect of rate of mass transfer on the optimum surface renewal rate.

Figure 4.89 shows that the faster the mass transport of sorbate to sorbent the more intense renewal of the surface should be, and thus the optimal rate of renewal of the surface is higher.

4.6 Membrane leaching – the new leaching process in membrane contactor

Membrane leaching is a process which can be classified as a separation process in membrane contactors in the system solid-liquid, as in the membrane sorption process The major advantage of this process is that raw materials (such as ores, milled grains etc.) in powder form can be used. The application of powders enables acceleration and improvement of the efficiency of each process in the system solid-liquid separation. These effects result from an increase in the interfacial area and a shortening of the diffusion path. The use of membranes allows faster separation of grains in comparison with the gravitational methods and those methods cheaper than the inertial methods of separation (such as centrifuges, cyclones).

It is assumed that leaching take place on the membrane surface during flow of the solvent through the layer of concentrated solid phase on membrane surface (see Fig. 4.90). The thickness of this layer is δ, and the concentration of the solid in suspension (X) is always higher in the vicinity of the membrane than in the bulk, because of the separation effect of the membrane. The idea of the process is to carry out the leaching process in the so-called concentration polarization layer (or cake) on the membrane surface. The main condition to be fulfilled during this process is to synchronize kinetics of the three different processes, i.e.,

1. kinetics of dissolution of the desired substance from the solid material;
2. flushing out this substance from the membrane vicinity to the permeate;
3. exchanging the solid material on the membrane surface for the fresh portion from the bulk.

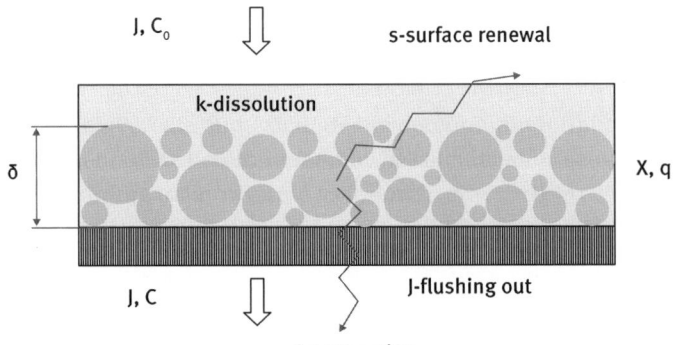

Fig. 4.90. The impact of surface renewal on the process of membrane leaching.

The kinetics of dissolution is a very complex phenomenon and may only be experimentally evaluated by determining the kinetics constant (k). The flushing rate is mainly dependent on the permeate flux and on the structure of the layer. The rate of exchange of solid material depends on the hydrodynamics inside the membrane module. It should be mentioned here that in membrane processes a lot of effort was devoted to the efficient removal of material accumulated on membrane surface. These are called depolarization models or depolarization techniques, and can be found in many publications.

The leaching process consists of dissolution and subsequent flushing away of the solute. The rates of both processes may be different, but they both decline with time. The net kinetics of the entire process can be determined as a superposition of both components. If dissolution is faster than the flushing, the concentration in the permeate increases. If flushing out is faster than dissolution, the concentration in the permeate decreases.

4.6.1 The kinetics of the dissolution of the substance extracted from solid grains

The grain of the solid has heterogeneous structure and consists of a substance that dissolves in the leaching process, and the part which is not dissolved and is a rigid skeleton of the solid such as a rock. The dissolved substance may have been in enclosed or open tubules (see Fig. 4.91). In the first of these cases, fragmentation only opens close to the surface, while those deeper remain closed. In the case of opened microchannels grinding accelerates both, i.e. the kinetics of mass transport, and it enhances the leaching efficiency. In general, the kinetics of dissolution must be determined experimentally.

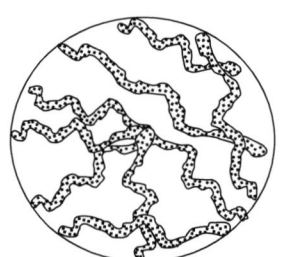

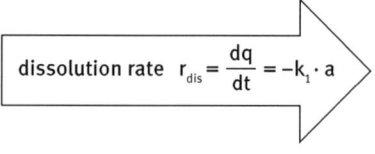

C(t) – solvent concentration

dissolution rate $r_{dis} = \dfrac{dq}{dt} = -k_1 \cdot a$

q(t) – concentration the solute

Fig. 4.91. Dissolution process in the pores of solid particle.

The dissolution rate decreases with time, but is proportional to the current and available content solute concentration in the solid material, and may be expressed as

$$\frac{dq}{dt} = -k \cdot (q - q_\infty).$$
(4.168)

Assuming initial condition in the solid phase,

$$q(0) = q_0.$$
(4.169)

Thus the solution of this equation can be expressed as

$$q = q_\infty + (q_0 - q_\infty) \cdot e^{-k \cdot t}.$$
(4.170)

The rate of dissolution can be described from the solid point of view as

$$\frac{dq}{dt} = -k \cdot (q_0 - q_\infty) \cdot e^{-kt}.$$
(4.171)

Now, assuming that the concentration of the solid surface is in thermodynamic equilibrium with the substance extracted in the solid,

$$C = m \cdot q,$$
(4.172)

we determine the concentration pattern of the change in the liquid under the influence of the change in the solid. Or from the liquid viewpoint,

$$C = m \cdot \left(q_\infty + (q_0 - q_\infty) \cdot e^{-k \cdot t}\right)$$
(4.173)

$$\frac{dC}{dt} = -m \cdot k \cdot (q_0 - q_\infty) \cdot e^{-k \cdot t}.$$
(4.174)

Dissolution ceases when the concentration in the solid reaches a minimal concentration q_∞, or if the incoming liquid is already saturated.

4.6.2 Flushing out of the substance extracted from solid grains

In the second leaching step after dissolution, the substance to be extracted must be flushed out by the solvent. The mechanism of flushing out on membranes is similar to diafiltration of the substance with concentration (C) from the given volume (V) which is

$$-V\frac{dC}{dt} = J \cdot F \cdot C, \tag{4.175}$$

where
V – is a volume of the system;
J – is a flux of the permeate;
F – is a surface of the flow;
C – is a concentration in the permeate;
δ – is a thickness of the layer of solid material on the membrane surface.

The problem lies in the fact that the volume of the filter cake consists of two phases, namely a liquid and a solid phase. The concentration of solid phase (X[g solid/l suspension]) at the membrane surface is very high and solid particles are closely packed. Thus the volume occupied by the liquid is some volumetric fraction of the entire system

$$\varphi_{liquid} + \varphi_{solid} = 1. \tag{4.176}$$

The volume subject to leaching concerns only a fraction of the solid structure which is accessible to the liquid. The volume fraction occupied by the liquid is a part of the whole system and can be determined based on the definition of mass concentration of suspension and its density:

$$\varphi_s = \frac{X}{\rho_s}. \tag{4.177}$$

Hence,

$$V_{liquid} = V\left(1 - \frac{X}{\rho_s}\right). \tag{4.178}$$

During flushing out of the extract, the fraction of the liquid volume should be taken into account solely during determination of the permeate concentration (C):

$$V_1 \cdot \frac{dC}{dt} = -J \cdot F \cdot C. \tag{4.179}$$

Hence,

$$\frac{dC}{dt} = -\frac{J}{\delta} \cdot \left(1 - \frac{X}{\rho_s}\right)^{-1} \cdot C. \tag{4.180}$$

The initial condition imposed on the concentration of the permeate can be written as

$$C(0) = C_0,$$
(4.181)

to obtain the solution

$$C = C_0 \cdot e^{-\frac{J}{\delta} \cdot \left(1 - \frac{X}{\rho_s}\right)^{-1} \cdot t}.$$
(4.182)

It should be noted that C_0 is the initial equilibrium concentration of the permeate with respect to the concentration in the fresh solid q_0, excluding the nonextractable portion q_∞:

$$C_0 = m \cdot (q_0 - q_\infty),$$
(4.183)

where q_0 is the initial concentration in solid material and q_∞ is the final concentration if it is not zero:

$$\frac{dC}{dt} = -\frac{J}{\delta} \cdot \left(1 - \frac{X}{\rho_s}\right) \cdot m \cdot (q_0 - q_\infty) \cdot e^{-\frac{J}{\delta} \cdot \left(1 - \frac{X}{\rho_s}\right)^{-1} \cdot t}.$$
(4.184)

4.6.3 The nonstationary (dead-end) membrane leaching process

4.6.3.1 Instantaneous mass flux during membrane leaching

When the leaching process takes place in a closed container, according to the law of conservation of mass, the concentrations in the liquid and solid are related by the equation

$$C_0 + X \cdot q_0 = C + X \cdot q = \text{const.}$$
(4.185)

The corresponding rates of concentration reduction in the solid and concentration increase in the solution are then equal, as is clear from the law of conservation of mass:

$$\left.\frac{dC}{dt}\right|_d = \left.\frac{dC}{dt}\right|_a + \left.\frac{dC}{dt}\right|_f.$$
(4.186)

When permeate flows through the membrane where the layer of solid material is leached, the two processes, i.e. dissolution and flushing out, occur simultaneously. The processes take place one after another. Therefore if they do not occur at the same rate accumulation occurs (positive or negative). In each case, the law of conservation of mass must be fulfilled. Therefore, the speed of these separate processes affects the final, the net effect of which is the final concentration of the permeate. The rate of the entire effect during the flow of the fluid through the solid material can thus be determined based on the superposition of these two effects, i.e. dissolution (positive)

$$\left.\frac{dC}{dt}\right|_d = m \cdot k \cdot (q_0 - q_\infty) \cdot e^{-k \cdot t}$$
(4.187)

and flushing out (negative)

$$\left.\frac{dC}{dt}\right|_f = -\frac{J}{\delta} \cdot \left(1 - \frac{X}{\rho_s}\right)^{-1} \cdot m \cdot (q_0 - q^*) \cdot e^{-\frac{J}{\delta} \cdot \left(1 - \frac{X}{\rho_s}\right)^{-1} t}.$$
(4.188)

The net rate of the process is the superposition of the dissolution and the rate of the flushing out as follows:

$$\frac{dC}{dt} = m \cdot k \cdot (q_0 - q^*) \cdot e^{-k \cdot t} - \frac{J}{\delta} \cdot \left(1 - \frac{X}{\rho_s}\right)^{-1} \cdot m \cdot (q_0 - q^*) \cdot e^{-\frac{J}{\delta} \cdot \left(1 - \frac{X}{\rho_s}\right)^{-1} \cdot t}. \quad (4.189)$$

Based on the differential equation, one can determine the function of concentration during the extraction of the solid material at the membrane vicinity, which is enhanced by the permeate flow is as follows:

$$C = m \cdot k \cdot (q_0 - q^*) \cdot \int e^{-k \cdot t} dt - \frac{J}{\delta} \cdot \left(1 - \frac{X}{\rho_s}\right) \cdot m \cdot (q_0 - q^*) \cdot \int e^{-\frac{J}{\delta} \cdot \left(1 - \frac{X}{\rho_s}\right) \cdot t} dt + \text{const.} \quad (4.190)$$

The initial condition before the dissolution and prior to leaching is obvious:

$$C(0) = 0. \quad (4.191)$$

Therefore a function has the form

$$C(t) = m \cdot (q_0 - q^*) \cdot \left(e^{-\frac{J}{\delta} \cdot \left(1 - \frac{X}{\rho_s}\right)^{-1} \cdot t} - e^{-k \cdot t}\right). \quad (4.192)$$

This model describes the change in concentration in the permeate which flows through the membrane through which the material is leached. It should be noted that the rate of elution of the substance extracted from the membrane layer depends on the thickness of the film, on the concentration of solids in the film X and on the rate of dissolution. Leaching is also dependent on the permeate flux passing through the membrane. Figure 4.92 shows the relationship of the concentration in the permeate flux (J) and kinetics of dissolution (k).

It should be noted that with increased permeate flux in the dead-end system, the concentration of eluted substance was reduced, and simultaneously the maximum of the concentration in time was shifted towards smaller values.

Also the values of these peaks shifted to the left with increasing dissolution kinetics (Fig. 4.92). Tab. 4.16 shows the values of the other parameters that were taken into account during the calculations.

4.6.3.2 The optimal time for backflushing

Since the concentration of the permeate versus time has a significant maximum, this can be practically applied. Backflushing can be used for periodical removal of the depleted material. The optimal interval of backflushing coincides with the time at which the maximum concentration occurs:

$$C_{max} = m \cdot (q_0 - q^*) \cdot \left(1 - e^{-k \cdot t_{opt}}\right) - m \cdot (q_0 - q^*) \cdot \left(1 - e^{-\frac{J}{\delta} \cdot \left(1 - \frac{X}{\rho_s}\right)^{-1} \cdot t_{opt}}\right). \quad (4.193)$$

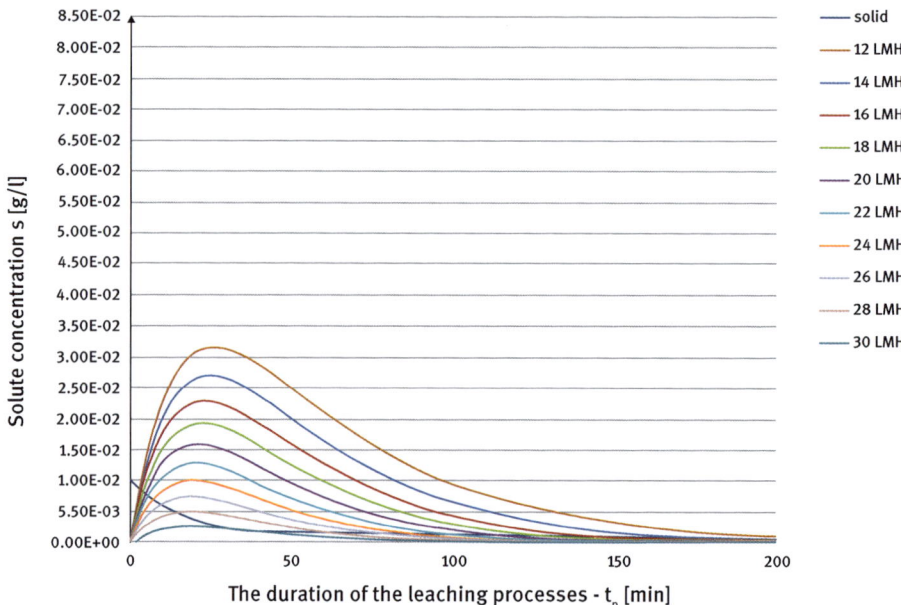

Fig. 4.92. Concentration in permeate as a result of permeate flux, in the dead-end membrane leaching process. A – higher fluxes.

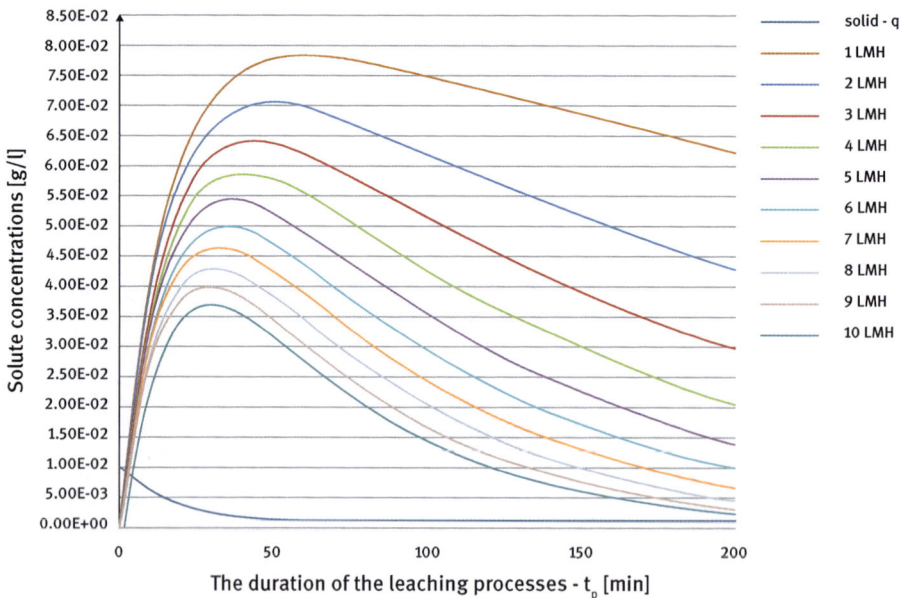

Fig. 4.93. Concentration in permeate as a result of permeate flux, in the dead-end membrane leaching process. B – lower fluxes.

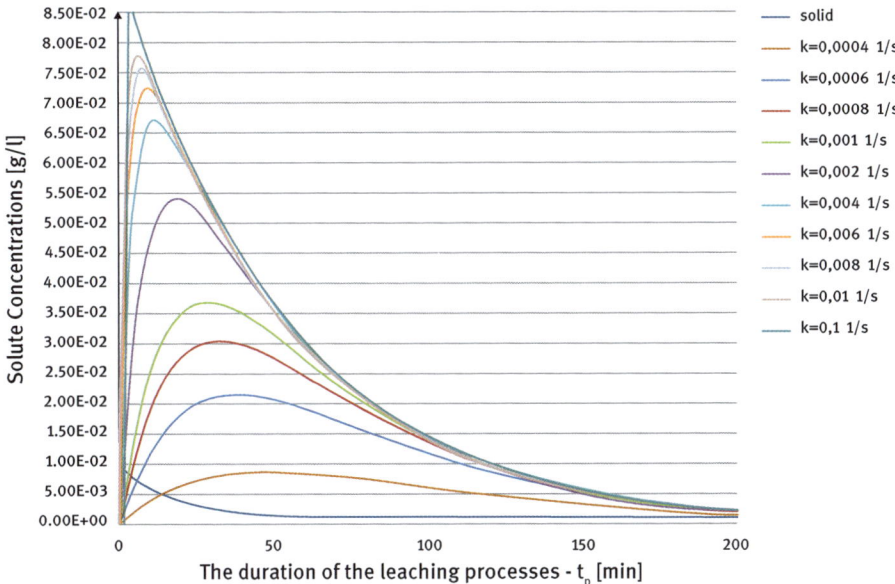

Fig. 4.94. Effect of dissolution rates on concentration in permeate during membrane leaching.

Table 4.16. Data taken from membrane leaching process simulations

Variable	Quantity	SI units
$k =$	1.00E-03	s^{-1}
$J =$	2.78E-06	m/s
$\delta =$	1.00E-02	m
$X_{max} =$	100	kg/m^3
$q_0 =$	0.01	kg/kg
$q^* =$	0.001	kg/kg
$m =$	10	kg/m^3
$\rho_s =$	1000	kg/m^3

To this end, the first derivative of the concentration must be likened to zero:

$$\frac{dC_{max}}{dt} = 0$$

$$= \left[m \cdot (q_0 - q^*) \cdot \left(1 - e^{-k \cdot t_{opt}}\right) - m \cdot (q_0 - q^*) \cdot \left(1 - e^{-\frac{J}{\delta} \cdot \left(1 - \frac{X}{\rho_s}\right)^{-1} \cdot t_{opt}}\right) \right]'. \quad (4.194)$$

Hence,

$$t_{opt} = -\frac{\ln \dfrac{J \cdot \rho_s}{k \cdot \delta \cdot (\rho_s - X)}}{\left(k - \dfrac{J \cdot \rho_s}{\delta \cdot (\rho_s - X)} \right)}. \quad (4.195)$$

It should be noted that the permeate stream must be adjusted to several parameters (see Tab. 4.16). Among which the most important are the thickness (δ), the degree of packing of solid film on the surface of the membrane (X) and the rate of dissolution (k). To find the right strategy for the leaching process it is convenient to use a 3D graph which shows the optimal period for backflushing (on the vertical axis) depending on the permeate flux and k (on the horizontal axis) (see Fig. 4.95). However, in order to be able to perform leaching on membrane surface the threshold defined by the following equation should not be exceeded:

$$J < \frac{k \cdot \delta \cdot (\rho_s - X)}{\rho_s}.$$

(4.196)

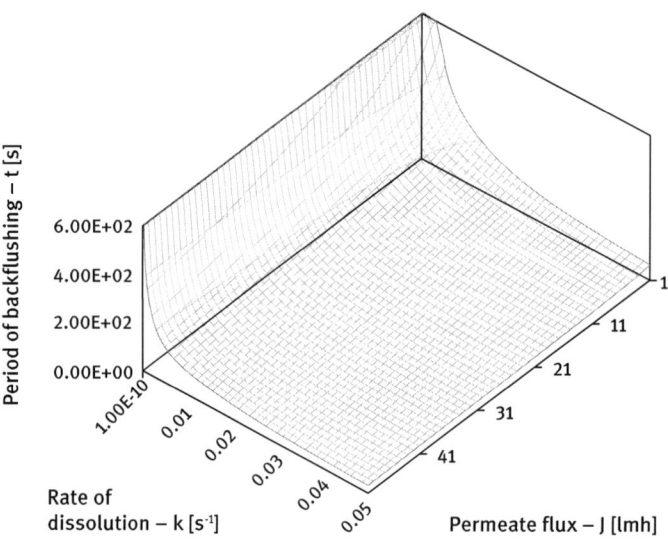

Fig. 4.95. Selection of backflushing frequency on the basis of the kinetics of leaching and the permeate flux.

4.6.4 The stationary process of membrane leaching

4.6.4.1 Using the surface renewal model for the membrane leaching process

Recharging material on the membrane surface can be also synchronized with the intensity of the dissolution (k) and the intensity of leaching (J) by means of proper hydrodynamics at membrane vicinity. Equally important effects can therefore be achieved by applying the depolarization methods widely known and used in membrane processes for the removal of the polarizing film. These methods were discussed earlier in Section 4.2 on membrane processes. The Danckwerts model of surface renewal is particularly well suited to the mathematical description of such synchronization aspects.

Danckwerts' model allows calculation of the average concentration of the permeate across the membrane, provided knowledge of concentration changes in time C(t) on the system dead-end:

$$C_P^m = \int\limits_0^\infty C(t) \cdot f(t) \cdot dt. \tag{4.197}$$

The correlations determined in this chapter can be applied to the Danckwerts model:

$$C(t) = m \cdot (q_0 - q^*) \cdot \left(e^{-\frac{J}{\delta} \cdot \left(1 - \frac{X}{\rho_s}\right)^{-1} \cdot t} - e^{-k \cdot t} \right). \tag{4.198}$$

Following the Danckwerts model surface renewal process is completely stochastic (with respect to space and time), which allows the application of a known age distribution function f(t):

$$f(t) = s \cdot e^{-s \cdot t} \tag{4.199}$$

$$C_P^m = \int\limits_0^\infty m \cdot (q_0 - q^*) \cdot \left(e^{-\frac{J}{\delta} \cdot \left(1 - \frac{X}{\rho_s}\right)^{-1} \cdot t} - e^{-k \cdot t} \right) \cdot s \cdot e^{-s \cdot t} \cdot dt, \tag{4.200}$$

after the mathematical integration of this equation to determine the mean concentration in the permeate. It should be noted that the established form of an age function of Danckwerts can be approached only after an infinitely long process. This is denoted by dashed lines on the graphs:

$$C_P^m = m \cdot (q_0 - q^*) \cdot s \cdot \left[\frac{\delta \cdot (\rho_s - X)}{J \cdot \rho_s + s \cdot \delta \cdot (\rho_s - X)} - \frac{1}{(k + s)} \right]. \tag{4.201}$$

4.6.4.2 Optimal operation conditions during membrane leaching process
In the cross-flow mode of membrane operation, the permeate concentration depends mainly on the rate of surface renewal controlled by hydrodynamic conditions, but is also a function of various process parameters. The rate of the surface renewal also has anoptimal value which can be determined as follows:

$$\frac{dC}{ds} = 0 = \frac{d}{ds} \left\{ m \cdot (q_0 - q^*) \cdot \left[\frac{s \cdot \delta \cdot (\rho_s - X)}{J \cdot \rho_s + s \cdot \delta \cdot (\rho_s - X)} - \frac{s}{(k + s)} \right] \right\}. \tag{4.202}$$

The rate of surface renewal can be controlled in different membrane modules by choosing suitable cross-flow velocity of the retentate over the membrane surface. It is important that it is synchronized with the flushing rate (dependent on permeate flux), the kinetics of dissolution (k), as well as with other parameters in the formula

$$s_{opt} = \frac{\sqrt{k \cdot J \cdot \rho_s \cdot \delta \cdot (\rho_s - X)} - J \cdot \rho_s}{\delta \cdot (\rho_s - X) - \sqrt{\dfrac{J \cdot \rho_s \cdot \delta \cdot (\rho_s - X)}{k}}}. \tag{4.203}$$

Figure 4.96 shows the dependence of the optimal rate of renewal (s, vertical axis) from the permeate flux and the rate of dissolution. This relationship demonstrates that the higher the permeate flux and the greater the rate of dissolution, the greater the surface renewal rate should be. It should be emphasized that the optimal rate of surface renewal has only been determined under the assumption of stabilization, which can

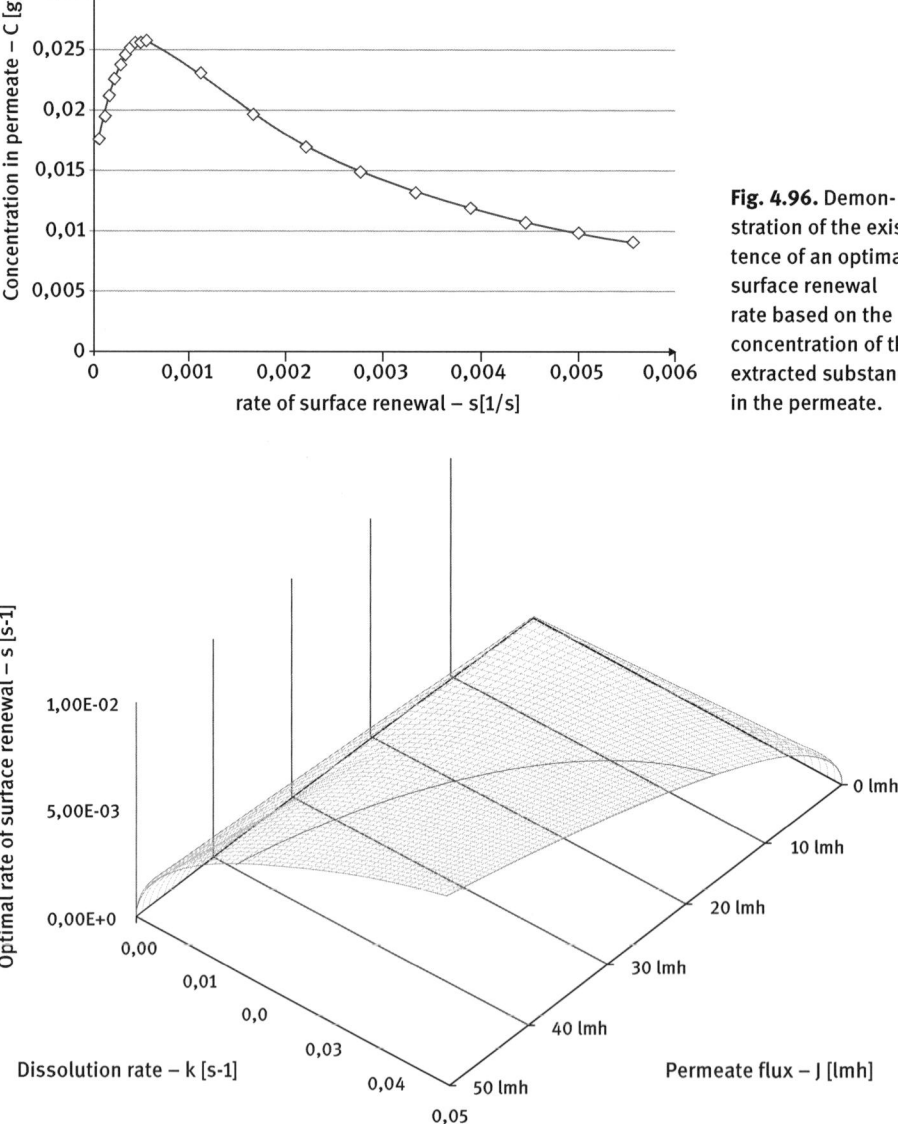

Fig. 4.96. Demonstration of the existence of an optimal surface renewal rate based on the concentration of the extracted substance in the permeate.

Fig. 4.97. Optimal rate of surface renewal based on the permeate flux and the rate of dissolution.

be achieved only after an infinitely long process duration, and therefore gives only the asymptotic values. The time factor must be taken into account in real processes, as described in the next Section 4.6.3.4 (see also the dashed lines in Figs. 4.98 and 4.99).

4.6.4.3 Concentration in the quasi-constant process of membrane leaching

Even under stationary conditions, the process is unsteady in the initial period until stabilization is achieved. In Danckwerts model it is assumed that the process has been going on for an infinitely long time after the age function is stabilized and this can be expressed by means of exponential distribution (see Eq. (2.273)). This equation was the result of assumptions which must be met by all statistical distributions, that the integral of the distribution of all the elements (from 0 to infinity) is equal to one. In other words, all the elements are within the range $\langle 0, \infty \rangle$. This assumption is of course only an approximation, but in fact one can only say that age of all the elements is within the finite range (0 etc.), i.e. up to the process duration:

$$\int_0^{t_p} f(t) \cdot dt = 1. \tag{4.204}$$

This obvious assumption completely alters the age functions, which now must lie in a finite range, and can be expressed as follows:

$$f(t) = \frac{s \cdot e^{-s \cdot t}}{1 - e^{-s \cdot t_p}}. \tag{4.205}$$

The change in concentration-time $C(t)$ is treated as before, i.e.,

$$C(t) = m \cdot (q_0 - q^*) \cdot \left(e^{-\frac{J}{\delta} \cdot \left(1 - \frac{X}{\rho_s}\right)^{-1} t} - e^{-k \cdot t} \right). \tag{4.206}$$

Thus, the formula for determining the average concentration of the permeate, as the average age distribution of the components on the membrane surface, will have modified only the upper limits of integration:

$$C_P^m = \int_0^{t_p} C(t) \cdot f(t) \cdot dt. \tag{4.207}$$

Hence,

$$C_P^m = \int_0^{t_p} m \cdot (q_0 - q^*) \cdot \left(e^{-\frac{J}{\delta} \cdot \left(1 - \frac{X}{\rho_s}\right)^{-1} \cdot t} - e^{-k \cdot t} \right) \cdot \frac{s \cdot e^{-s \cdot t}}{1 - e^{-s \cdot t_p}} \cdot dt. \tag{4.208}$$

In this way, an important formula was obtained after integration in which the duration of the process is an important parameter. The formula allows determination of the

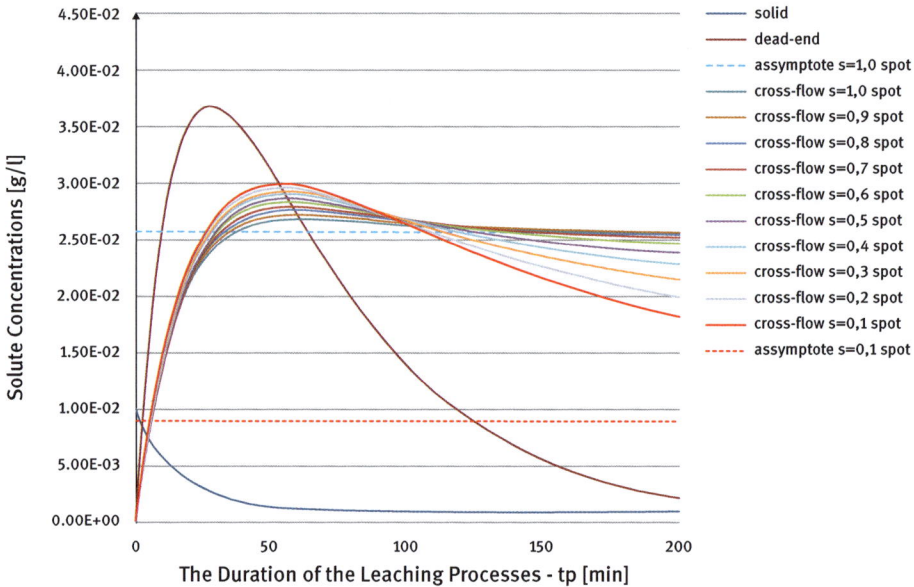

Fig. 4.98. Effect of surface renewal on the steady state of membrane leaching for the lesser than optimum values.

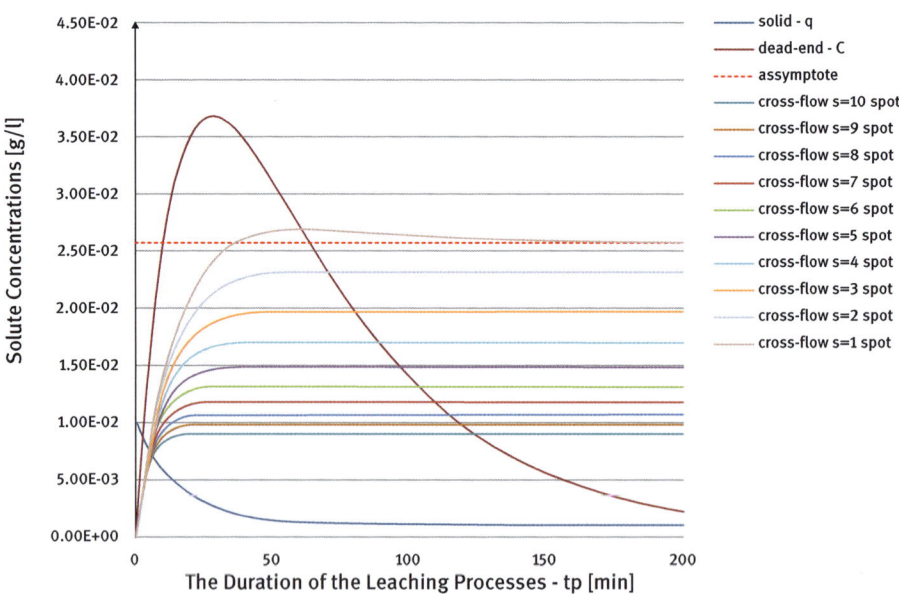

Fig. 4.99. Effect of surface renewal on the steady state of membrane leaching for the higher than optimum values.

dynamic properties of the new process, i.e. "membrane leaching":

$$
C_P^m = \frac{m \cdot (q_0 - q^*) \cdot s}{1 - e^{-s \cdot t_p}} \cdot \left[\frac{1 - e^{-\left(\frac{J}{\delta} \cdot \left(1 - \frac{X}{\rho_s}\right)^{-1} + s\right) \cdot t_p}}{\left(\frac{J}{\delta} \cdot \left(1 - \frac{X}{\rho_s}\right)^{-1} + s\right)} - \frac{1 - e^{-(k+s) \cdot t_p}}{(k + s)} \right]. \tag{4.209}
$$

The graphs have shown how the changes to concentration levels (in the permeate) occur in relation to different surface renewal rates. Pay attention to the investigation of the optimum value of smaller values as well as higher values. As in the biosorption process, the synchronization of the three speeds is essential. Thus, the rate of surface renewal must be adapted to the rate of the permeate flux and mass transport kinetics (here the rate of leaching).

5 Bioprocesses

5.1 Biotechnology

The semantic meaning of the word biotechnology is very simple because the prefix "bio" means life and the core "technology" means knowledge about targeted and cost-effective transformation processes of natural resources into useful goods. Biotechnology is as old as human civilization and has been an integral part of human life. For many centuries a range of goods has been produced such as food, beverages, clothes, beer, wine, curd, cheese, bread and numerous other goods. Similarly, other aspects of biology in breeding and other activities have been known since ancient times. Modern biotechnology includes new genetic manipulations such as gene modifications (GM), marker assisted selection (MAS), cloning, transgenic and cisgenic. The establishment of a new journal entitled Biotechnology and Bioengineering in 1961 appears to be a good starting point for modern biotechnology on the time axis.

The European Federation of Biotechnology, established in 1978, defines biotechnology as an 'integration of natural sciences and organisms, cells, parts thereof, and molecular analogues for products and services'. In 1981 the International Union of Pure and Applied Chemistry (IUPAC) defined biotechnology as "The application of biochemistry, biology, microbiology and chemical engineering to industrial process and products and on environment" [215]. Again in 1981, the OECD defined biotechnology as "... the application of scientific and engineering principles to the processing of materials by biological agents to provide goods and service". The Convention on Biological Diversity which was agreed upon at the UN Earth Summit mentioned Biotechnology twice as "... the use of living systems and organisms to develop or make useful products", or "any technological application that uses biological systems, living organisms or derivatives thereof, to make or modify products or processes for specific use". Similar definitions of biotechnology were presented by several scientific national organizations such as the US National Science Foundation, who stated that "... biotechnology is the controlled use of biological agents, such as microorganisms or cellular components, for beneficial use". The American Chemical Society defined biotechnology as the application of biological organisms, systems, or processes by various industries to learning about the science of life and the improvement of the value of materials and organisms such as pharmaceuticals, crops and livestock. Japanese biotechnologists defined biotechnology as "using biological phenomena for copying and manufacturing various kinds of useful substances". British Biotechnologists have provided the following definition: "Biotechnology is the application of biological organisms, system or processes to manufacturing and service industries". In a study made under the auspices of the National Research Council entitled Frontiers in Chemical Engineering (the Amundson report) biotechnology is defined as "... an excellent example of the creative use of Process Engineering methods for biological processes to enable the technical implementation of the prospects offered by modern biology and biochemistry" [216]. The major biotechnology products identified below in Figure 5.1 are the following:

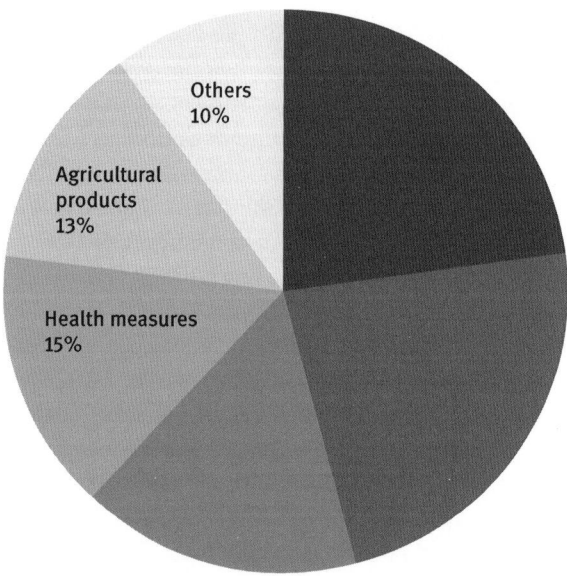

Fig. 5.1. Biotechnology products: 1 – energy carriers, 2 – foodstuffs, 3 – chemicals, 4 – health measures, 5 – agricultural products, 6 – others.

In more detail, the individual products in those sectors are presently as follows:
1. *energy carriers:* biodiesel from different crops and wastes, ethanol from biomass and waste paper (waste liquors), methane from agricultural waste and waste water treatment, production of hydrogen and hydrocarbons by photosynthesis of algae, production of hydrogen and electricity from wastewater;
2. *foodstuffs:* protein foods, amino acids, sugars, organic acids, fragrances and flavors, sweeteners;
3. *chemicals:* organic products (butanol, acetic acid, alcohols, diols), enzymes, pesticides, higher organic acids, bulk chemicals, bioplastics; functional foods and nutraceuticals. Microbial protein production of inorganic substances, textile and fabric finishing and fabric coatings, surfactants;
4. *health procedures:* antibiotics, vitamins, amino acids, hormones, steroids, immune proteins (vaccines, interferon, antibodies), insulin, human growth hormone, monoclonal antibodies. pharmaceuticals, citric acid, polysaccharides (dextrans, xanthan, alginate);
5. *agricultural products:* feed additives, organic manures, plant protection products, veterinary medicines;

6. *biohydrometalurgy:* of copper, uranium, (10 % of these metals), nickel, cobalt, silver, gold;
7. *environmental protection:* desulfurization of coal, waste water treatment, treatment of industrial waste water, treatment of waste gases, use of enhanced microorganisms in bioremediation, use of biomining techniques.

All of these will be produced in biorefineries, i.e. future factories (see Fig. 5.2). Biorefineries are factories, in which the products are all from biomass, which is an analogous concept to today's petroleum refineries, which produce a variety of products from only one raw material: crude oil. Biorefineries have already existed in our economic reality for several years. It is fair to say that not all technologies to convert biomass can be considered sustainable (as burning), although they are better than conventional, such as using a biomass as fuel. Such technologies can be used for a transition period, but they should be gradually and definitely reduced in order to prevent them taking root in the global economy. Only those which do not affect the ecological balance, do not emit pollutants, and do not consume fossil fuels can be considered to be sustainable technologies. Future biorefineries should play an exclusive role in producing chemicals, bioplastics and biofuels, including biohydrogen.

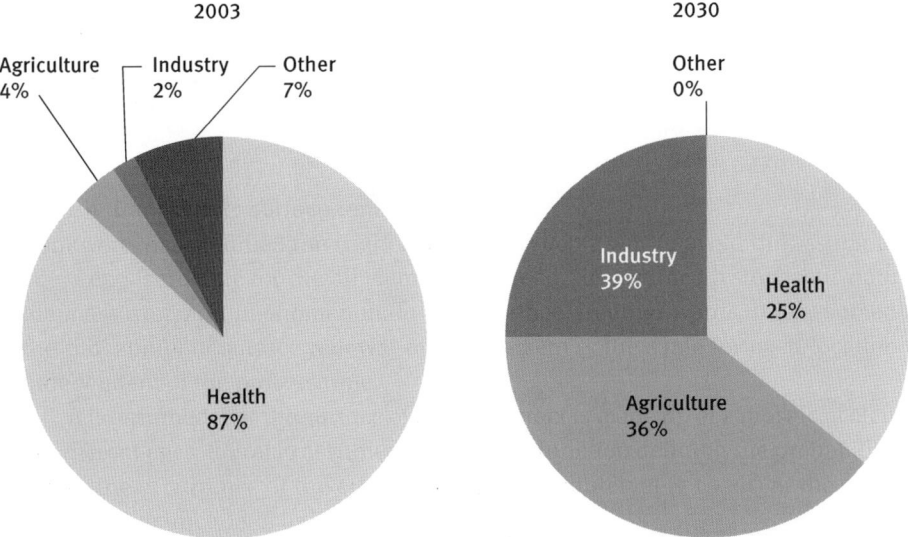

Fig. 5.2. Expected development of such sectors as: 1. medicine, 2. agriculture and 3. industry.

5.2 Bioprocess engineering

Bioprocess engineering deals with the design and development of biological processes using enzymes, living cells or tissues. Bioprocess engineering plays exactly the same role in biotechnology as process engineering plays in industry (Fig. 5.3).

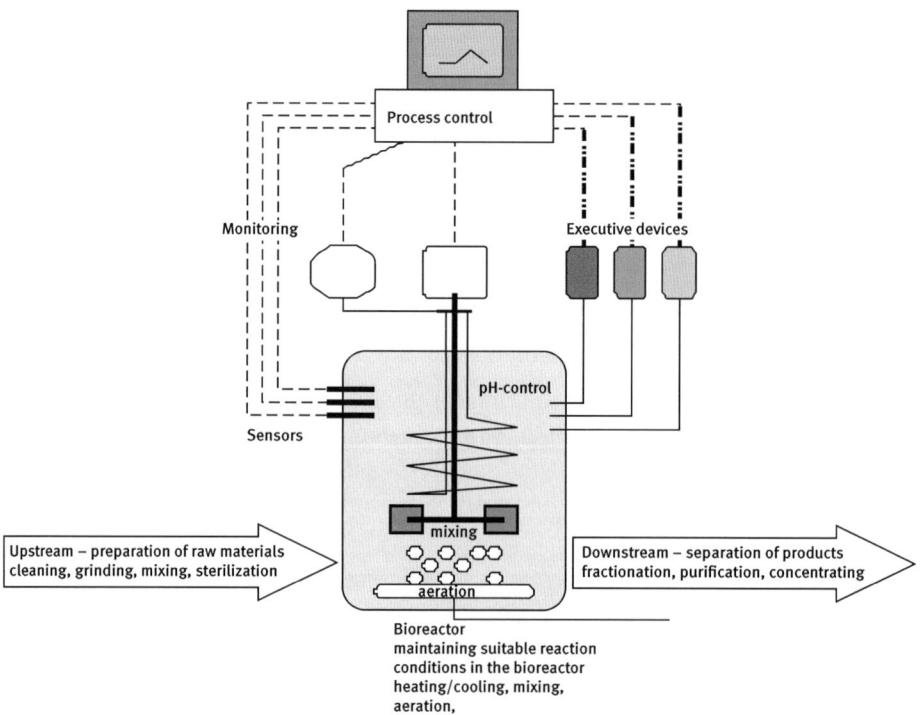

Fig. 5.3. The generalized conceptual diagram of any biotechnological process.

5.2.1 Bioprocess simplifications in biotechnology

Biochemical reactions are usually very complex, and may be consequential, sequential, parallel or mixed. The course of a reaction depends on process conditions such as temperature, pH and a very narrow range of concentration. In addition, the presence of trace amounts of certain substances which may be important for the catalytic reactions is also important. Bioprocess engineering introduces a number of simplifications in order to determine the basic products which originate from the main feedstock. The first simplification is to assume that biotechnological reactions involve only four main elements: carbon, hydrogen, oxygen and nitrogen. These four elements make up about 95 percent of the participation of all elements involved in biochemical reactions. Carbon is the basic building block of the whole our ecosphere. Thus, it may be regarded as basis of mass balances or the reference element, as certain kind of weight. All equations can therefore be expressed by the elemental formulas [217]

$$C_\delta H_a O_b N_c \quad \text{where} \quad \delta = 1 \text{ or } 0. \tag{5.1}$$

Providing all the elements involved in biochemical reactions would be very difficult. In some cases it is necessary to balance the phosphorus and sulfur elements also. If the substance has no carbon atoms, the pattern takes the form of a molecular formula of the substance, and is equivalent to c-mol mole. The average elemental composition of biomass was determined experimentally and is

$$CH_{1,79}O_{0,5}N_{0.2}. \tag{5.2}$$

Therefore the C-mol molecular mass is

$$M_{C,X} = \frac{1 \cdot 12 + 1,8 \cdot 1 + 0,5 \cdot 16 + 0,2 \cdot 14}{0,95} = 25,9 \frac{g}{C - mol}. \tag{5.3}$$

Table 5.1. The C-mol formulas for various microorganisms

Microorganism	C-mol formula
Candida utilis	$CH_{1,83}O_{0,53}N_{0,1}$
Candida utilis	$CH_{1,87}O_{0,56}N_{0,2}$
Saccharomyces cerevisiae	$CH_{1,83}O_{0,56}N_{0,17}$
Klebsiella aerogenes	$CH_{1,75}O_{0,43}N_{0,22}$
Paracoccus denitrificans	$CH_{1,81}O_{0,46}N_{0,19}$
Escherichia coli	$CH_{1,77}O_{0,49}N_{0,24}$
Aerobacter aerogenes	$CH_{1,83}O_{0,55}N_{0,25}$

5.2.2 Stoichiometry of bioprocesses

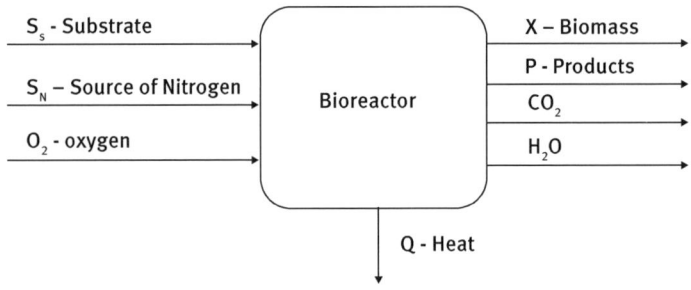

Fig. 5.4. The symbolic notation of reaction in the bioreactor.

Simplified model reaction of heterotrophic growth is assumed in a bioreactor [218, 219]:

$$S_S + S_N + O_2 \rightarrow X + P + CO_2 + H_2O + Q. \tag{5.4}$$

The equation for the reaction in a bioreactor is

$$\nu_S CH_{a_S} O_{b_S} N_{c_S} + \nu_N C_{\delta_N} H_{a_N} O_{b_N} N_{c_N} + \nu_O O_2 \rightarrow$$
$$\nu_X CH_{a_X} O_{b_X} N_{c_X} + \nu_P CH_{a_P} O_{b_P} N_{c_P} + \nu_C CO_2 + \nu_W H_2O, \tag{5.5}$$

where

a – stoichiometric coefficient for hydrogen;
b – stoichiometric coefficient for oxygen;
c – stoichiometric coefficient of nitrogen;
δ – stoichiometric coefficient for carbon;
ν – balance coefficients for S, N, O, X, P, C, and W.

The balance for all four elements on both sides of the reaction equation leads to the following four algebraic equations:

$$C: \quad 1 \cdot \nu_S + \delta_N \cdot \nu_N + 0 \cdot \nu_O = 1 \cdot \nu_X + 1 \cdot \nu_P + 1 \cdot \nu_C + 0 \cdot \nu_W \tag{5.6}$$

$$H: \quad a_S \cdot \nu_S + a_N \cdot \nu_N + 0 \cdot \nu_O = a_X \cdot \nu_X + a_P \cdot \nu_P + 0 \cdot \nu_C + 2 \cdot \nu_W \tag{5.7}$$

$$O: \quad b_S \cdot \nu_S + b_N \cdot \nu_N + 2 \cdot \nu_O = b_X \cdot \nu_X + b_P \cdot \nu_P + 2 \cdot \nu_C + 1 \cdot \nu_W \tag{5.8}$$

$$N: \quad c_S \cdot \nu_S + c_N \cdot \nu_N + 0 \cdot \nu_O = c_X \cdot \nu_X + c_P \cdot \nu_P + 0 \cdot \nu_C + 0 \cdot \nu_W. \tag{5.9}$$

After arranging we obtain four equations with seven unknowns:

$$1 \cdot \nu_S + \delta_N \cdot \nu_N + 0 \cdot \nu_O - 1 \cdot \nu_X - 1 \cdot \nu_P - 1 \cdot \nu_C - 0 \cdot \nu_W = 0 \tag{5.10}$$

$$a_S \cdot \nu_S + a_N \cdot \nu_N + 0 \cdot \nu_O - a_X \cdot \nu_X - a_P \cdot \nu_P - 0 \cdot \nu_C - 2 \cdot \nu_W = 0 \tag{5.11}$$

$$b_S \cdot \nu_S + b_N \cdot \nu_N + 2 \cdot \nu_O - b_X \cdot \nu_X - b_P \cdot \nu_P - 2 \cdot \nu_C - 1 \cdot \nu_W = 0 \tag{5.12}$$

$$c_S \cdot \nu_S + c_N \cdot \nu_N + 0 \cdot \nu_O - c_X \cdot \nu_X - c_P \cdot \nu_P - 0 \cdot \nu_C - 0 \cdot \nu_W = 0. \tag{5.13}$$

The vector notation for this system of equations is

$$B \times N = 0, \tag{5.14}$$

where B is the matrix of elementary coefficients:

$$B = \begin{bmatrix} 1 & \delta_N & 0 & -1 & -1 & -1 & 0 \\ a_S & a_N & 0 & -a_X & -a_P & 0 & -2 \\ b_S & b_N & 2 & -b_X & -b_P & -2 & -1 \\ c_S & c_N & 0 & -c_X & -c_P & 0 & 0 \end{bmatrix}, \tag{5.15}$$

and N is a vector of balance coefficients:

$$N = [\nu_S \nu_N \nu_O \nu_X \nu_P \nu_C \nu_W]. \tag{5.16}$$

Because there are four equations and seven unknowns, we need to introduce three additional relationships to solve. To this end, the three so-called yield coefficients may be introduced, whose values are determined experimentally.

The biomass yield coefficient relative to the substrate:

$$y_{XS} = \frac{v_X}{v_S}. \tag{5.17}$$

The product yield coefficient relative to the substrate:

$$y_{PS} = \frac{v_P}{v_S}. \tag{5.18}$$

The product yield coefficient relative to the biomass:

$$y_{PX} = \frac{v_P}{v_X}. \tag{5.19}$$

It must be distinguished between the molar yield coefficients and the mass yield coefficients, however they can be mutually converted according to the definition:

$$Y_{XS} = \frac{m_X}{m_S} = y_{XS}\frac{M_{C,X}}{M_{C,S}}. \tag{5.20}$$

Example 5.1. Calculate the molar ratio of biomass yield relative to the substrate Y_{XS} if we know the mass yield coefficient $Y_X = 0.5$ g/g, for yeast grown on glucose.

Solution:
stoichiometric formula of the glucose: $C_6H_{12}O_6$;
C-mol elemental formula: CH_2O;
molecular mass of the glucose: $M_{CS} = 12 + 2 + 16 = 30$ g/C-mol;
elemental formula of yeast according to Roels: $CH_{1,8}O_{0,5}N_{0,2}$;
molecular mass of yeast according to Roels:

$$M_{CX} = 12 + 1 \times 1.8 + 16 \times 0.5 + 14 \times 0.2 = 12 + 1.8 + 8 + 2.8 = 25.9$$

$$y_{XS} = Y_{XS}\frac{M_{C,S}}{M_{C,X}} = 0.5\frac{30}{25.9} = 0.58\frac{\text{C-mol}}{\text{C-mol}}.$$

5.2.2.1 Calculation of the consumption of nitrogen in biochemical reactions

The amount of nitrogen consumed can be calculated from the general reaction equation:

$$v_S CH_{a_S}O_{b_S}N_{c_S} + v_N C_{\delta_N}H_{a_N}O_{b_N}N_{c_N} + v_O O_2 \rightarrow$$
$$v_X CH_{a_X}O_{b_X}N_{c_X} + v_P CH_{a_P}O_{b_P}N_{c_P} + v_C CO_2 + v_W H_2O. \tag{5.21}$$

The balance of stoichiometric coefficients for nitrogen is

$$c_S \cdot v_S + c_N \cdot v_N + 0 \cdot v_O = c_X \cdot v_X + c_P \cdot v_P + 0 \cdot v_C + 0 \cdot v_W. \tag{5.22}$$

After rearranging this equation we can arrive at the following formula:

$$\frac{v_N}{v_X} = \frac{1}{c_N}\left(c_X - c_S \cdot \frac{v_S}{v_X} + c_P \cdot \frac{v_P}{v_X}\cdot\right) \tag{5.23}$$

Taking the yield coefficients which were introduced earlier into account, the next equation can be expressed as

$$\frac{v_N}{v_X} = \frac{1}{c_N}\left(c_X - \frac{c_S}{y_{XS}} + c_P y_{PX}\right), \tag{5.24}$$

where:

y_{XS} – yield coefficient of the biomass with respect to the substrate:
y_{PX} – yield coefficient of the product with respect to the biomass.

5.2.2.2 Calculation of the consumption of oxygen in biochemical reactions

Complete oxidation of biochemical compounds can be expressed by the following reaction:

$$C_\delta H_a O_b N_c + nO_2 \rightarrow \delta \cdot CO_2 + \frac{1}{2}a \cdot H_2O + \frac{1}{2}c \cdot N_2. \tag{5.25}$$

The balance of the oxygen atoms in the reaction leads to a simple algebraic equation:

$$b + 2n = 2\delta + \frac{1}{2}a. \tag{5.26}$$

This shows the number of molecules of oxygen required to oxidize any substance:

$$n = \frac{1}{4}\left(4\delta + a - 2b\right) \tag{5.27}$$

Each oxygen molecule consists of two atoms and each atom takes two electrons during the oxidation reaction (and its reduction). Thus the number of electrons passing to 1 molecule of oxygen as a result of the complete oxidation of a given substance "i" are shown in parentheses. The number of electrons is called the reduction factor:

$$\Gamma_i = \left(4\delta_i + a_i - 2b_i\right). \tag{5.28}$$

The exchange of electrons between particles participating in the reaction can be noted in each biochemical reaction. Based on the overall biochemical reaction, one can write the balance of electrons thus:

$$v_S \cdot \Gamma_S + v_N \cdot \Gamma_N + v_O \cdot \Gamma_O = v_X \cdot \Gamma_X + v_P \cdot \Gamma_S + v_C \cdot \Gamma_{CO_2} + v_W \cdot \Gamma_{H_2O}. \tag{5.29}$$

For those substances identified in our biochemical reaction, i.e. oxygen, carbon dioxide, and water, the reduction coefficients can be calculated at once:

$$\Gamma_{O_2} = \left(4\delta_{O_2} + a_{O_2} - 2b_{O_2}\right) = 4 \cdot 0 + 0 - 2 \cdot 2 = -4, \tag{5.30}$$

$$\Gamma_{CO_2} = \left(4\delta_{CO_2} + a_{CO_2} - 2b_{CO_2}\right) = 4 \cdot 1 + 0 - 2 \cdot 2 = 0, \tag{5.31}$$

$$\Gamma_{H_2O} = \left(4\delta_{H_2O} + a_{H_2O} - 2b_{H_2O}\right) = 4 \cdot 0 + 2 - 2 \cdot 1 = 0. \tag{5.32}$$

The new balance equation of the electrons of our biochemical reaction is now simpler:

$$v_S \cdot \Gamma_S + v_N \cdot \Gamma_N + v_O \cdot (-4) = v_X \cdot \Gamma_X + v_P \cdot \Gamma_P. \tag{5.33}$$

From this equation, we can calculate the oxygen consumption in the reaction as follows:

$$v_O = \frac{1}{4} \left(v_S \Gamma_S + v_N \Gamma_N - v_X \Gamma_X - v_P \Gamma_P \right). \tag{5.34}$$

To sum up, the balance yield coefficient of oxygen for any biochemical reaction can be calculated on the basis of the reduction factors of the: substrate, nitrogen sources, biomass and product. Nitrogen is also an electron acceptor like oxygen, and therefore a second type of reduction factors, relative only to oxygen, was introduced. These reduction factors are called relative reduction coefficients. In order to remove electrons originating from nitrogen, the relation on the use of nitrogen consumption (which was previously derived) can be substituted to the last formula:

$$v_N = \frac{v_X}{c_N} \left(c_X + c_P \cdot \frac{v_P}{v_X} - c_S \cdot \frac{v_S}{v_X} \right). \tag{5.35}$$

As a result of this substitution one could obtain

$$v_O = \frac{1}{4} \left(v_S \Gamma_S + \frac{v_X}{c_N} \left(c_X + c_P \cdot \frac{v_P}{v_X} - c_S \cdot \frac{v_S}{v_X} \right) \Gamma_N - v_X \Gamma_X - v_P \Gamma_P \right). \tag{5.36}$$

And, after further arrangement and grouping, one can arrive at the equation of the form

$$v_O = \frac{1}{4} \left(v_S \cdot \left(\Gamma_S - \frac{c_S}{c_N} \cdot \Gamma_N \right) - v_X \cdot \left(\Gamma_X + \frac{c_X}{c_N} \Gamma_N \right) - v_P \cdot \left(\Gamma_P + \frac{c_P}{c_N} \cdot \Gamma_N \right) \right). \tag{5.37}$$

Now, in this equation is repeated the expression, called the relative reduction factor:

$$\gamma_i = \Gamma_i - \frac{c_i}{c_N} \Gamma_N \quad \text{for} \quad i = S, X, P. \tag{5.38}$$

This enables the equation of the oxygen demand to be expressed, including nitrogen as the electron acceptor, as

$$v_O = \frac{1}{4} \left(v_S \gamma_S - v_X \gamma_X - v_P \gamma_P \right). \tag{5.39}$$

The relative reduction coefficient for any substance can be calculated based on stoichiometric coefficients:

$$\gamma_i = 4\delta_i + a_i - 2b_i - \frac{c_i}{c_N} \Gamma_N. \tag{5.40}$$

For ammonium salts the formula for the relative reduction coefficient is even simpler:

$$\gamma_i = 4 + a_i - 2b_i - 3c_i. \tag{5.41}$$

It should be noted that the reduction factors testify to the ease with which a substance can be oxidized. Tab. 5.2 shows the reduction factors for selected substances.

Table 5.2. Reduction factors

Substance	Elemental formula	Γ	γ
Biomass	$CH_{1.8}O_{0.5}N_{0.2}$	4.8	4.2
Glucose	CH_2O	4	4
Ethanol	$CH_3O_{0.5}$	6	6
Citric acid	$CH_{1.33}O_{1.17}$	3	3
Methane	CH_4	8	8

5.2.2.3 Limitation of heterotrophic growth resulting from the balance of oxygen

The yield coefficient of oxygen consumption relative to biomass Y_{OX} can be expressed based on the balance coefficient which was previously derived, to arrive at

$$y_{OX} = \frac{v_O}{v_x} = \frac{1}{4}\left(\frac{v_S}{v_X}\gamma_S - \gamma_X - \frac{v_P}{v_X}\gamma_P\right) \tag{5.42}$$

After taking the yield factors into account (y_{XS} – biomass relative to the substrate and y_{PX} – the biomass product), one can arrive at a working form of the equation:

$$y_{OX} = \frac{v_O}{v_x} = \frac{1}{4}\left(\frac{\gamma_S}{\gamma_{XS}} - \gamma_X - y_{PX}\cdot\gamma_P\right). \tag{5.43}$$

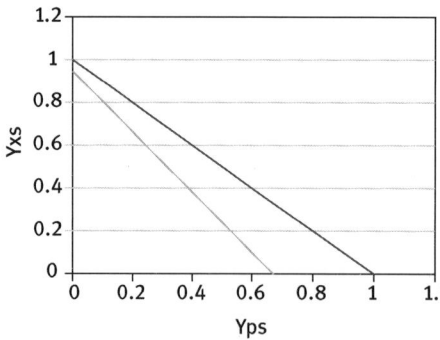

Fig. 5.5. Limitation of heterotrophic growth of the yeast Saccharomyces Cerevisiae (glucose-ethanol) resulting from lack of oxygen (lighter line), andlimitation of carbon (darker line).

5.2.2.4 Limitation of heterotrophic growth resulting from the balance of carbon

The balance of stoichiometric coefficients at carbon atoms gives

$$1 \cdot v_S + \delta_N \cdot v_N + 0 \cdot v_O = 1 \cdot v_X + 1 \cdot v_P + 1 \cdot v_C + 0 \cdot v_W, \tag{5.44}$$

and after simplifications,

$$v_S + \delta_N \cdot v_N = v_X + v_P + v_C. \tag{5.45}$$

Dividing by the coefficient of the substrate balance gives the equation

$$1 = \frac{v_X}{v_s} + \frac{v_P}{v_s} + \frac{v_C}{v_s}. \tag{5.46}$$

Since $\frac{v_C}{v_S} \rangle 0$, this equation can be replaced by an equivalent inequality

$$y_{XS} \leq 1 - y_{PS}. \tag{5.47}$$

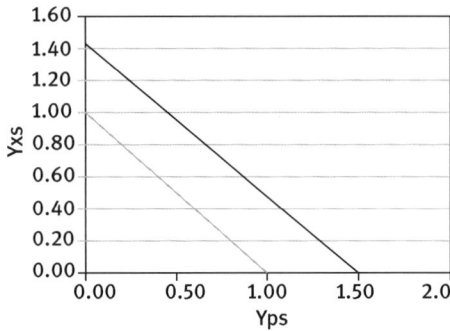

Fig. 5.6. Limitation of heterotrophic growth of the bacteria Acetobacter on glucose resulting from lack of carbon (lighter line) and limitation of the oxygen (darker line).

5.2.3 The energy issues of microbial growth

Assuming that enthalpy at standard conditions (25 °C, 1 bar) is equal to zero, the thermodynamic functions of combustion are proportional to the reduction coefficients:

Heat of combustion:

$$\Delta h_C^0 = 115 \cdot \Gamma \quad [\text{J/mol}]. \tag{5.48}$$

The amount of heat emitted during the growth of a substance can be estimated based on the enthalpy of combustion:

$$Q = v_S \cdot \Delta h_{C,S}^0 + v_N \cdot \Delta h_{C,N}^0 - v_X \cdot \Delta h_{C,X}^0 - v_P \cdot \Delta h_{C,P}^0 \quad [\text{J}] \tag{5.49}$$

$$Q = 115 \cdot (v_S \cdot \Gamma_S + v_N \cdot \Gamma_N - v_X \cdot \Gamma_X - v_P \cdot \Gamma_P) \quad [\text{J}] \tag{5.50}$$

The demand for oxygen is equal to the contents in brackets, which has previously been established (see Eq. (5.34)), and thus

$$Q = 115 \cdot 4 \cdot v_0 = 460 \cdot v_0. \tag{5.51}$$

The experimentally determined enthalpy of combustion of various substances accurately confirms the linear relationship between the molar enthalpy of combustion and the reduction coefficient.

The amount of heat generated on the amount of oxygen used can be expressed as another yield coefficient:

$$y_{QO} = \frac{Q}{v_O} = 460 \left[\frac{\text{kJ}}{\text{mol } O_2} \right]. \tag{5.52}$$

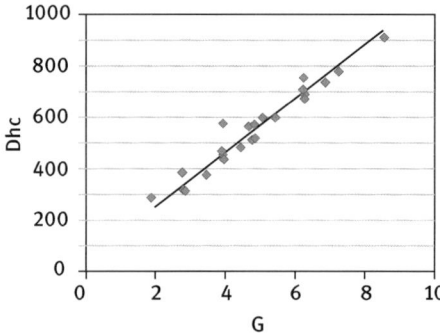

Fig. 5.7. The experimental relationship between the heat of combustion and reduction factors for different substances.

The amount of heat generated based on the amount of substrate used may be expressed thus:

$$y_{QS} = \frac{Q}{v_S} = \frac{Q}{v_O \cdot \frac{v_S}{v_O}} = 460 \cdot \frac{v_O}{v_S} = 460 \cdot y_{OS} \left[\frac{kJ}{mol\,S} \right]. \tag{5.53}$$

The amount of heat generated in relation to the resulting biomass can be described as follows:

$$y_{QX} = \frac{Q}{v_S \cdot \frac{v_X}{v_S}} = 460 \cdot \frac{v_O}{v_S} \cdot \frac{v_S}{v_X} = \frac{Q}{v_X} = 460 \cdot y_{OX} \left[\frac{kJ}{mol_X} \right]. \tag{5.54}$$

Example 5.2. Calculate the thermal productivity of the biomass of yeast cultured on glucose assuming that there is no product. The yield coefficient of the biomass relative to substrate is $Y_{XS} = 0.55$ g/g. The elemental mass of yeast $M_{C,X} = 25.9$ and glucose $M_{C,S} = 30$ (as previously determined).

Solution: First, calculate the molar yield coefficient of biomass relative to the substrate

$$y_{XS} = Y_{XS} \frac{M_{C,S}}{M_{C,X}} = 0.55 \frac{30}{25.9} = 0.64. \tag{5.55}$$

Calculate the coefficient of oxygen consumption of biomass:

$$y_{OX} = \frac{1}{4} \left(\frac{y_S}{y_{XS}} - y_X \right) = \frac{1}{4} \left(\frac{4}{0.64} - 4.2 \right) = 0.51 \frac{mol\,O_2}{C\text{-}mol}. \tag{5.56}$$

Then calculate the coefficient thermal performance of biomass:

$$y_{QX} = \frac{Q}{v_X} = 460 \cdot y_{OX} = 0,51 \cdot 460 = 235 \frac{kJ}{C\text{-}mol}. \tag{5.57}$$

5.3 Enzymes

Enzymes are catalysts which increase the rate of all biochemical reactions. Nearly all known enzymes are proteins, although some ribonucleic acids (found in the cell nucleus and cytoplasm) may also be used as catalysts. Enzymes play an increasingly important role in industry because they enable biotransformations to be integrated in production cycles and work in mild operation conditions, which is important for sustainable technologies [220]. Enzymes are catalysts for metabolic and catabolic natural processes. Many factors affect the activity of enzymes, including temperature, pH, and the concentration of the enzyme, substrate and product. Water is a particularly important component of the enzyme reaction, which is the product of many new bond-forming reactions or the substrate for the reaction, resulting in the breakage of bonds. Enzymes can increase the reaction rate or selectivity, or allow the reaction at lower temperatures. The unit for measuring the activity of the catalyst is Cat = mol/s, the biochemical equivalent is the enzyme unit. The General Conference on Weights and Measures and other international organizations recommend using Cat units in the SI system. For example, one Cat trypsin is the amount of trypsin which causes the breaking of 1 mole of peptide bond within seconds under defined conditions.

The mechanism of the enzymatic reaction can be divided into several stages. First, enzymes are combined with substrates by using the so-called active sites of the substrate molecules to form active complexes.

The active site of the enzyme catalyst combines with the binding site. The structure and chemical properties of the active site allow detection of the type of substrate bond. This is normally a "pocket" or gap surrounded by amino acids or side chains on the surface of the enzyme, which comprises residua responsible for substrate specificity such as charge, hydrophobicity, steric specificity. Residues often operate as proton donorsor acceptors, or are frequently responsible for binding cofactors. Many enzymes require additional components to be activated or achieve full activity. Such non-protein optional ingredients are called co-factors.

The concept of step, which determines the rate of reaction, is very important for the optimization and understanding of many chemical processes including catalysis. This is the slowest step among the mechanisms leading to a chemical reaction, namely "the bottleneck". The formation of the active complex is the most common reaction rate determining step. However, the bottleneck may also be the rate of diffusion. This occurs especially where the reaction rate is very high and the substrates are slowed down by diffusion. Reactions where the active complex and the products are developed quickly are always controlled by mass transport. Knowledge of process engineering is particularly useful in such cases. The limiting step may be evaluated based on the mixing effect. If the impact of mixing is distinct, then the rate-limiting step (of enzymatic reaction) is mass transport, otherwise it is the reaction rate.

Knowledge of enzyme structure is helpful in understanding and interpreting data on reaction kinetics. The structure of the enzyme may even suggest how the reactants and products are bonded during catalysis, which changes occur during the reaction and the role of the particular amino acid residues. Some enzymes change shape in a very significant way. It is possible to recognize the construction of enzymes capable of forming a complex with substrates and without them through observation.

Daniel Koshland proposed modifying the "lock and key" model in 1958. Since enzymes are normally quite flexible structurally, the active center is subjected to a continuous spatial rearrangement during interaction with the substrate [221]. Enzymatic catalysis can only proceed if the side groups of amino acids are subjected to rearrangement through accurate fitting to the spatial position of the substrate. The active site continues rearrangement until the final matching stage with the substrate. The initial substrate-enzyme reaction based on a weak bond leads to the formation of the active complex. However, this leads to poor bonding of the conformational changes which results in the strengthening of the bond. There are two types of substrate-enzyme bonds:
1. homogeneous binding, which has a strong substrate bond;
2. differential binding which has a strong bond transition state.

The stabilizing effect of homogeneous binding increases the affinity for both the substrate and the intermediate state. These effects lead to the reduction of ΔG, and therefore most of the proteins have a high affinity to the enzyme in the transition state.

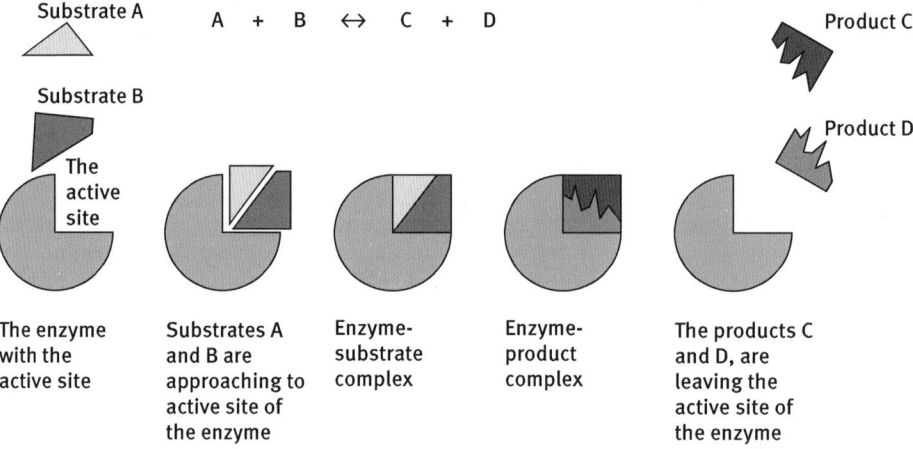

Fig. 5.8. Conceptual diagram of the enzymatic reaction.

After the reaction when bonding has already occurred, one or more of the active catalyst mechanisms reduce the energy of the transition state of the reaction by providing an alternative reaction. There are six possible mechanisms to overcome the energy barrier:
1. catalysis by tightening bonds (forcing) [222–224];
2. catalysis by approximation and direction setting [225–227];
3. acid-base catalysis [228];
4. electrostatic catalysis [229, 230];

5. covalent catalysis [231, 232];
6. quantum tunneling [233–240].

Enzymes must be immobilized in order to fulfill their role economically. There are several ways of keeping the enzymes in the reactor. Usually membranes are used for this purpose. Membranes can be used outside the reactor to separate and recycle the free enzyme from the product. They may also be integrated with the enzymes in various ways for immobilization. In such cases the enzymes may be immobilized in the pores of microfiltration or ultrafiltration membranes, in the gel layer, or may be linked by a chemical method. They may also be entrapped in microcapsules with the possibility of supplying the substrates and discharging the products. Most important in this case is to determine where the rate of the enzymatic reaction is controlled. If the reaction is fast, the bottleneck is the rate of mass transport. A convenient measure is to use the Theile modulus [248] (the Thiele modulus expresses the relation between diffusion time to the reaction time):

$$\phi = L \cdot \sqrt{\frac{r_m}{D_e \cdot K_m}},$$
(5.58)

where: r_m – maximum rate of the reaction, K_m – Michaelis–Merten constants, and D_e – effective diffusivity. When $\phi < 1$, diffusion time is shorter than the reaction time, and therefore diffusion can be omitted in considerations about the rate of enzymatic reactions. Otherwise, the mass transport plays an important role.

5.3.1 Michaelis–Menten equation of enzymatic reaction equilibrium

The Michaelis–Mentren empirical equation describes the rate of enzymatic reaction based on its concentration in the substrate. Waveforms of various enzymatic reactions are associated with changes in concentrations of reactants during the enzymatic reaction. At the beginning, the rate of enzymatic reaction is linearly dependent on the amount of product produced. As the concentration of the product increases and that of the substrate decreases, the reaction rate is reduced asymptotically to a constant value. At this time the reaction becomes slower, until exhaustion of the substrates or saturation of the product. Particular attention should be paid to the accuracy of the measurement of the concentration in the initial period where it varies rapidly. There are many methods of measurement for determining the concentration. One of these is spectrophotometry, i.e. observation of the absorbance of samples at different concentrations. This method allows continuous measurement of concentrations of various substances and is characterized by varying the wavelength of light at which light absorption takes place. Similar measurements can be performed using mass spectrometry, a method by which stable isotopes are determines during the conversion of substrate to product. The most accurate method of measurement is the observation of individual molecules of the catalyst under special microscopes in the light of the laser.

Of particular interest is the observation of the fluorescence of specific enzymes and co-factors and their movement during the enzymatic reaction. These studies provide new information on the kinetics and dynamics of a single molecule of the enzyme which is an extension of the traditional study of the behavior of the average performance of the population of many millions of particles.

The Michaelis–Menten equation describes the simple enzymatic reaction as follows:

$$S + E \underset{k_{-1}}{\overset{k_1}{\rightleftarrows}} ES \underset{k_{-2}}{\overset{k_2}{\rightleftarrows}} E + P. \tag{5.59}$$

Initially, the reaction takes place between the enzyme (E) and the substrate (S), thereby forming a complex (ES). The second stage is the decomposition of the complex, which can be very complicated but is generally one rate-limiting step of the reaction and therefore can be treated as a simple catalytic reaction with the rate constant K_{cat}. If the reaction path leads through several intermediate products then the value of the rate constant is a function of many constants corresponding to the intermediate steps. This apparent reaction rate is often called the number of rotations of the cycle. The Michaelis–Menten equation describes the initial reaction rate r_0 depending on the rate constant k_2:

$$r_s = \frac{r_m \cdot C_s}{K_m + C_s}, \tag{5.60}$$

where

$$K_m = \frac{k_2 + k_{-1}}{k_1} \approx K_D, \tag{5.61}$$

$$r_m = k_{cat} C_{E0}. \tag{5.62}$$

The Michaelis–Menten equation is the basis of the kinetics of the enzymatic reaction with a single substrate. This equation is based on two fundamental assumptions, regarding the mechanism of inhibition of the product and the lack of allostericity. The term allostericity comes from the Greek *allos* (other) and *stereos* (object), which means that the central regulatory protein is physically different from the active site. These effectors may slow down or speed up the action of enzymes, and they are referred to allosteric activators or inhibitors.

The concentration of the enzyme-substrate complex, and thus the concentration of the unbonded enzyme vary much slower than the concentration of product and the substrate. Therefore the first assumption is that the change in concentration of the enzyme complex during time is equal to zero:

$$\frac{dC_{ES}}{dt} = 0. \tag{5.63}$$

The second assumption is that the total enzyme concentration does not change with time:

$$C_{E0} = C_E + C_{ES} = \text{const.} \tag{5.64}$$

The constant K_M must be determined experimentally and is defined as the concentration at which the reaction rate is half the maximum speed. This can be verified by substituting $[S] = K_M$ to the Michaelis–Menten equation. If the controlling step of the reaction rate is slower than the degree of dissociation constant ($k_2 \ll k_{-1}$), then the constant (K_m) is simply the degree of dissociation (K_{D0}) of the ES complex. For the case where the substrate concentration C_s is small compared with the constant K_m, then

$$\frac{C_s}{K_m + C_s} \approx \frac{C_s}{K_m}. \tag{5.65}$$

When the amount of an enzymatic complex (ES) produced is small, the enzyme concentration is approximately equal to the initial concentration. In this case, the rate of formation of the product is equal to

$$\text{if} \quad C_s \ll K_m, \quad \text{then} \quad r_s = \frac{k_{cat}}{K_m} C_E \cdot C_s. \tag{5.66}$$

In cases where the production rate depends on the concentrations of enzyme and substrate, the rate constant (k_2/Km), corresponding to pseudo second order reaction can be introduced into the equation. This constant is a measure of enzyme reaction rate. The most efficient enzymes reach a rate of 10^8–10^{10} [$M^{-1}s^{-1}$]. These enzymes are so efficient that they catalyze the reaction each time they encounter a substrate molecule and achieve the highest theoretical limit where diffusion is a limiting step. They are commonly known as perfect enzymes.

Let us consider the enzymatic reaction of the decomposition of urea with urease based on experimental data. The parameters of the reaction obtained by means of best fitting curves are $K_M = 20, 14$ and $r_M = 4.25$ (see Fig. 5.9).

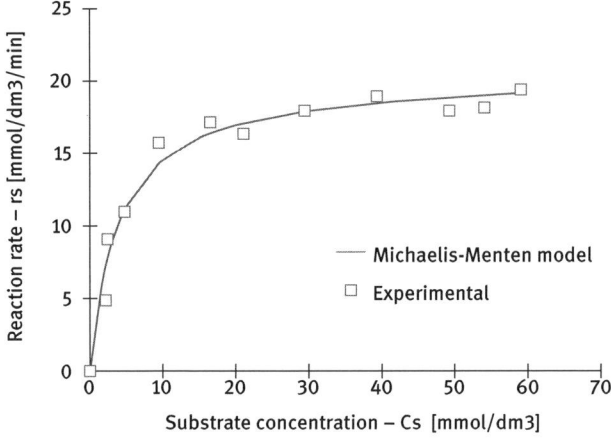

Fig. 5.9. The graphical representation of the Michaelis–Menten equation for decomposition of urea by means of urease.

More information on the mechanisms of enzymatic reactions can be interpreted on the basis of the relevant graph of the enzymatic reaction rate on the concentration of the substrate in the form of straight lines (linearized). Researchers have proposed various methods of linearization function for R_S vs C_S, which would help increase the accuracy of fixed r_m and K_m. All of these methods can be useful for visualizing the experimental data and verification of the assumptions and simplifications. The following are the most common methods of linearization of the Michaelis–Menten equation:
1. Lineweaver–Burk;
2. Eadie–Hofstee;
3. Hanes–Woolf.

The Lineweaver–Burk method (see Fig. 5.10) is the most common method of introducing the inverse to both sides of the Michaelis–Menten equation:

$$\frac{1}{r_s} = \frac{K_m}{r_m \cdot C_S} + \frac{1}{r_m}. \tag{5.67}$$

In this equation, one can easily show parameters $1/r_m$ and $-1/K_m$ as the intersection points of a straight line graph with respective axes graph y = mx + c. During the experiment, there are no zero values of $1/C_s$ or negative values of concentration. The value of $1/C_s = 0$ refers only to very large (approximately infinite) values of concentration of the substrate where $1/r_s = 1/r_m$ and the intersection point with the x-axis is only extrapolated concentrations. The Lineweaver–Burk plot moves importance of the measurement to concentrations of low values.

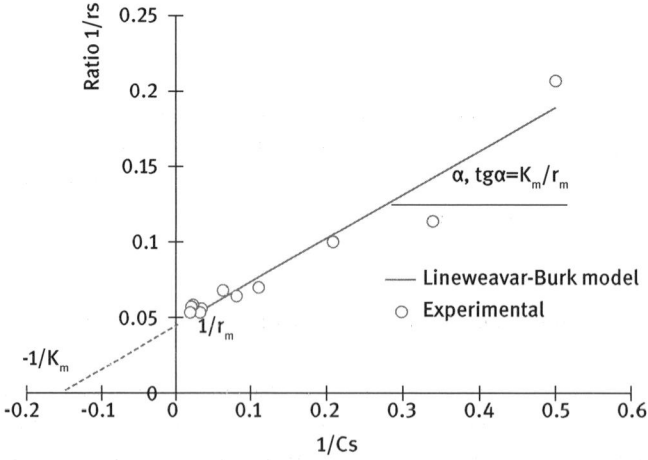

Fig. 5.10. Lineweaver–Burke, or double-reciprocal plots of kinetic data, showing the significance of the axis intercepts and gradient.

Another method is the Eadie–Hofstee plot (see Fig. 5.11), which draws up the value of r_s with respect to r_s/C_s:

$$r_S = r_m - K_m \cdot \frac{r_S}{C_S}. \tag{5.68}$$

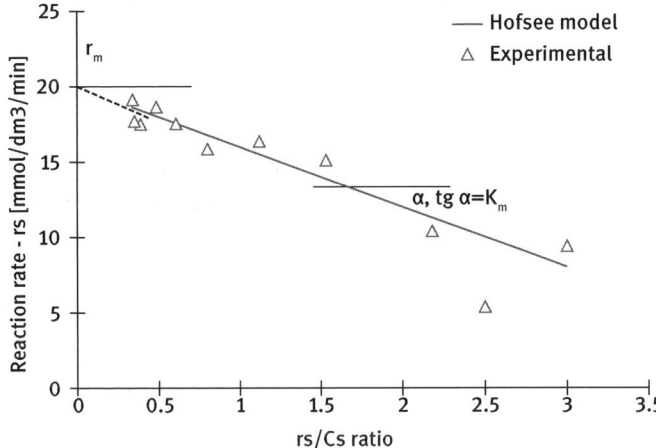

Fig. 5.11. Hofstee linearization.

The Hanes–Woolf plot (see Fig. 5.12) makes the quotient of C_s/r_s of the substrate concentration C_s as follows:

$$\frac{C_S}{r_S} = \frac{1}{r_m} \cdot C_S + \frac{K_m}{r_m}. \tag{5.69}$$

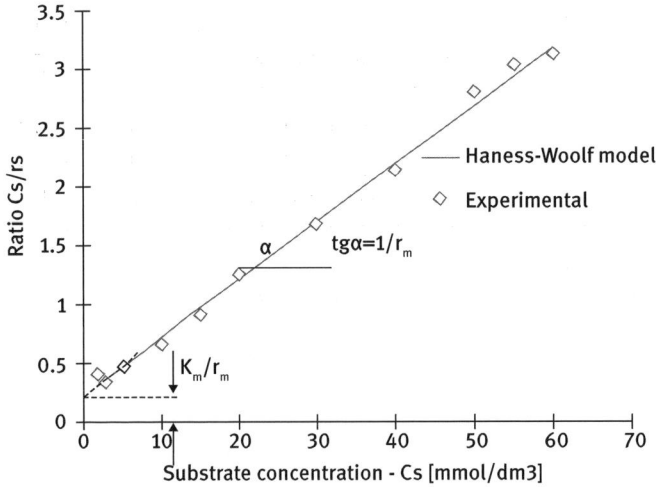

Fig. 5.12. Hanes–Woolf linearization.

In general, all experimental data linearization methods help reduce the workload while increasing accuracy.

The kinetics of enzymatic reactions is important for two reasons. Firstly, it helps to explain the principles of enzyme reaction mechanisms, and secondly to help predict the behavior of the enzymes in living organisms. Kinetic constants are the size of K_m and r_m, which are important to understand the control of the metabolism of living organisms. The Michaelis–Menten equation can also include other more complex mechanisms. Consider the more complicated case of enzymatic reactions in which there is an intermediate product (EI):

$$E + S \underset{k_{-1}}{\overset{k_1}{\rightleftharpoons}} ES \overset{k_2}{\longrightarrow} EI \overset{k_3}{\longrightarrow} E + P. \tag{5.70}$$

In this case, the catalytic reaction equation can be written as

$$r_s = k_{kat} \cdot \frac{C_s \cdot C_{E0}}{K'_M + C_s}, \tag{5.71}$$

where the appropriate constants are different compared to the simple Michaelis–Menten reaction:

$$K'_M = \frac{k_3}{k_2 + k_3} K_M = \frac{k_3}{k_2 + k_3} \cdot \frac{k_2 + k_{-1}}{k_1} \tag{5.72}$$

$$k_{cat} = \frac{k_3 \cdot k_2}{k_2 + k_3}. \tag{5.73}$$

The Michaelis–Menten equation is important again for the limit case, in which the decomposition of the intermediate product of the enzyme complex is very fast ($k_3 \gg k_2$):

$$EI \rightarrow E + P. \tag{5.74}$$

Thus, if the last stage of the decomposition reaction is many times faster than the previous stage, we obtain the previous Michaelis-Menten equation with the corresponding constants:

$$K'_m \approx K_m \quad \text{and} \quad k_{cat} \approx k_2. \tag{5.75}$$

5.3.2 The role of the enzymes in the processes of life

All forms of life i.e. nutrition, multiplication etc. are based on biochemical reactions involving enzymes. A particularly interesting feature of living organisms is that they pass their features on to future generations through a very complex enzymatic catalysis including transcription, translation and replication. The DNA sequence of a gene (through an intermediary RNA) is used for the production of sequence specific proteins. The production of RNA molecules from DNA and gene sequence is called transcription. Relay RNA molecule is then used to produce the appropriate amino acid sequence by a process called translation. Transcription is the rewriting of the information contained in the DNA (matrix) for RNA. DNA replication is a process in which the DNA double strand (double helix) is copied. Replication is semi-conservative, i.e. each of the two resulting double stranded DNA contains one strand of a stem and one new strand. There is a small probability of error in replication (about 1 error per 10^9 nucleotides, compared with transcription error of 1 out of 10^4), and the two DNA molecules are identical. RNA, or ribonucleic acid, is a very important biological protein molecule which consists of a long-chain nucleotide. Nucleotides are the organic chemical compounds from the group of phosphate esters of nucleoside and phosphoric acid. They are the basic structural components of nucleic acids (DNA and RNA). In addition, nucleotides play an important role in metabolism and signal transduction in the cell. ATP (adenosine triphosphate) and GTP (guanosine triphosphate) are the main sources of chemical energy used in chemical reactions in the body. RNA synthesis occurs in the process of enzymatic reactions involving RNA polymerase. RNA polymerase builds a new RNA molecule by combining single nucleotide code contained in the DNA strands.

In order to control all life processes and allow the cells to operate efficiently in both directions, enzymes must be activated under certain conditions and inhibited in others. Inhibitors are molecules which bond with enzymes, reducing their activity. Thus, they can both regulate metabolism as well as kill pathogens, e.g. biocides.

The cofactors are (nonprotein) chemical components which attach to the protein molecules in order to allow biological activity. These active proteins are typically enzymes, and the cofactors are only auxiliary molecules which support biological transformations. Cofactors may be organic or inorganic. If cofactors are associated with weakly bonding enzymes, they are called coenzymes. In the case of strong bonds, they are known as prosthetic groups. Enzymes without cofactors are called apoenzymes, and those with cofactors are called holoenzymes. Vitamins are often organic cofactors.

5.3.3 Industrial importance of enzymes

Enzymes are particularly useful for sustainable applications because of low energy consumption, higher reaction rates, milder reaction conditions and greater stereospecifity. Enzymes may be used in several forms, i.e. suspended in solution and immobilized by membrane or immobilized within the membrane itself.

Protease hydrolysis of proteins can be used to remove organic stains and in the food industry to intensify flavor and aging process.

Lipase is an enzyme which catalyzes the hydrolysis of fats (lipids). Lipase assists in digestion, transport and processing of lipids in all living organisms. Lipase is used in detergents for removing greasy stains, in the fat industry for hydrolysis of fat, in the preparation of polysaturated fatty acids and glycerol. Lipase also has numerous applications in the cosmetic, pharmaceutical and food industries in the preparation of fragrances and flavors. In the chemical industry lipases, amylases and nitrilases are used for the separation of enantiomers of intermediates for pharmaceuticals and agrochemicals. They are also used for the hydrolysis of esters, amides, nitriles.

Amylases, such as glucoamylase, glucose isomerase, and pullulanase are used to remove residues of starch, to liquefy starch, for fragmentation of gelatinized starch, and for saccharification and complete degradation of starch. In the sugar industry amylases are used to convert starch to saccharose, glucose, and fructose and for isomerisation. In the textile industry amylases are used to produce fibers from less valuable raw materials.

Cellulases are used in the detergent industry to restore smooth surfaces and colors of clothes during laundering. In the wine industry they may be used for the breakdown of cell walls, to improve clarification and storage stability, and in the fruit juice industry to improve juice yield and color.

Lactase is mainly used to produce low lactose content milk for special dietary requirements. β-Glucanase may be used to help the clarification process and to reduce glucans and pentosans in animal feed.

Pectinase is used to reduce fruit viscosity and to improve fruit extraction.

Practically, enzymes are used for the following purposes on a large scale:
1. production of L-aspartic acid by Escherichia coli immobilized in polyacrylamides [241];
2. synthesis of dipeptide aspartame using thermolysin [242];
3. production of L-alanine using Pseudomonas dacunhae [243];
4. production of L-amino acids from racemic mixtures [244];
5. hydrolysis of lactose in whey and milk by means of β-galactosidase [245];
6. production of fructose syrups by means of glucose reticulate [246];
7. production of L-malic acid which is a dicarboxylic acid made by living organisms, contributes to the pleasantly sour taste of fruits, and is used as a food additive [247];
8. production of pharmaceuticals.

5.3.4 The industrial importance of microorganisms

The economic importance of microorganisms derives from the fact that bacteria are exploited by humans in biotechnology in a number of beneficial ways. Despite the fact that some bacteria play harmful roles, such as causing disease and spoiling food, the economic importance of bacteria includes both their useful and harmful aspects, including production of chemicals such as ethanol, acetone, organic acid, enzymes, perfumes etc. Among other things, the pure enantiomers used in the chemical, pharmaceutical and agrochemical industries may be produced by bacteria.

Different types of microorganisms are used on an industrial scale for the production of biofuels, alcohols, organic acids, antibiotics, hormones, enzymes, vitamins or amino acids. They have always been used in households (old biotechnology) to coagulate milk, which is the basis for the production of yogurt and cheese, to pickle cucumbers and cabbage and for olive oil. Cyanobacteria are used as a natural fertilizer for their ability to assimilate nitrogen from the air. Biogas is currently produced during the anaerobic digestion of organic matter. The processes of aerobic or anaerobic oxidation of wastewater by microorganisms are necessary for the functioning of the biological treatment plant.

There are two types of microorganism cells: eukaryotes and prokaryotes. Eukaryotic cells are about ten times larger than prokaryotic cells. Unlike prokaryotic cells, the walls of eukaryotic cells do not contain peptidoglycan. Eukaryotes undergo mitosis while prokaryotes reproduce by binary division or simple cell division. The domain eukaryotes Eukaryotes has more than 8.7 million species, but only 2,000,000 are described. Eukaryotes contain chromosomes and organelles in the cell nucleus. Prokaryotes are mostly unicellular microorganisms without a nucleus and organelles. The organelles of eukaryotes enable them to perform intracellular division of labor at a higher level than prokaryotic cells. The genetic material (DNA) is in the nucleus, while the DNA of prokaryotes floats freely in the cell.

Prokaryotic cells use sulfur compounds or simple organic compounds as electron donors, as in this case oxygen is not emitted. This process is known as anaerobic-type photosynthesis. Prokaryotes are the main source of food for the majority of organisms and therefore they take part in the circulation of matter in ecosystems, of all elements throughout the biosphere. A long time ago, prokaryotes were involved in the creation of oil, sulfur, iron ore deposits and the natural deposits of ammonium nitrate.

5.3.5 The industrial importance of photosynthesis

In the discussion of the model of the ecosphere (Chapter 1) it was shown that photosynthesis supports a wide variety of food chains and generally life on Earth in its present form. Photosynthesis is a biochemical process for the production of organic compounds from inorganic matter by cells containing chlorophyll or bacteriochlorophyll, with active participation of light. The process of photosynthesis is very beneficial because it maintains a high level of oxygen in the atmosphere and contributes to the increase in the amount of organic carbon at the expense of inorganic matter. Carbohydrates convert light energy into chemical bond energy with an efficiency rate of 22–33 %. There are only two sources of energy available to living organisms, i.e. sunlight and reduction-oxidation (redox) reactions. Photosynthesis is carried out mainly by plant tissues and cells of cyanobacteria, which can be found in almost every terrestrial and aquatic habitat, both in sweetwater and in the ocean, moist soil, temporarily damp rocks and deserts.

Plant cells are eukaryotic cells, which are different to the cells of other organisms in some key aspects. In plants the leaves are the main organs capable of photosynthesis because they contain cell chloroplasts. Also, prokaryotic cyanobacteria are able to use the photosynthesis process typical for eukaryotes, which allows the degradation of water and releases oxygen. Chloroplasts contain chlorophyll, a green pigment that absorbs sunlight. It allows plants to make their own food through photosynthesis. Other types of plastids are amyloplasts, designed for storing starch, elaioplasts specialize in storing fat, and chromoplasts specialize in the synthesis and storage of pigments. Chloroplasts are composed of lamellar systems, which are granums and intergranums.

Photosynthesis takes place in two stages – the phase of light, which is referred to as the phase of energy conversion. In this stage light is absorbed and its energy is converted into the energy of chemical bonds and oxygen is emitted to the atmosphere as a byproduct. The dark stage is defined as the transformation of the compounds produced in the previous (light) stage and subsequently used for the synthesis of organic compounds. Both the direct products of photosynthesis and some of their derivatives (such as starch and sucrose) are defined as assimilates.

These two stages occur simultaneously. In a simplified form, the synthesis of carbon dioxide and water into sugars and oxygen on exposure to light is as follows [249, 250]:

$$6H_2O + 6CO_2 \xrightarrow{h\nu} C_6H_{12}O_6 + 6O_2 \quad \Delta E = -2872 \text{ kJ/mol}. \tag{5.76}$$

A granum is a flattened stack of thylakoids, in which the "light phase" of photosynthesis occurs. A thylakoid is a membrane-bound compartment inside chloroplasts and cyanobacteria. They are the sites of the reactions of photosynthesis. Photons energize the electrons that come out of the water, thereby by obtaining O_2, with the ultimate purpose of generating the reduced form of nicotinamide adenine dinucleotide phosphate (NADPH) for CO_2 fixation as organic carbon. Nicotinamide adenine dinucleotide phosphate, ($NADP^+$), is a coenzyme used in anabolic reactions, such as lipid and nucleic acid synthesis, which require NADPH as a reducing agent. Nicotinamide adenine dinucleotide phosphate (NADP) is reduced to NADPH. A certain substrate must charge phototrophs to generate adenosine triphosphate (ATP) [251]:

$$2H_2O + 2NADP^+ + 3ADP + 3Pi \rightarrow 2NADPH + 2H + 3ATP + O_2. \tag{5.77}$$

In the dark phase (also known as the Calvin-Benson cycle) the energy stored in ATP and NADPH is used to convert carbon dioxide into simple organic compounds, which can be represented as

$$3CO_2 + 9ATP + 6NADPH + 6H \rightarrow C_3H_6O_3 + 9ADP + 8Pi + 6NADP + 3H_2O. \tag{5.78}$$

Adenosine-5'-diphosphate (ADP) is an organic compound, which is formed from ATP hydrolysis as a result of the transfer of phosphorus residues from ATP to an acceptor (for example glucose or protein).

Adenosine-5'-triphosphate (ATP) is an organic compound which is the universal energy carrier for all organisms. Thus electrons move spontaneously from donor to acceptor through electron transport chains. This process allows additional pumping protons, through the thylakoid membrane, contributing to the chemiosmotic gradient, which is the "driving force" of ATP production. The Pi in Eq. (5.77) denotes the dikinases, i.e. the kind of enzymes (phosphotransferases) that catalyze the chemical reaction.

Thus photosynthesis is mainly used by aerobic phototrophs (plants, algae and cyanobacteria) which are able to produce oxygen. However, anoxygenic phototrophs can also use photosynthesis without oxygen production. Anoxygenic photosynthesis is a process in which light energy is captured and stored in ATP but water is not an electron donor. Anoxygenic phototrophs occur in 10 % of all ocean organisms which are exclusively prokaryotic, in contrast to aerobic phototrophs. Aerobic anoxygenic phototropic bacteria (AAPB) are aerobes which are able to capture energy from light by anoxygenic photosynthesis.

In the case of photosynthetic bacteria such as aerobic phototrophs, for the storage of light energy corresponds to the bacteriochlorophyll, which is analogous to chlorophyll, but with the strongest absorption of infrared light at a wavelength (700–1000 nm). In this case, the electron transfer in photosynthesis is connected with the generation of an electrochemical potential [252]. Energy efficiency is defined as the ratio of energy produced to the resources consumed [253]. For photochemical processes, Einstein's fundamental law of photochemistry establishes that every photon absorbed causes a elementary reaction. The reaction may consist of the chemical transformation of molecules of the substance or in their physical excitation and the emission of energy. The number (N) of reacted molecules is related to the energy (E) absorbed by the system by the equation, where v is the frequency of radiation, C is the velocity of light, λ is the wavelength of light, and h is Planck's constant:

$$E = Nhv = \frac{C}{\lambda}.$$

(5.79)

For example, in order to ionize hydrogen, photons need an energy greater than 13.6 electron-volts, which corresponds to a wavelength of 91.2 nm [254]. For photons with greater energy than this, the energy of the emitted photoelectron is given by

$$\frac{mv^2}{2} = hv.$$

(5.80)

The next important parameter is the photochemical quantum yield, Φ, defined as the ratio of the number of photochemical products to the number of absorbed protons. The quantum yield is a measure of the efficiency of the photochemical process. Its value is equal to unity, which means that every absorbed photon leads to the creation of the products, in contrast to zero value, which means that no products of photosynthesis are formed. The efficiency of energy conversion is estimated to be between 10 and 24 %, whereas a typical steam turbine generator which produces an electrical current has about 30 % efficiency, equal to photovoltaic cells, and 20 % the efficiency of automobiles, whereas the bicycle has about 75 % energy conversion [255–260]. It should be mentioned here that an important limitation of the conversion efficiency of photosynthesis lies in the difficulty of providing appropriate conditions for cultures, which is usually less than 1 %, and is a huge research potential for future biorefinery development.

The problem of low efficiency of photosynthesis is therefore that too much of the light energy converted by the cells in other than the photosynthesis processes is lost. Wild algae use only about 10 % of the photons in full sun for photosynthesis, and thus lose the rest of the photon energy for heat and fluorescence. The second reason for the low energy efficiency of photosynthesis is the loss of dehydrogenase activity under the influence of the oxygen produced [261].

5.4 Biorefineries

5.4.1 Biorefinery principles

According to the International Energy Agency (IEA) definitions,

"Biorefinery is the sustainable processing of biomass into a spectrum of marketable products (food, feed, materials, chemicals) and energy (fuels, power, heat)" [262, 263].

The IEA is an autonomous intergovernmental organization based in Paris. It was established within the framework of the organization for Economic Cooperation and Development (OECD) in 1974 after the oil crisis of 1973. The role of the IEA is responding to physical disruptions in oil supply, as well as a source of information on the statistics of the international oil market and other energy sectors.

Biorefineries are the future in the industry and energy sectors, as they embody all of the conditions for sustainable development. First of all, they do not use fossil fuels, and do not emit of any wastes to the environment because they utilize all wastes as substrates for new products. Biorefineries are safe for the environment by further converting solar energy which is stored in the plants due to photosynthesis. The significance of biorefineries will be justified entirely when they become even more competitive in relation to conventional production methods. This will be achieved upon fulfillment of a wide variety of conditions which have been described in many chapters of this book. It should be noted, however, that the developments and latest achievements in green chemistry and clean technologies are only the prerequisites.

Thus biorefineries produce multiple fuels, polymers and various commodities from biomass resources. These are possible as a result of advances in the field of bioprocess engineering, which formed advanced processes and pathways allowing the extremely precise separation (such as racemic mixtures) and conversion of various raw materials of biological origin to a wide variety of final products. The spectacular development of biorefineries became a reality as a result of the possibility of obtain-

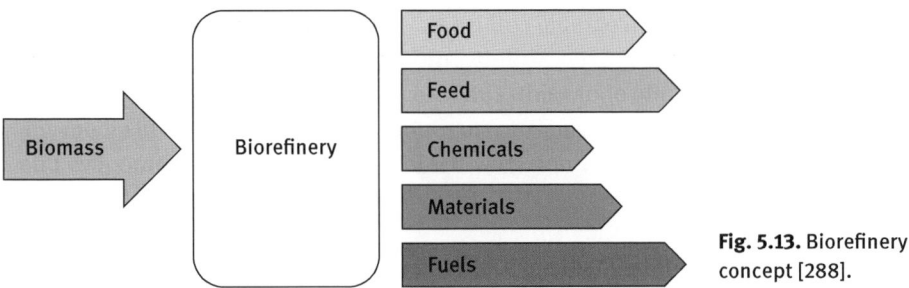

Fig. 5.13. Biorefinery concept [288].

ing many different products ranging from energy and feed products to sophisticated molecular products (see Fig. 5.13).

The design principles of the ideal biorefinery are the following:
1. all fossil raw materials are prohibited;
2. all products must be biodegradable;
3. all processes must apply to green chemistry and to clean technologies;
4. all wastes must be converted into valuable products or energy;
5. maximum flexibility of products.

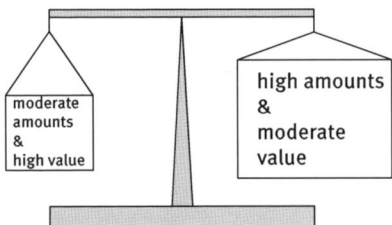

Fig. 5.14. One of the main principles of biorefineries: low price-high volume and high value-low volume.

High production flexibility can be achieved by using a wide range of products of varying value simultaneously (see Fig. 5.14). It is important to apply the principle that a similar benefit can be achieved with low price products produced in larger quantities, as with smaller but highly processed quantities. Biologically renewable raw materials usually contain a lot of very useful substances. In conventional technologies these are introduced to the waste waters, increasing their nuisance value. This is usually the fate of valuable substances such as polyphenols, proteins, sugars and fats in food production, or metals, solvents, coolants. There are no absolutely worthless substances in the world. Only some mixtures are wastes, and the use of separation processes may exponentially increase the values of particular constituents.

Molecular farming uses plants to produce therapeutic proteins in animal feed byproducts and for nonfood production. Hitherto such systems were known based on microbial cells, animal cells or transgenic animals. Plants provide cheap and convenient products on a large scale in the production of valuable recombinant proteins without compromising product quality and safety. The advantage is that they are known plant pathogens capable of infecting humans. Biopharmaceutical proteins include antibodies, vaccines, blood products, human growth hormones and other valuable intermediates such as itaconic acid, lysine and others. Large-scale process integration will maximize profits by diversifying production of bulk materials such as fuel, fodder and energy to materials with high added value as chemicals and pharmaceuticals.

Different levels of biorefineries are possible, from small biorefineries, located directly at rural biomass producers, to large manufacturing companies, located near large cities. Biorefineries should always adjust their range and scale of production to meet local needs, continually diversifying their production. There is great scientific potential for process engineering regarding such issues as optimization of photobioreactors, optimization of separation processes for recycling, diversification of products in the production of biofuels, diversification of raw materials, the use of various types of wastes, and the search for new technologies for manufacturing biofuels. For example large amounts of glycerol are produced during the production of biodiesel, which could be a raw material for other industries such as the pharmaceutical industry, where glycerin is used to prepare medicines for the extraction of herbs for lowering osmotic pressure in the cells, in the manufacture of syrups and elixirs. In the food industry glycerin is used to retain moisture or as a solvent for aromatic substances. It can also be used to soften casings for meats, cheeses, cakes and emulsifiers. Glycerin is used in the textile industry to soften fibers. In the chemical and cosmetic industries glycerin is used in the manufacture of anhydrous alcohol, paints, coatings and polymers. In cosmetics, glycerin is used as a moisturizing and softening component of soaps, toothpaste and hair care products. The addition of glycerin to soap increases its foaming properties.

Esters of fatty acids can also be important for other applications, such as
1. solvents, paints;
2. degreasing substances;
3. cleaning agents;
4. pesticides;
5. drilling muds;
6. modifiers, lubricants, greases and fuels;
7. surfactants, used in the cosmetic, food and fuel industries;
8. fuel additives [284].

So far, biorefineries have been used for different purposes. One of them was management of various types of waste from various sectors of the economy including the municipal waste (garbage), waste from crops and animal production and processing in the food industry [264]. The main motive for the development of biorefineries has been advances in biotechnology, which use the various elements such as enzymes, bacteria, algae, fungi and tissue cells for production. Biotechnology has contributed to major advances in various areas of production in terms of energy savings because most biological processes are carried out at moderate temperature conditions. At the same time, biotechnology reported success in protecting the environment in general use for wastewater treatment. However, it seems that the primary use of biorefineries should be the need for sustainable development in production. The priority should be the use of modern solutions which are affordable, energy-saving, efficient and consume cheap raw materials. They should be waste-free because they do not emit waste but recycle all as a result of the precise separation of material streams (including water), and thus will always be safe for the environment. Reduction of energy and resources should of course be taken into account when choosing a technology, but they are not the only criteria for selection. Low cost of the production can also be achieved by integrating production systems where waste generated in one line, can be the raw material for another production line at the same time. Such production requires diversification of final products; their number should be as large as possible, to allow for the elimination of waste management. Instead, an increased number of operations and processes should be expected. It is difficult to implement because it complicates the manufacturing process, but also profitable because of the greater flexibility of production portfolio and scales. Flexibility relates to the possibility of easy changes in the amounts of individual products, for instance certain products can be produced in larger quantities, whereas production of others may be temporarily reduced or even abandoned. For example, the value of potato starch is comparable to the value of unique proteins present just under the skin of the potato. In traditional technology, this small amount of protein is discarded as waste and very burdensome in wastewaters. The use of ultrafiltration in this case could significantly increase potato processing profitability. Similar results can be obtained in the food and the wine industry in the processing of fruits and olives. On the one hand, polyphenols are very problematic, resulting in inhibition of fermentation in biological wastewater treatment plants; on the other hand they are very valuable food ingredients. It has been found on several occasions that polyphenols have a very good effect on human health, affecting many diseases such as cancer, bacterial infections and hypertension. It is only necessary to use the appropriate method of separation (ion exchangers for example). Another example is lysozyme, one of the enzymes of egg white, which has the ability to damage the walls of bacteria and can be easily isolated. It may therefore be successfully used as a natural food preservative or aseptic agent for many applications. When eggs are dried industrially, water is removed using membranes. During ultrafiltration the small lysozyme molecules (about 14 kDa) pass easily through the membrane into the permeate. This then makes it difficult to recycle the water. The recovery of lysozyme is easy and significantly increases the profitability of the process. The use of microreactors to produce fine chemicals has increased potential flexibility. The application of hybrid processes which use in parallel the several similar separation processes simultaneously has similar potential. Many of are both more efficient and cheaper. Combining distillation and pervaporation processes is much more efficient compared to the individual processes.

Elasticity can be realized by different production routes. For instance the bioconversion of organic waste can be achieved by transforming it into methane in anaerobic digestion. Hydrogen, the cleanest source of energy, can be also produced from the methane. However it is easier to get the hydrogen through algae photo-bioreactors for aerobic and anaerobic bioreactors. Hydrogen can also be obtained during the photo-catalytic reaction.

5.4.2 Biorefinery products

Biorefinery products are: food, feed, chemicals, materials and biofuels (see Figs. 5.15 and 5.16). Nevertheless today in biorefineries primarily only sugars are produced from biomass [268, 269]. The most common intermediates are: hemicellulose (pentosans-xylan) and cellulose (hexosans-glucan). As a result of various enzymatic reactions, these sugars are subjected to further processing in order to obtain other products. The enzymes used for this purpose are xylose isomerases, belonging to the family of oxidoreductases, aldoses and ketoses. A hemicellulose (also known as polyose) is any of several heteropolymers (matrix polysaccharides), such as arabinoxylans, present along with cellulose in almost all plant cell walls. While cellulose is crystalline, strong, and resistant to hydrolysis, hemicellulose has a random, amorphous structure with little strength. It is easily hydrolyzed by dilute acid or base as well as hemicellulase enzymes. Lignin is one of the most abundant organic polymers on Earth, exceeded only by cellulose, employing 30 % of nonfossil organic carbon and constituting from a quarter to a third of the dry mass of wood. As a biopolymer, lignin is unusual because of its heterogeneity and lack of a defined primary structure. In the process of bio-refining, specific components such as proteins, mainly in the various forms of carbo-hydrate, are recovered from plants and then undergo further processing. A biorefinery should either produce the products in small quantities, but with a high value or low-value products on a large scale. Products with a high value but in small quantities allow for additional profits due to the reduction of energy and production costs. By producing a plurality of products, biorefineries use various biomass components and their intermediates. Biorefinery products should eliminate the products made from minerals completely, such as gas, oil, coal and ore (see Tab. 5.3).

The bio-based products, such as amino acids, lipids, organic acids, have been used for a long time in the production of pharmaceuticals, vitamins, cosmetics, food and feed, and detergents. Currently, the newest substrates for the production of biopolymers are organic materials such as the following acids: fumaric, malic, succinic, itaconic acid, and the 1,3-propanediol [272]. Although biobased polymers and plastics are still in their infancy, this industry has been growing rapidly, driven by new synergies and collaborations. The global production capacity for biobased polymers was estimated to be 360 million tons in 2007, with an annual market growth rate of 48 % in Europe and 38 % globally. Recent developments in processes combining biotechnology and chemical synthesis are leading to the emergence of materials such as ethylene from bio-ethanol, but their higher production cost currently limits their commercial viability to some specific sectors such as packaging [270, 271].

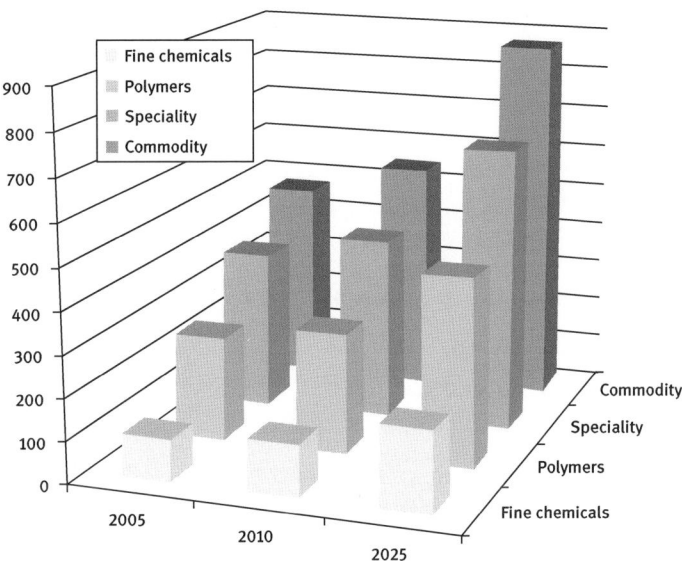

Fig. 5.15. Predicted global conventional production of chemicals.

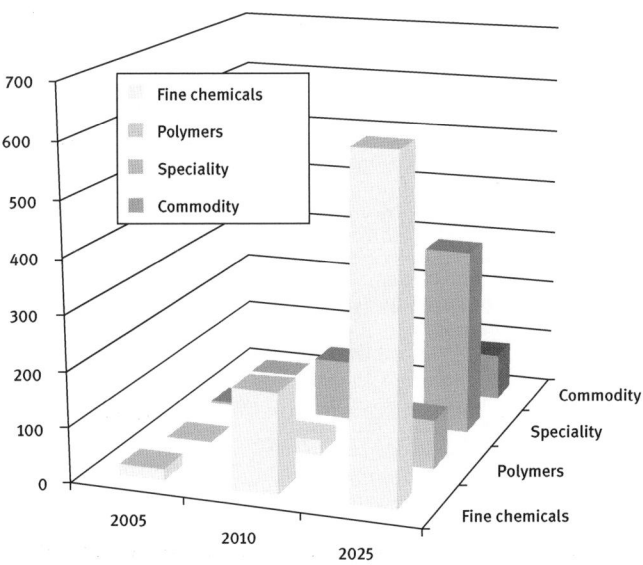

Fig. 5.16. Predicted global biobased production of chemicals in billions of USD [272].

Table 5.3. Fermentation products

Fermentation product	Typical organism used	(kg/year)
Ethanol (nonbeverage)	Saccharomyces cerevisiae	2×10^{10}
Acetone/butanol	Clostridium acetobuticilum	2×10^{6}
Yeast for food and agriculture	Lactic acid bacteria/ yeast	5×10^{8}
Single cell protein	Pseudomonas methylotrophus	$0.5–1 \times 10^{8}$
Single cell protein	Candida utilis	$0.5–1 \times 10^{8}$
Citric acid	Aspergillus niger	$2–3 \times 10^{8}$
Gluconic acid	Aspergillus niger	5×10^{7}
Lactic acid	Lactobacillus delbrueckii	2×10^{7}
Itaconic acid	Aspergillus itaconicus	
L-glutamic acid	Corynebacterium glutamicum	3×10^{8}
L-lysine	Brevibacterium flavum	3×10^{7}
L-phenylalanine	Corynebacterium glutamicum	2×10^{6}
L-arginine	Brevibacterium flavum	2×10^{6}
Steroids	Rhisopus arrhisus	
D-sorbitol to l-sorbose (vit. C)	Acetobacter suboxydans	4×10^{7}
Penicillins	Penicillinum chrysogenum	$3–4 \times 10^{7}$
Cephalosporins	Cephalosporinum acremonium	1×10^{7}
Tetracyclines (7-chlortetracycline)	Streptomyces aureofaciens	1×10^{7}
Macrolide (erythromycin)	Streptomyces erytheus	2×10^{6}
Polypeptide (gramicidin)	Bacillus brevis	1×10^{6}
Aminoglycoside (streptomycin)	Streptomyces griseus	
Xanthan gum	Xantamonas campestris	5×10^{6}
Dextran	Leuconostos mesenteroides	small
5'-guanosine monophosphate	Brevibacterium ammoniagenes	1×10^{5}
Proteases	Bacillus spp.	6×10^{5}
α–amylase	Bacillus amyloliquefaciens	4×10^{5}
Glucoamylase	Aspergillus niger	4×10^{5}
Glucose isomerase	Bacillus coagulans	1×10^{5}
Pectinase	Aspergillus niger	1×10^{4}
Rennin	Mucor miehei or recombinant yeast	1×10^{4}
Vitamin B_{12}	Propionibacterium shermanii	1×10^{4}
Vitamin B_{12}	Pseudomonas denitrificans	
Riboflavin	Eremothecium ashbyii	
Ergot alkaloids	Claviceps paspali	5×10^{3}
Shikonin pigment	Lithospermum erythrorhizon (plant)	60
β-carotene pigment	Blakesela trispora	
Diphtheria (vaccine)	Corynebacterium diphtheriae	< 50
Tetanus (vaccine)	Clostridium tetani	
Pertussis (whooping cough)	Bordetella pertussis	
Poliomyelitis virus	Monkey kidney or human diploid cells	
Rubella	Baby-hamster kidney cells	
Hepatitis B	Surface antigen in recombinant yeast	
Insulin (therapeutic protein)	Escherichia coli recombinant	< 20
Growth hormone	Escherichia coli recombinant	
Erythropoietin	Recombinant mammalian cells	
Factor VII-C	Recombinant mammalian cells	
Tissue plasminogen activator	Recombinant mammalian cells	
Interferon-α₂	Recombinant Escherichia coli	
Monoclonal antibodies	Hybridoma cells	< 20
Bacterial spores insecticides	Bacillus thuringiensis	
Fungal spores insecticides	Hirsutella thompsonii	

Ethanol is a substrate for biodiesel and is used for fuels and fuel oxygenate additives. Ethanol can form ethyl acetate with acetic acid, which is a green solvent.

Ferulic acid is produced from lignin, which is a byproduct of the paper industry. It is found in the seeds of coffee, apple, artichoke, peanut and orange, as well as in both seeds and cell walls of plants such as rice, wheat, oats, the Chinese water chestnut (Eleocharis dulcis) and pineapple. Ferulic acid glucoside can be found in flaxseed. Ferulic acid, also known as hydroxycinnamic acid, is a precursor of vanillin and other aromatic compounds.

3-Hydroxypropionic acid is used for the production of polymers, biodegradable polyester known as poly (3-hydroxypropionic acid). This is used for contact lenses, absorbing polymers and Sorronna fibers. Sorronna is DuPont's brand, which was named and commercialized in 2000. The fibers are claimed to be soft and extremely stain resistant, plus high in strength and stiffness. Sorronna has been used in the manufacture of clothing, diapers, residential carpets, automotive fabrics and plastic parts. Mohawk Industries is currently the exclusive carpet manufacturer currently making carpets using DuPont Sorronna fiber. Furfural is an organic compound derived from a variety of agricultural byproducts, including corncobs, oat, wheat bran and sawdust. Furfural is an important renewable, nonpetroleum based, chemical feedstock. The hydrogenation of furfural leads to furfuryl alcohol (FA), which is a useful chemical intermediate and which may be further hydrogenated to tetrahydrofurfuryl alcohol (THFA). THFA is used as a nonhazardous solvent in agricultural formulations and as an adjuvant to help herbicides penetrate the leaf structure. Furfural is used to make thermoplastic polymers, solvents and other furan chemicals, such as furoic acid via oxidation [273], and furan itself via palladium catalyzed vapor phase decarbonylation [274]. Furfural is also an important chemical solvent.

Levulinic acid is classified as a keto acid. This is soluble in water and polar organic solvents. It is derived from the degradation of cellulose and is a potential precursor to biofuels [275]. Levulinic acid is the precursor to pharmaceuticals, plasticizers,and various other additives such as pesticides, oxygenates, solvents, polycarbonate resins [276]. Potential biofuels can be prepared from levulinic acid including methyltetrahydrofuran, valero lactone, and ethyl levulinate. Dehydration of levulinic acid gives angelica lactone [277].

Lactic acid is of great importance because it is used in the production of biodegradable polymers used mainly for packaging. In 2006, the global production of lactic acid reached 275,000 tons with an average annual growth of 10 % [278]. Lactic acid has chiral properties because it has two optical isomers. It is used for the production of biodegradable polymers and in pharmacy and medicine. It is most commonly used for fluid resuscitation after blood loss due to trauma, surgery or burns.

Succinic acid is a precursor to some specialized polyesters. Succinic acid was known as the spirit of amber because it was originally obtained from amber by distillation. It is also a component of some alkyd resins [279]. Global production is estimated at 16,000 to 30,000 tons a year, with an annual growth rate of 10 % [280]. In nutraceutical form as a food additive and dietary supplement it is safe and approved by the US Food and Drug Administration [281]. Succinic acid is used in the food and beverage industry, primarily as an acidity regulator. As an excipient in pharmaceutical products it is used to control acidity and, more rarely, in effervescent tablets [282]. Succinic acid is used for production of fibers Lycra and water soluble polymers. Succinic acid is created as a byproduct of the fermentation of sugar. It lends to fermented beverages such as wine and beer a common taste that is a combination of saltiness, bitterness and acidity.

Sorbitol, also known as glucitol, is a sugar alcohol, which the human body metabolizes slowly. It can be obtained by reduction of glucose, changing the aldehyde group to a hydroxyl group. Most sorbitol is made from corn syrup, but it is also found in apples, pears, peaches, and prunes [283]. It is synthesized by sorbitol-6-phosphate dehydrogenase, and converted to fructose by succinate dehydrogenase and sorbitol dehydrogenase. Sorbitol is used for water soluble polymers, antifreeze additives, production of fuels from biomass resources and as amateur rocket fuels.

Xylitol (food additive E967) is used mainly in the food industry to sweeten; primarily chewing gums and candies due to its anti-caries action. It is recommended for diabetics because it is metabolized by a small percentage of insulin. Derivatives of unsaturated resins are used as antifreeze.

5.4.3 Biorefinery substrates

Plant biomass is formed as a result of photosynthesis, in which carbon dioxide contained in the atmosphere is converted into sugar, and by complex metabolic pathways results in other more complex compounds such as proteins, fats and carbohydrates. All forms of biomass have the same basic components – cellulose, hemicellulose and lignin. Agriculture and forests are the potential biomass resources. In Europe fields occupy 42 % of the area and the forests 40 %, which is 170 million hectares. The raw materials for biorefineries can also be obtained from agricultural wastelands, forests and meadows which will also contribute to the reduction of CO_2 in our atmosphere. Aquacultures such as algae and seaweeds are also quite promising. The total potential biomass supply is expected to be between 10 and 60 EJ in 2050. The forestry potential could be 0.9–2.4 EJ/yr. in 2030, while the potential for agricultural residues and other waste could be around 3 EJ/yr.

Agricultural crops containing starch (see Fig. 5.17), sugars and oil are traditionally the main source of biomass. These are cereals such as corn, wheat, barley, oilseeds, potatoes, sugar beets, agricultural wastes, wood clearings, forest trimming and dedicated energy crops such as switchgrass, willow or hybrid poplar. The category "dedicated energy crops" includes agricultural biomass resources such as sugar- and starch-derived biomass (e.g. sugar beet, wheat and maize), oilseed derived biomass (e.g. rapeseed, sunflower) and lignocellulosic crops (C4-grasses or short-rotation coppice, for example willows). In contrast, estimates of waste streams from forestry, agriculture and other organic sources cover a narrower range.

Wood biomass from forests could range from 625 to 898 million m^3 per year in 2030. This is the total production capacity of the EU Forestry, regardless of the application. According to the 2010 final report, in 2030 the energy potential of forests [265] will be from 0.79 to 2.74 EJ after deduction of wood for construction materials. The estimated potential of total biomass production from European woodland in 2010 was 1,277 billion m^3 including bark. Approximately half of this is stem wood (i.e., derived from the main trunk) and the rest consists of logging residues, stumps and woody biomass from early thinning in young forests. The potential is, however, reduced to about 747 million m^3 by various environmental, technical and social constraints. After subtracting the wood used directly or processed to produce fiberboard and other materials, the EU wood report estimates that the remaining biomass has an energy potential of 2.56 EJ (Exajoule, 10^{18} joules). The final report of the BEE project (Biomass Energy Europe) gives a similar estimate (2.6 EJ) for potential energy. The BEE figure takes account of stem wood, primary above-ground residues and primary stump wood. If secondary forestry residues are included, the potential energy content rises to 3.3 PJ. Biomass production is not equally distributed across all EU Member States. Five countries (Sweden, Germany, France, Finland and Italy) produce about 62 % of the total.

Lignocellulosic annual crops. Several other studies [249, 272] clearly show the advantages of a C4 photosynthetic system, where CO_2 forms the four-carbon compound (hence the name C4; for example, miscanthus-bamboo and switchgrass) in terms of dry biomass yield (up to 25–30 t/ha with averages of around 15 t/ha 20). Switchgrass has dominated the scene as a perennial grass suitable for cellulosic ethanol production in the United States. Switchgrass uses C4 carbon fixation, giving it an advantage in conditions of drought and high temperature. Miscanthus is used commercially for small-scale heat and power in Denmark and in the United Kingdom. Other countries working with biomass crops include Hungary, Germany, Austria, Italy and Sweden. In 2006, the European Environment Agency (EEA) [266] estimated additional land availability from arable areas (not including woodland) for the 22 member states of the EU to be from 13–15 Mha in 2010 to 19–25 Mha in 2030. However one study [267] looks as far as 2040, with an estimate of 108.2 Mha of land available for biomass production in the EU-27.

In the European Union, starch is mainly produced from maize, wheat and potatoes, from more than 21 million tons of agricultural raw materials. In Europe, the starch-based biorefinery concept has focused on integrated production for food, feed and nonfood applications, while aiming for zero waste and getting maximum added values from co-products.

By 2030, the global demand is expected to be more than 100 million tons, meaning an average annual growth of 2 %. Further growth is expected through the development of additional biobased products including specialty materials and chemicals. The major bottlenecks hindering the achievement of this by 2030 will be the cost competi-

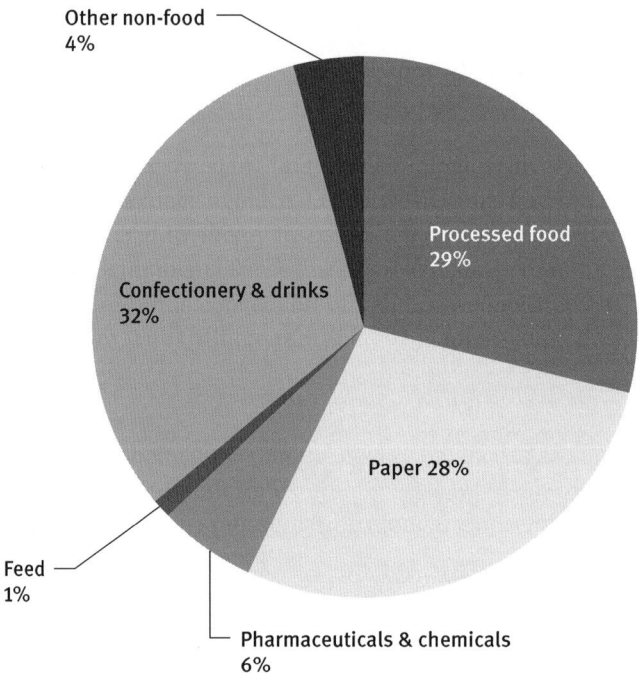

Fig. 5.17. The actual starch use.

tiveness of biobased products made from starch, and scale-up and implementation of the necessary processes. Development will therefore focus on: the efficiency of industrial processes, reduction of energy and water use, the use of starch as a polymer for biobased products

Biorefinery processes of starch (see Fig. 5.18) and sugar (see Fig. 5.19) run using such resources as cereals (e.g. wheat or maize) and potatoes, sugar crops, sugar beet or sugar cane. In Europe, the main application of this biorefinery is currently the production of starch derivatives, ethanol and organic acids, with the protein being used for food and animal feed, according to the section on biorefinery concept. Figure 5.18 shows the processes schematically.

The annual sugar production in the EU is 19–20 million tons. It is produced in almost all member states, although France, Germany and Poland together account for more than half of the total 2.1 million hectares of sugar beet currently cultivated in the EU. Currently, only 2 % of total sugar production is used in nonfood applications. By 2030, the global demand is expected to be around 260 million tons, requiring a 90 million ton increase in global production. Brazil alone (a major sugar cane grower) already accounts for 60 % of the global market.

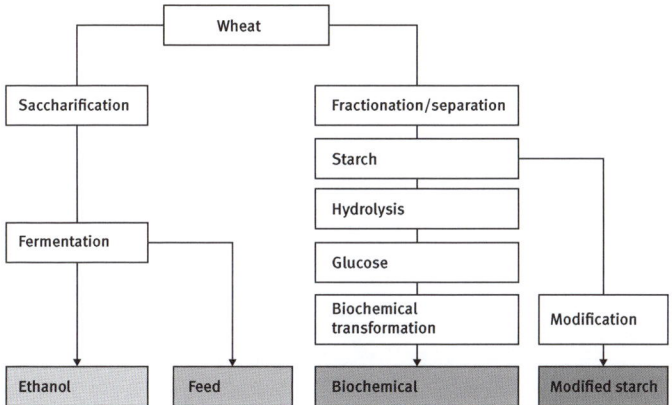

Fig. 5.18. Simplified diagram of starch processing in biorefineries.

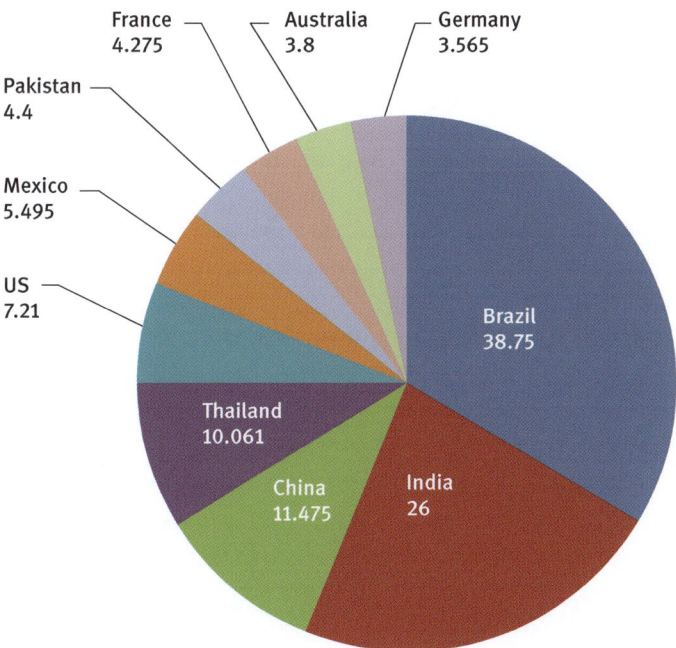

Fig. 5.19. Sugar production in leading countries in millions of tons.

According to the biorefinery concept, biomass products should be diversified to the maximum extent. The main product of the sugar beet– is sucrose, which is a component of food and other products such as a protein mainly suitable for animal feed (see Fig. 5.20). Ethanol is obtained from sucrose as the primary product, but also organic acids, surfactants and other chemicals.

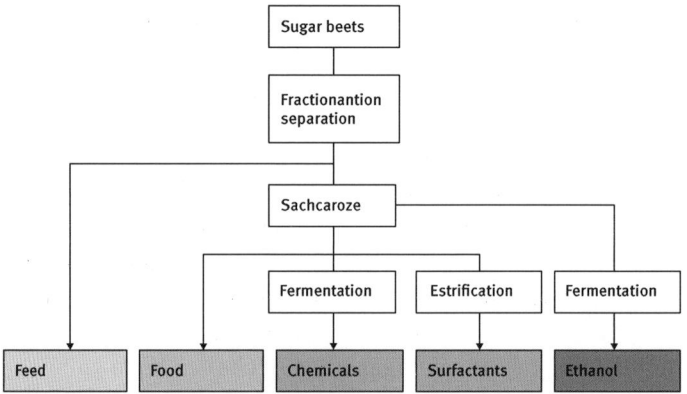

Fig. 5.20. Sugar beet processing scheme [288].

The world production of vegetable oil is expected to increase by almost 40 % from 2010 to 2019. The contribution of vegetables to oils for biodiesel production is predicted to be around 15 %.

Currently, oilseed biorefineries (see Figs. 5.21–5.23) produce mainly food and feed ingredients, biodiesel and oleochemicals from oilseeds including rape, sunflower, soybean and olives. In the first step, oilseeds are crushed, mechanically pressed and then usually treated with chemical solvents to separate the vegetable oil from the protein fraction. The protein fraction is mainly used as animal feed (Biorefinery Euro view [289]). Vegetable oil (or animal fat) is increasingly being used to make biodiesel (fatty acid methyl ester or FAME), via the transesterification process, which produces

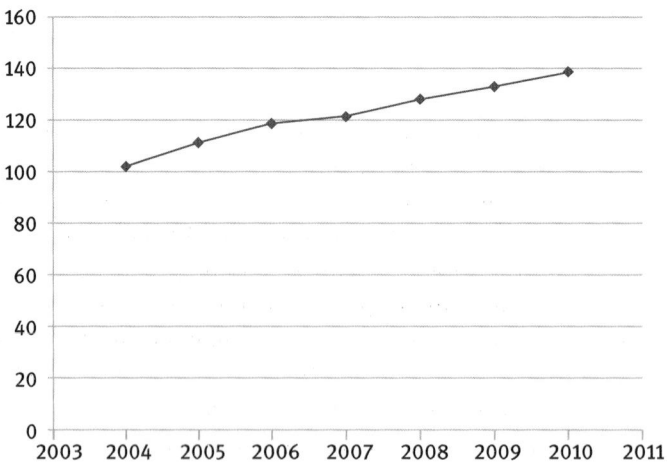

Fig. 5.21. Global vegetable oil production in Mt (millions of tons).

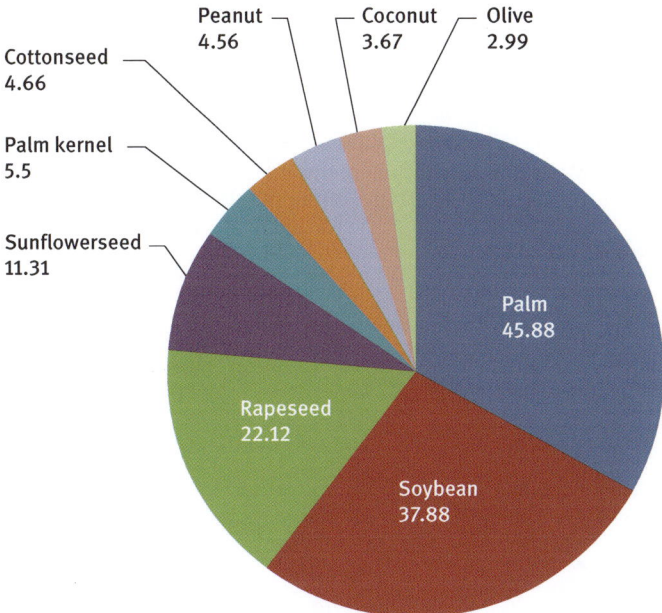

Fig. 5.22. The percentage of resources in the global vegetable oil production (138.57 = 100 %) in 2010.

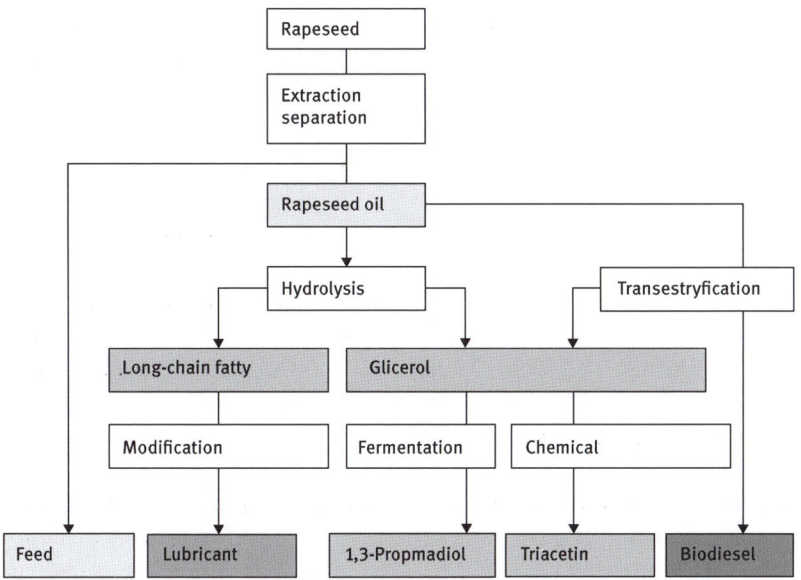

Fig. 5.23. Example of rapeseed processing biorefinery of the German Agency for Renewable Resources (FNR) [288].

glycerol as a by-product. The chemical and enzymatic modification of vegetable oil produces oleochemicals such as fatty acids, alcohols, fatty esters, ketones, dimer acids and glycerol. Oleochemicals are primarily used in personal care products or as raw materials and additives in industrial applications [290] such as textiles, lubricants, household cleaners and detergents, plastic and rubber.

Small-scale biorefineries which produce ethanol, biogas and protein for animal feed or human food are being developed at a scale of 10,000–50,000 tons of primary dry weight inputs [291].

Green biorefineries (see Fig. 5.24) are multi-product systems which handle their refinery cuts, fractions and products in accordance with the physiology of the corresponding plant material, i.e. the maintenance and utilization of the diversity of syntheses achieved by nature [292].

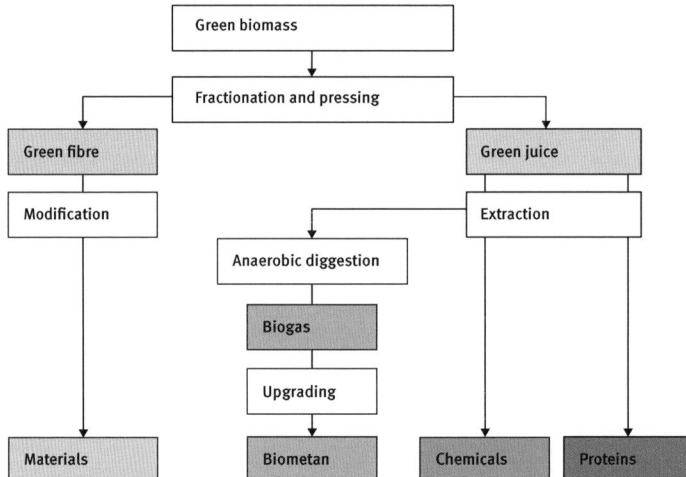

Fig. 5.24. Scheme of green biorefinery.

Green biomass for example includes grass from the cultivation of permanent grassland, closure fields, nature preserves and green crops, such as lucerne, clover and immature cereals from extensive land cultivation. Thus, green plants represent a natural chemical factory and food plant. Green crops are separated into a fiber-rich press cake and a nutrient-rich green juice. Besides cellulose and starch, the press cake contains valuable dyes and pigments, crude drugs and other organics. The green juice contains proteins, free amino acids, organic acids, dyes, enzymes, hormones, other organic substances and minerals. In particular, the application of biotechnological methods is predestined for conversions. Lignin–cellulose composites are not as strong as those in LCF materials. Starting from green juice, the main focus is directed to products such as lactic acid and the corresponding derivatives, amino acids, ethanol and proteins. The press cake can be used for the production of green feed pellets, as a raw material for the production of chemicals, such as levulinic acid, and

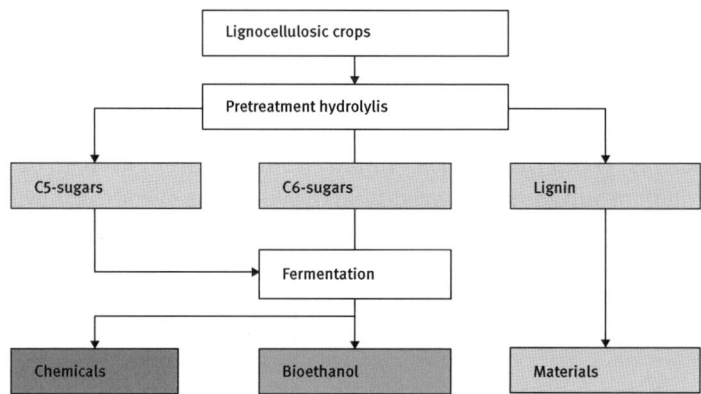

Fig. 5.25. Scheme of biorefinery based on lignocellulosic crops.

for conversion to syngas and hydrocarbons (synthetic biofuels). The residues of a substantial conversion are suitable for the production of biogas, combined with the generation of heat and electricity [288, 293–295].

Lignocellulosic biorefineries have two primary process routes: thermochemical and biochemical (see Figs. 5.25 and 5.26). The biochemical approach (Fig. 5.25) is used primarily for biochemical fractionation refining lignocellulosic raw material into three streams, such as: cellulose, hemicellulose and lignin. These fractions are then subject to separate processes. Cellulose is hydrolyzed to produce sugars, which are then used as the fermentation substrate for the production of alcohols, organic acids and solvents. The hemicellulose fraction is converted to xylose, gelling agents, barrier, furfural, and nylon. Lignins can be used to produce adhesives and glues (glyoxalised lignin), formaldehyde resins, as well as fuels or carbon fibers.

The thermochemical approach involves the gasification of lignocellulosic material, which is then used for the production of fuels and chemicals. Lignocellulosic raw materials are dry agricultural residues such as straw, peels and husks, wood waste, waste paper and lignin.

5.4.4 Algae aquacultures

Microalgae are microorganisms which are able to photosynthesise. They can grow rapidly even in difficult conditions due to their simple structure, being unicellular or multicellular. Solar energy could be transformed very efficiently and directly into valuable biological products by microalga according to the formula:

$$CO_2 + H_2O \rightarrow C_n (H_2O)_n + O_2. \tag{5.81}$$

Biomass (see Fig. 5.27) in the form of microalgae and seaweeds can also be obtained from water. Aquaculture can be characterized by a high yield and a high content of valuable components, including fats, proteins, polysaccharides and other biological

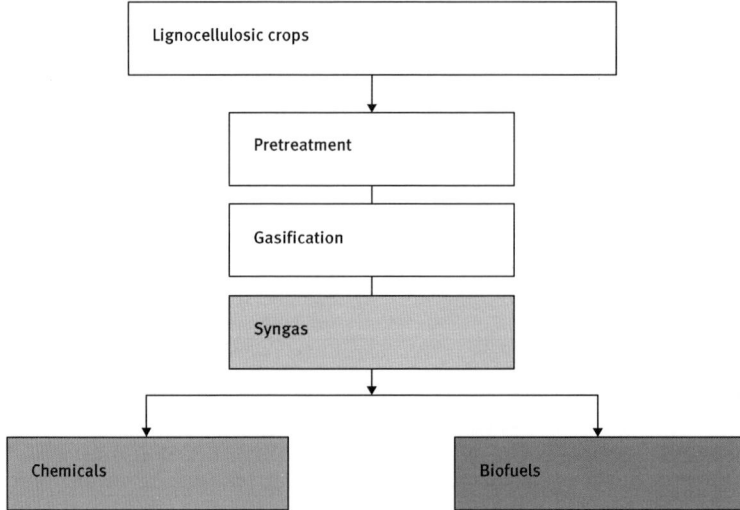

Fig. 5.26. Scheme of biorefinery based on thermochemical approach to lignocellulosic biorefinery.

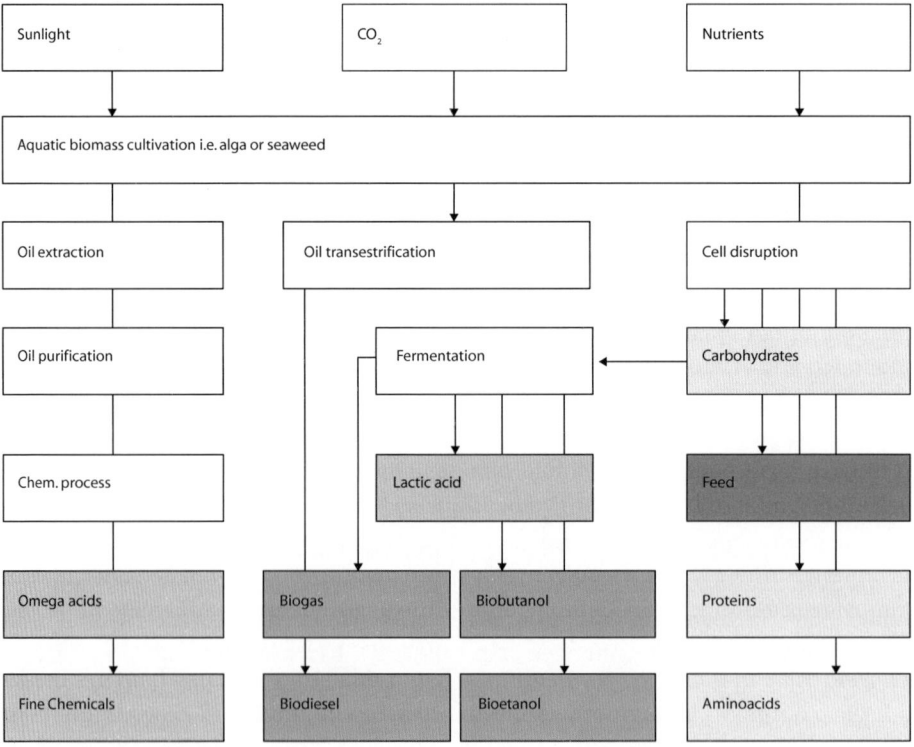

Fig. 5.27. Scheme of biorefinery based on aquaculture.

molecules. The diverse composition of microalgae and seaweed makes them very suitable raw material for biorefining, fuels and chemicals from bulk specialty chemicals and food ingredients and feed.

Microalgae and other microorganisms and plants produce triacyglycerides. These can be used for the synthesis of fatty acid methyl esters, as biodiesel. Microalgae are a very attractive alternative to land-based oil crops due to their superior performance and not being grown on arable land. Furthermore, they have a unique composition, including the ability to gather large amounts of oils, and a great diversity of species and end products. Crop-growing algae may also be combined with water treatment systems, and the use of CO_2 from combustion gases and sources of low temperature waste heat.

Cultivation of seaweed could be a significant source of biomass in the future.

Aquaculture can have a very positive impact on the environment by removing carbon dioxide from the air and excess nutrients which cause eutrophication. To do this, the integration of different technologies is required, which are now treated separately. The use of exhaust gases for cultivation of algae, or the use of excess fertilizers from the surface-waters or wastes from agricultural production is a hope for the future. Feed for aquaculture should ensure the recycling of nutrients and byproducts from food and biofuel production.

The condition is access to water, some minerals, nutrients and CO_2. It should be noted that the efficiency of photosynthetic algae to convert carbon dioxide into oils, proteins and carbohydrates is significantly higher than that of agricultural crops in the soil, without the need for agricultural land [296]. Algae can produce 2,400 commercial products for industrial use [297, 298], such as biofuels [299], nutritional products [300, 301], feed, cosmetics, pharmaceuticals and even resins.

Algae holds great potential for future sustainable technologies because its use may contribute simultaneously to the production of energy, chemical products and purifying the environment. Carbon dioxide (CO_2) is considered the largest contributor to the greenhouse gas effect. The many countries which are signatories of the Kyoto Protocol will demand clean technologies which may be based on algae in the near future. Of the 30,000 species of algae which have hitherto been identified, only a few thousand strains have been tested for chemical content, and only a few tons per year are used in industrial practice.

Algae are typically autotrophic organisms, ranging in form from unicellular to multicellular, such as the giant kelp (large brown alga), that may grow up to 50 meters in length. The largest and most complex marine forms are called seaweeds.

Algae may be cultivated in open ponds and in bioreactors. Open pond systems are the most common method of algae cultivation, already used commercially to produce nutritional products and treat wastewater. Open pond systems use shallow ponds, in which the algae are exposed to natural solar radiation (sunlight) which they convert into biomass.

Fig. 5.28. Algae cultivation in open ponds (photo courtesy of Joel Cuello) .

Algae farming (see Fig. 5.28) can be an effective countermeasure to reduce the harmful emissions from agricultural sources. Currently agriculture is the biggest waste emitter and the main culprit of natural environment degradation. As a rule, i.e. deliberately, intentionally, and commonly chemicals like fertilizers and biocides are dosed excessively, in order to fulfill their role in a particular stage of the growing season without consideration of the weather. Emission sources are scattered over a large area iand it is difficult to control and clean such places. Gas production from agricultural waste is not widely implemented and does not solve the problem completely. The conventional methods of treatment are costly and cannot be economically justified in this case. Phytoremediation is a very promising and sustainable treatment method used to protect the environment. Microalgae, macroalgae and biofilters have long been used to remove nutrients and heavy metals as exemplified by the so-called trickling filters,used for the last stages of wastewater treatment at the plants. The long-lasting effect is biological, since wastewater flows down the stones that are overgrown with algae and are placed in open tanks with air. Algae grow rapidly under such conditions, consuming nutrients from the wastewater. The biomass excess is eaten by animals such as snails, which are then eaten by birds and rodents. In this way, contaminants are placed in the natural cycle. Therefore, in order to prevent environmental pollution caused by farming, especially in the protection of people, plants and animals and to prevent undesirable ecological changes, the EU has set strict limits on the quality of water which can be removed from agricultural waste.

International studies showed that macroalgae biofilters can be used to clean wastewater streams by removing inorganic nutrients and heavy metals mostly from aquaculture waste streams and secondarily from treated effluent, while producing a natural, useful and valuable biomass. Algae can be a source of energy in many ways, through direct combustion of biomass, and production of methane, ethanol and biodiesel. Algae and microalgae can be used to produce fuel.

Photobioreactors (see Fig. 5.29) are closed vessels, with controlled environmental parameters. They can be used for efficient cultivation of algae. Photobioreactors allow the use of solar energy, while CO_2 may be supplied from flue gases, from steel mills, cement plants or simply the atmospheric air. Thus the algae could contribute to the recirculation of carbon dioxide in clean technologies. The concentration of cells in photobioreactors can be up to one million times greater than the concentration of biomass in seawater. One milliliter of seawater contains on average 200 cells of microalgae.

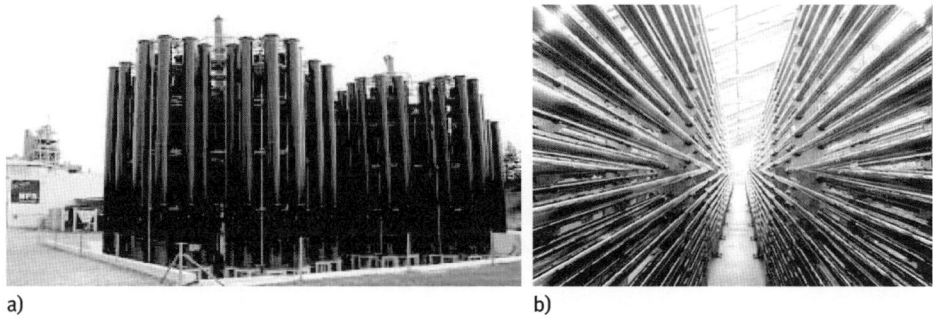

a) b)

Fig. 5.29. The tubular photobioreactors: (a) a view of BFS vertical photo-bioreactors in Alicante (Spain), (b) horizontal photobioreactor installed in a greenhouse (courtesy of Bio Fuel Systems BFS).

Production of microalgae biomass is so far limited to a few thousand tons per year, mostly produced in open ponds [302]. Only a few hundred tons are produced in closed photobioreactors. However, these open systems exhibit plenty of defects, for example poor control of environmental conditions such as temperature, weather causing the risk of contamination, low light intensity. Moreover, prevention of water loss by evaporation is possible only in closed systems. Basic process engineering principles regarding light distribution, mass transfer and hydrodynamics have been set up [303]. As a condition for the profitability of the installation to produce energy, the cost of investments may not exceed 40 €/m² [304]. The main parameters for photobioreactor design are the following:

1. total working volume;
2. total surface area of the light acceptor (transparent part): A_L;
3. total surface area of the ground occupied: A_G;

4. volumetric productivity P_V:

$$P_V = \frac{1}{V}\frac{dm}{dt}; \tag{5.82}$$

5. areal productivity P_A:

$$P_A = \frac{1}{A}\frac{dm}{dt}; \tag{5.83}$$

6. I_0 the intensity of radiation of photons PFD = $\mu E/(m^2\, s)$ (microEinstein).

In practice, only a range is usable for photosynthesis, i.e. photosynthetic active radiation range (PAR 400 nm–700nm). PCE, the photoconversion efficiency, measures the fraction of solar energy converted to chemical energy in a photobioprocess. The maximum theoretical value has been estimated to be 9 % [305] for full sunlight. For the calculation of PCE the energy content of the biomass has to be measured. It can range from 20 MJ/kg to 30 MJ/kg for oil-rich algae. According to thermodynamics oil-rich algae could show lower areal biomass productivities than other algae [306], but this can nevertheless mean a high PCE, which is at the end the decisive value. The three most popular types of photo-bioreactor design are flat-plate, annular and tubular (see Fig. 5.30). The vertical alignment of the reactor used is shown in Figure 5.30.

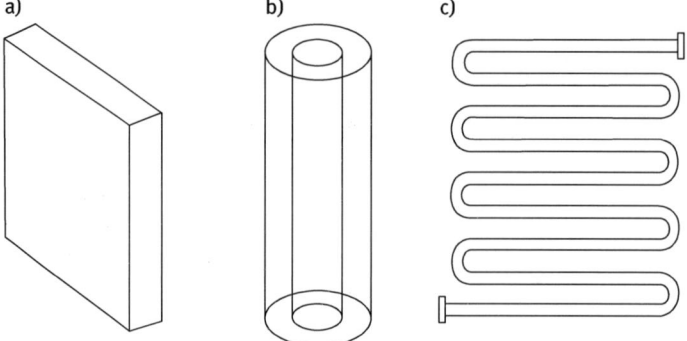

Fig. 5.30. The three main configurations of photo-bioreactors i.e.: (a) plate-frame, (b) annular and (c) tubular.

The CO_2 supplied at the bottom is within the range of 0.25–2 v/v/min, which consumes about 50 W/m^3 energy [307–309]. The annular photobioreactors are similar in operation to the bubble columns; a hollow space in the middle allows for better light transmission into the reactor. A diameter of about 20 cm of light was insufficient for proper photosynthesis. The diameters of tubular photobioreactors do not exceed 40 mm, but the length of the tube may be 100 m and more. A drawback of tubular configuration is large energy consumption, but the efficiency is also much higher [310].

Tubular and flat-plate photo-bioreactors have become very popular in the last decade [311, 312]. They use much higher concentrations of biomass [313]. Although they are more expensive to build, operate and maintain [314], they also present additional advantages [308]. Recovery of biomass is easier, the culture operates under sterile conditions, making it possible to keep the strains that provide high quality products exclusively, reduced dependence on the weather (rain, wind, temperature, evaporation of water) and better growth conditions, such as temperature, pH, nutrient concentrations and the possibility of using CO_2 from industrial waste. Moreover, in the design of closed bioreactors suitable solar access, temperature control, heat and mass transfer, as well as the reaction kinetics must be taken into account [315, 316]. The following photobioreactors are described in the literature: tubular, spiral cascade reactors, bubble, horizontal or vertical flat panels [317], vesicular. The vertical location reduces the land area employed, benefits the sun-light transfer process, CO_2 dispersion by controlling the bubbling system of CO_2 which can kill the algae and the cleaning process [318]. The concentric layout in vertical photobioreactors permits control of the light/shadow ratio.

Microalgae are useful for the preparation of biologically active compounds [319], such as antibiotics, toxins, algaecides, pharmaceutically active as a source of vitamins and vitamin precursors, in cosmetics (e.g. in skin and hair care products) and fine chemicals [320] including carotenoids, vitamins, polysaccharides, and essential fatty acids. Aquaflow [321] has identified a number of other interesting potential products, including bio-derived chemicals, as specialty industry solvents, precursors for polymer manufacture, surfactants (detergents) and pharmaceutical components.

In cosmetics the use of extracts of algae act preventively as thickening agents, water scavengers and antioxidants [322]. In addition, microalgae extracts are used as bioactive substances which accelerate the healing process. Multifunctional algae extracts contain active substances such as vitamins, enzymes, phytochemicals, antioxidants and essential oils. They can be used in products such as creams, lotions and ointments [323–325]. The algal polysaccharides, algal proteins or lipids, vitamin A, vitamin B1 may be produced from algae. Microalgae diversity promises to provide new and diverse enzymes and biocatalysts and has the potential to make industrial biotechnology [326] as alternatives to conventionally produced enzymes from animals and bacteria [327].

Algae can produce enzymes with the ability of capturing free radicals and thereby preventing cancer or other diseases [328]. The most promising enzyme in this case is superoxide dismutase, which is an important antioxidant enzyme which is currently showing powerful anti-inflammatory activity in clinical trials [329]. Peroxidase is able to reduce UV-induced erythema [328]. Some algae show the presence of unique haloperoxidases or vanadium bromoperoxidase with a high resistance to thermal and chemical denaturation [330, 331].

The European industry for flavors and fragrances is estimated to be worth hundreds of millions of Euros. Flavorings and fragrances are used in many applications, such as in food products like beverages, dairy products, sweet goods and in personal care applications, household cleaners and detergents. There is a growing demand among consumers for natural ingredients [332]. Among others, algae contain a wide number of molecules with a high potential for the fragrance and flavor industry such as: pyroles, cyclopentenones [333, 334], phenols [335], furans, pyrrolidones, indoles [336–338] and skatol.

The global production of surfactants has been estimated to be 7 M tons per year. The major surfactant is a synthetic, linear alkylbenzene sulfonate (LAS), which is non-biodegradable and thus problematic for the environment because it is retained in soil and water for a long time [339–341]. In contrast, biodegradable surfactants are based on natural fatty acids derivatives or on sugar (alkylglycosides). The bio-oil of microalgae is exceptionally rich in triamines, and thus can be easily modified to behave as novel bio-based surfactants. Pyrrolidone derivatives are easily biodegradable [342], and of relatively low toxicity, and they are environmentally friendly.

Inks and coatings are made from resin with appropriately selected physical properties, such as rheologic, high-gloss, hardness and scratch resistance. The current cost of resin ranges from € 1.50 to € 3.00/kg. BASF recently developed copolymers of vinyl-lactams to achieve these desired features at relatively low cost [343]. Professor Huber of the University of Massachusetts-Amherst recently patented the development of a bio-derivative of benzene [344, 345]. He developed as an alternative method for butadiene, which comes from the thermal processing of biomass [346].

Table 5.4. Fatty acid composition of different vegetable oils

	C14-18:0	C16-18:1	C18-22:n>1	Others
Microalgae	28	31	35	6
Codliver oil	20	24	31	35
Soybean oil	16	23	61	3
Corn oil	13	28	59	–
Sunflower oil	12	19	69	–

Fish oil is the main source of omega3 fatty acids, but it has recently been suspected that small children and pregnant or lactating women may experience negative symptoms due to the potential presence of trace amounts of heavy metals. Algae can be an alternative source of omega3 (see Tab . 5.4) because they do not contain heavy metals such as fish oils.

5.5 Biopolymers

5.5.1 Background and context

Some biopolymers have been known since time immemorial. Galalith or caseinate is the first known bioplastic, and is derived from milk casein. In ancient times it was probably used for ivory forgery. The Mayans, harvested natural latex from the Hevea tree in 1600 BC [347] to make a ball. Natural rubber is a complex emulsion consisting of proteins, alkaloids, starches, sugars, oils, tannins, resins and gums.

Today polymers are widely used in various sectors and are the one of the best-developed fields of chemistry. However, these materials require the use of fossil fuels and large amounts of energy (see Fig. 5.31). In Europe, the assessment of the potential of bioplastics in different sectors of the European economy is performed by the Committee of Professional Agricultural Organizations in the European Union (COPA) and the General Committee for the Agricultural Cooperation in the European Union (COGECA). In the United States in 2002 $ 310 billion was paid for bioplastics.

Unfortunately polymers are made primarily from petroleum, which is completely contrary to the principles of sustainable development, but in accordance with the laws of economics. The polymers cost more than fuel, and this is already a source of big profits. The cost of gasoline is $ 0.35/kg, but plastic can cost up to 20 times more,as much as $ 7/kg. This will only become worse because oil prices are on the increase, as are the prices of all products derived from petroleum. Plastic materials are durable and have a long life so spontaneous biodegradation is difficult. A temporary preventive solution is the landfill mining, i.e. extraction of different polymers, and their re-use such as in construction materials, railway sleepers etc.

According to the United States Environmental Protection Agency, only 6 % of the plastic made in the United States was recycled in 2005, compared with 50 % recycling of paper, 37 % recycling of metals or 22% glass [348]. The National Non-Food Crops Centre (NNFCC) forecast that the global annual capacity of bioplastics would grow to 2.1 million tons by 2013. At this stage it is still too small, however, and the area of use of bioplastics is also limited to a few niche areas as shown in Figure 5.31.

Currently the best companies in the world are interested in the production of biodegradable polymers indicating an imminent reversal of negative trends. The production of starch, sugar and cellulose, currently the three most important raw materials, has increased by 600 %. Moreover, sustainable technology polymers can be based on waste materials such as sugar-cane bagasse, castor oil, cellulose from carrots in the cosmetic industry, sticky water from fish reloading and fatty acids from wastes which have an advantage, compared to the production of triglycerides from plants because there is no competition with food production [349, 350]. Algae require much less land to produce the same amount of oil. Triglycerides, which are converted into fatty acid,s are not chemically different from those derived from vegetable oils (see Table 5.4), although the monounsaturated fatty acid content is higher, which has

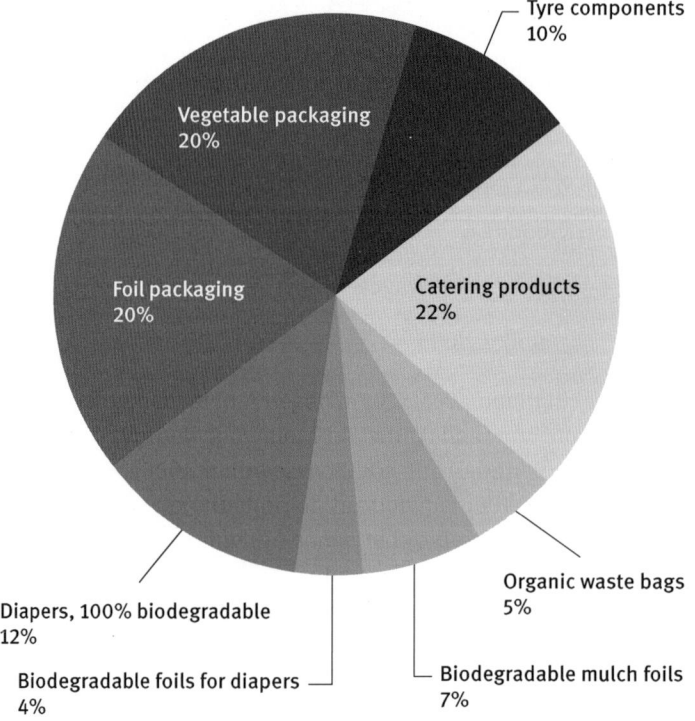

Fig. 5.31. Contemporary uses of bioplastics.

a positive effect on product stability. Conducting polymers are now very widely used in capacities such as anti-corrosion coatings, biomolecular electronics and telecommunications devices for display. The most commonly used conductive polymers are polyaniline (PANI), polyacetylenes and polypyrroles (PPy). The use of biomass can reduce the cost of production, as DuPont recently did with the conductive polymer consisting of polyaniline chains grafted to lignin, called Ligno-Pani, which was originally developed by NASA [546].

Some biodegradable polymers are even compostable, and can be marked with a 'compostable' symbol, under European Standard EN 13432 (2000). This means that they will break down within six months or less.

Polymers should be able to participate in the metabolism of nature. The prospect of producing bioplastics in transgenic plants may completely alter the environmental consequences of production. The International Organization for Standardization (ISO) has been developing Life-cycle Impact Assessment (LCIA) since 1993, which concerns material flow, environmental impact and resource consumption. All polymers designed now and in the future should be biodegradable in order to meet the demands of sustainable development. Through skillful application of process engineering, the production of bioplastics can be feasible in a number of commercially produced materials. Bioplastics are derived from renewable biomass sources, such as vegetable fats and oils, starch and biomass [351]. At present the following bioplastic types are manufactured [351–354]:

1. starch-based plastics;
2. cellulose-based plastics [352];
3. some aliphatic polyesters [353];
4. polylactic acid (PLA) plastics;
5. polyhydroxyalkanoates (PHA);
6. poly-3-hydroxybutyrate (PHB);
7. polyamide 11 (PA 11);
8. bio-derived polyethylene;
9. genetically modified bioplastics.

5.5.2 Production of biopolymers

Obtaining polymers produced in plant cells directly is most promising and suitable from the point of view of sustainable development, but it is still a future prospect for economic reasons. Biopolymers may be produced from natural substrates such as simple sugars, polysaccharides such as starch and cellulose, and fatty acids. A typical production method is creation of the monomer as a result of fermentation and the appropriate polymerization of said monomer.

Starch is the main raw material for about 50 % of thermoplastic biopolymers. It has many valuable properties due to its natural origin [354]. The main feature of the polymers derived from starch is their ability to absorb moisture. They are therefore used in the production of capsules in the pharmaceutical industry. Thermoplastics made of starch are obtained thanks to sorbitol and glycerin. Note that glycerin is formed as a byproduct of biodiesel production, and thus is widely available in large quantities. Starch was originally used as an admixture with other polymers to induce biological degradation. Such compounds are already produced in large quantities by BASF under the trade name ECOFLEX, in amounts of 60,000 tonnes per year. These mixes contain mostly starch with polycaprolactone or Polybutylene adipate-co-terephthalate. ECOFLEX conventional polyethylene has properties in accordance with DIN EN 13432. However, it can not be considered biodegradable, because it contains a biopolymer additive derived from petroleum. The Roquette company has developed the polyolefins (based on starch), which are similar to the previous biopolymers, i.e. are not 100 % biodegradable, but exhibit lower emissions of carbon dioxide in relation to plastics based on petroleum [355]. Many useful biopolymers such as biopolyesters are made out of vegetable oils [356] or sugars [357, 358].

Polyhydroxyalkanoate (PHA) and polylactic acids (PLA) are produced by starch hydrolysis. The production of PHA and PLA by microbial fermentation, is less energy consuming. The bioproduction of polyactides is performed according to the following scheme:

$$CO_2 + H_2O \xrightarrow{\text{plants photosynthesis}} \text{sugars, oils} \xrightarrow{\text{bacteria-fermentation}} \text{Biopolyesters} \qquad (5.84)$$

$$CO_2 + H_2O \xrightarrow{\text{Plants Bacteria, Algae-photosynthesis}} \text{Biopolyesters} \qquad (5.85)$$

$$CO_2 + H_2O \xrightarrow{\text{plants photosynthesis}} \text{sugars} \xrightarrow{\text{bacteria-fermentation}} \text{lactic acid}$$
$$\xrightarrow{\text{catalyst-chemosynthesis}} \text{Polyactide} \qquad (5.86)$$

The development of PLA began in 1932 [359], but in 1954 the DuPont Corporation introduced the synthetic polymer PLA, which has very good properties (clean, high molecular weight) [360]. In the early 1990s, the Cargill Corporation developed a method for the continuous production of this polymer [361–364]. PLA is currently manufactured from corn starch which is converted to dextrose. Lactic acid is obtained as a result of anaerobic microbial fermentation. The lactic acid becomes cyclic lactide monomer, with dimerization of the reactive distillation. Lactide is eventually polymerized to PLA (see Fig. 5.32).

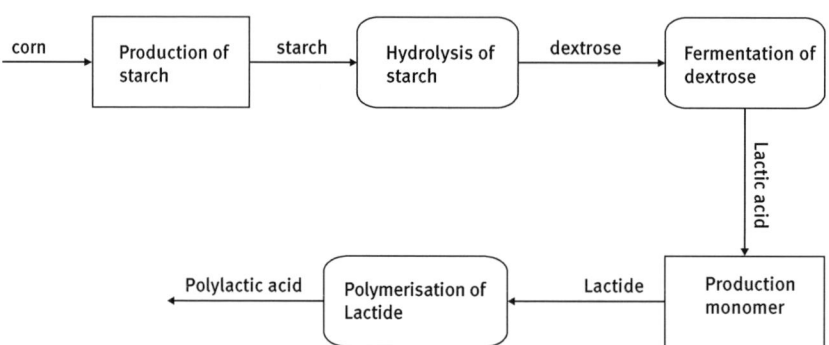

Fig. 5.32. Schematic diagram of simplified technology for the production of biopolymer of lactic acid PLA.

PLA can be recovered because more than 70 million tons of this material ends up in landfills each year [365, 366]. Breaking down waste PLA to lactic acid is possible. It may then be converted back to the lactide and again subjected to polymerization [367]. PLA has unique properties which predispose it to medical applications. The degradation of this polymer leaves no residue in the body [368]. PLA is therefore used for surgical sutures [369] studs [370, 371] plates, screws [372], suture anchors [373] and intra-vascular stents [374, 375]. Biodegradable PLA [376, 377] and its copolymers [378, 379], are also often used for controlled release of drugs [380], substrates to skin grafts [370] as well as in various dental applications [374].

Polyhydroxyalkanoates are linear polyesters produced in nature by bacterial fermentation of sugar or lipids. PHAs are thermoplastic and are resistant to UV radiation, in contrast to other biopolymers like polyactides, and have low moisture permeability. In order to produce PHA, bacterial culture of Alcaligenes eutrophus is grown, until the population reaches an acceptable level. Then the composition of the medium is changed, which forces the bacteria to synthesize PHA. The biosynthesis of PHA usually occurs with a deficiency of elements (e.g. phosphorus, nitrogen, trace elements, oxygen), or an excess of carbon. The resulting PHA may be up to 80 % dry matter.

More than 150 different monomers can be combined within this family to give materials with extremely different properties [382]. They can be either thermoplastic or elastomeric materials, with melting points ranging from 40 to 180 °C. PHA polymers can be processed using conventional processing equipment and are soluble in halogenated solvents such chloroform, dichloromethane or dichloroethane [383]. PHAs are benign materials and are highly biodegradable.

Starch and whey are also used to produce poly (3-hydroxybutyrate, P3HB) in fedbatch cultures of Azotobacter chroococcum and recombinant Escherichia coli respectively [384].

PHB is a biodegradable polymer from the group of aliphatic polyesters produced and stored by the bacterium Ralstonia eutropha H16. Microbial biosynthesis of PHB starts with the condensation of two molecules of acetyl-CoA to give acetoacetyl-CoA which is subsequently reduced to hydroxybutyryl-CoA. This latter compound is then used as a monomer to polymerize PHB. This is the first thermoplastic polymer with properties originating in biosynthesis which has been introduced effectively to the plastics market. PHB is biocompatible and therefore is frequently used in medicine. Moreover, it has high mechanical strength (40 MPa, close to that of polypropylene, PP), has good moisture resistance, good oxygen permeability, good ultra-violet resistance but poor resistance to acids and bases. However, the production of PHB from microorganisms is very expensive and therefore unprofitable. Among the various possible structural isomers of this polymer only poly (3-hydroxybutyrate, P3HB) is of practical importance [385]. In 2001 Metabolix closed all bioreactors, and in 2004 began to focus on the production of P3HB from plants [386, 387].

Cellulose bioplastics [381] are cellulose esters and their derivatives, including celluloid, cellulose acetate butyrate, cellulose propionate and ethyl cellulose. Cellulose is the most abundant organic polymer on Earth and therefore is mainly used to produce a number of commodities such as paper, paperboard, and bioplastics such as cellophane, rayon and artificial silk. Cellulose is mainly obtained for industrial use from wood pulp and cotton. Artificial silk was developed in the 1890s from cellulose fiber. Rayon was invented in 1893 by Arthur D. Little. In 1910 Camille and Henry Dreyfus were making acetate motion picture films. Cellulose acetate was used as lacquer and for other commercial products during World War I in England. The first commercial textiles from cellulose acetate in fiber form were developed by the Celanese Company in 1924. Cellulosic plastics are characterized by good strength, toughness, transparency and high surface gloss. Cellulosic compounds are available for extrusion, injection molding, blow molding and rotational molding. They also are widely used in the form of film and sheet. Cellulose acetate (CA) is a cheap cellulosic material of good mechanical properties and is commonly used in the production of membranes for reverse osmosis. Cellulose acetate butyrate (CAB), although a little more expensive than acetate, is somewhat tougher, and has lower moisture absorption and excellent transparency. CA is still one of the best materials for reverse osmosis.

Bio-derived polyethylene is chemically and physically identical to traditional polyethylene. Although it is not biodegradable it can be recycled. It can also considerably reduce greenhouse gas emissions because the technology of using its route from sugar cane ethanol to produce one ton of polyethylene captures, and removes from the environment, 2.5 tons of carbon dioxide, while the traditional petrochemical route results in emissions of close to 3.5 tons. It is used in packaging manufacturing such as bottles and containers. A new technology was developed in 2010 to produce bio-butene, which is required for the production of low-density polyethylene. Oil from canola, corn, soybean, and linseeds were used as substrates for monomers to produce elastomers [388].

5.6 Renewable energy

5.6.1 Global energy policy

Energy and mass are fundamental concepts, common to the various physical processes and phenomena. On Earth, energy occurs in different forms but always comes to us as solar energy. The demand for energy is increasing dramatically and it is expected that by the end of this century the total energy consumption could amount to 47 gigatons [389] (see Fig. 5.33). It should be mentioned that a ton of oil equivalent, i.e. 1 toe = 42 GJ, and global energy consumption in 2000 from all sources amounted to 9963 Mtoe. The total amount of energy consists of the following shares of energy sources. (see Fig. 5.33.) As shown in the graph (Fig. 5.33), the global renewable energy consumption in 2000 did not exceed 13.4 %, and there is therefore still a lot to do on this issue.

The first diesel engine was demonstrated at the World Exhibition in Paris in 1900. It was powered by peanut oil, the first biofuel. It was used in South Africa to supply to heavy equipment long before the Second World War. In the US in 1940 Colgate patented the transesterification biolipids as technology for new explosives. Research was also performed on the use and production of biodiesel from sunflower oil for transestrification in South Africa in 1979. In 1987, the Austrian company Gaskoks acquired biodiesel technology and built the first industrial plant for the production

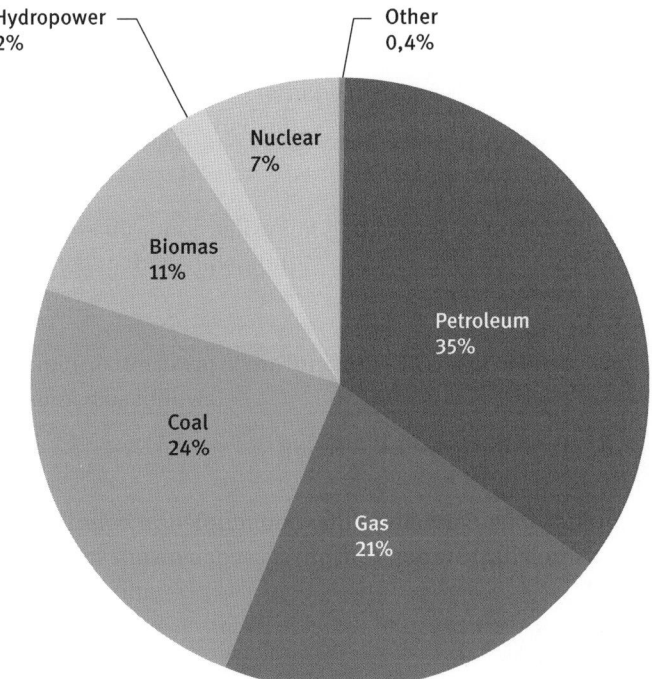

Fig. 5.33. Contribution of energy sources to global energy consumption in 2000.

of biofuels from rapeseed oil in 1989, with a capacity of 30,000 tons/year. Soon factories also appeared in Czechoslovakia, France, Germany and Sweden, and by 1998 biodiesel was already being produced in 21 countries. Local production of biodiesel by rapeseed oil transestrification was launched in France in 1990. In the years 1950–1994 the number of vehicles has increased from 70 to 630 million, and every day the number of new cars increases by 137,000. Thus by the year 2025 more than one billion will be made. In 2010, worldwide biofuel production reached 105 billion liters (i.e. 28 billion gallons US), up 17 % from 2009, and biofuels provided 2.7 % of the world's fuels for road transport, a contribution largely made up of ethanol and biodiesel. Global ethanol fuel production reached 86 billion liters (23 billion gallons US) in 2010, with the United States and Brazil the world's top producers, accounting together for 90 % of global production. The world's largest biodiesel producer is the European Union, accounting for 53 % of all biodiesel production in 2010. The International Energy Agency has a goal for biofuels to meet more than a quarter of world demand for transportation fuels by 2050 to reduce dependence on petroleum and coal.

The European demand for electricity is projected to grow by 0.9 % annually, from 244 Mtoe to 300 Mtoe by 2030.

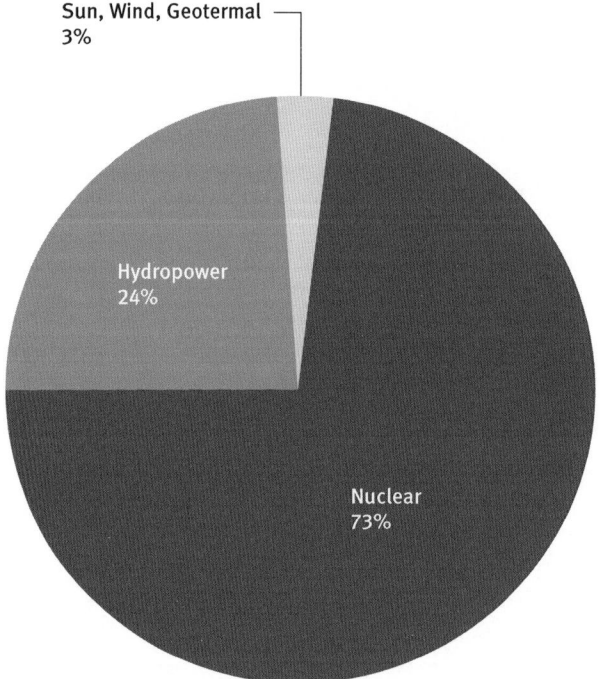

Fig. 5.34. Contribution of emission-free energies in 2000 in global production.

Figure 5.34 shows that in the year 2000, of all types of emission-free energy, nuclear energy (73 %) has the largest share, hydropower has 24 % and all other energies, i.e. wind energy, geothermal energy and solar only have a total share of 3 %.

Although the advancement and interests of different energies vary from country to country, globally the proportion of emission-free energy is still very small.

The most significant changes in patterns of electricity generation by 2030 will be in the shares of coal, nuclear and wind power. Coal-fired and nuclear electricity, which currently dominate the market, will each see declines of 9 %, from 31 % to 22 % for coal-fired stations and from 28 % to 19 % for nuclear. Their current contribution will be replaced mostly by wind power, followed by solar and gas. Wind power's share grows from 3 % to 15 % during this time-frame, while the share of solar electricity rises from essentially zero to 3 %, which is far too little growth. Gas-fired capacity meanwhile rises from 22 % to 25 % and becomes the dominant source of power. Hydro power's share of total generation will rise from 9 % to 10 %, with biomass and waste together increasing their share from 3 % to 5 %.

In the Renewable Energy Directive 2009/28/EC, second generation biofuels, i. e., biofuels made out of ligno-cellulosic, nonfood cellulosic, waste and residue materials will get double credit towards the 10 % target for renewable energy in transport in 2020.

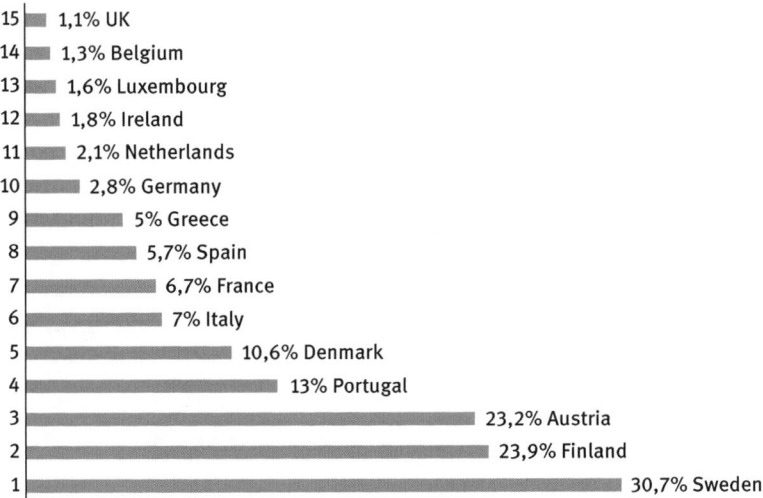

Fig. 5.35. Contribution of clean energies in various European countries.

The Council of Europe in July 2001 in Gothenburg agreed on the European Community strategies for sustainable development in the provisions of this package regarding biofuels. It has been declared that biofuels, i.e. bioethanol and biodiesel, will be used for vehicles in their pure forms or as a blend.

Renewable liquid fuels can be produced on the basis of photosynthesis, which is the real way to store the energy of sun. Examples of this carbon fixation are plants and microalgae. Intentional biomass cultivation (e.g. willow), which is then used for direct combustion, results in the emission of harmful VOCs. In this way the atmosphere produces the same amount of carbon dioxide and other toxic gases as in the case of combustion of timber. However, burning the biomass as the terminal waste material from the different types of production, wastewater treatment plants, as a method of final management of sewage deposits is obviously correct. However, in such cases appropriate gas emission cleaning of the waste incineration plant must be ensured.

Biofuels have increased in popularity because of the rising oil prices and the need for energy security, but the most important reason for the promotion of biofuels should be the need for sustainable development. The full consequences of environmental change, climate change, changes in the diversity of plants and animals resulting in the damage of some food chains are rarely taken into account. We forget that the ecosphere can be easily destabilized (or even derailed) from a state of equilibrium. For example, very small changes in the concentration of all substances in the earth can lead to disastrous consequences, much more and much faster than expected. Thus, maintaining the steady state condition on Earth is the most important factor of sustainable development. A very good example is the oxygen content of the atmosphere, which is about 20 % and is stable. This stability results from the oxygen concentration equilibrium, in which the various food-webs participate. To date, there is no absolute certainty as to who dominated as a supplier of oxygen on Earth, tropical forests or algae in the ocean. But we know for sure that the violation of this balance (in both directions), by as little as 5 % would be catastrophic. If there were only about 5 % more oxygen, then there would be a spontaneous ignition, and if there were only 5 % less oxygen than there is now, it would not be possible to start a fire at all on earth. This is the same for all the elements components and conditions. The basic principle of sustainable development is that it satisfies present needs, without taking away the opportunities for future generations to meet their own needs [392].

The production of biofuels will soon become an obvious necessity due to the decreasing resources of fossil fuels. But will it be too late for any action on our part then? Biofuels may be used as substitutes for diesel oil. Some biofuels may be used as substitutes for gasoline. These are mainly alcohol-derived, primarily from the fermentation of various sugars.

Increased profitability of biofuel production can be achieved through the introduction of new, cheaper and more productive varieties of plants, including adaptations of known varieties, genetic modification and algae. There is a large potential for increasing the yield even in the manufacturing processes including new methods of enzymatic transesterification in the reactors, modern methods of separation, in order to simplify the technology, and the use of all byproducts such as biosurfactants, solvents, glycerin, fertilizers, fatty acids and solvents.

The criteria to be taken into account in the selection of raw material for biofuel (bioethanol and biodiesels) production include not only the price of raw materials, but also all additional associated costs, logistics, and the value of byproducts obtained in the process, which should be diversified. The diversification of biofuels can also include various fuel additives, organic solvents and chemicals used in other industries. In this way profitability and flexibility are increased, and the principles of clean technologies and sustainable development are implemented in practice.

The regulations which impact the EU biofuels market are the Biofuels Directive (2003/30), the EU Climate and Energy Package and the Fuel Quality Directive (FQD, 2009/30). The Package was adopted as a white paper by the European Council on April 6, 2009 (0147/2009). The Package includes the "20/20/20" mandatory goals for 2020, one of which is a 20 % share of renewable energy in the EU total energy mix. Part of this 20 % share is a 10 % minimum target for renewable energy consumed in transport to be achieved by all Member States (MS).

Fuel additives in the United States are regulated under Section 211 of the Clean Air Act (as amended in January 1995). The Environmental Protection Agency (EPA) requires the registration of all fuel additives which are commercially distributed for use in highway motor vehicles in the United States, and may require testing and banning of harmful additives. The EPA also regularly reviews the health and net economic benefits of Clean Air Act policies [2].

The act also requires deposit control additives (DCAs) be added to all gasolines. This type of additive is a detergent additive that acts as a cleansing agent in small passages in the carburetor or fuel injectors.

The most commonly used additives for fuels include ethanol, MTB, ETB and TAME. Ethyl tert-butyl ether (ETBE) is commonly used as an oxygenate gasoline additive in the production of gasoline from crude oil. ETBE offers equal or greater air quality benefits than ethanol. Methyl tert-butyl ether (MTBE) is a gasoline additive used as an oxygenator to raise the octane number. Its use is controversial because of its occurrence in groundwater and legislation favoring ethanol. However, worldwide production of MTBE has been constant at about 18 million tons/y (2005). Tertiary amyl methyl ether (TAME) is mostly used as an oxygenator in gasoline. It is added for three reasons: to increase octane enhancement, to replace banned tetraethyl lead, and to raise the oxygen content in gasoline. It is known that TAME in fuel reduces exhaust emissions of some volatile organic compounds, but also it is suspected of toxicity [285].

Biofuel can also be produced as a result of enzymatic reactions (see Tab. 5.5), instead of KOH and NaOH. This allows the introduction of many improvements in the technology of biofuels. Firstly, biofuels are cheaper and of better quality because biotechnology is more energy efficient compared to conventional technology (see Tab. 5.6). Enzymatic reactions run at lower temperatures and therefore there is no problem with the separation of hydroxides.

Table 5.5. The degree of conversion for the different transesterification of various biodiesels

Oil	Alcohol	Lipase	Conversion
Rapeseed oil	Methanol	Lipozyme IM-60	19.4
Rapeseed oil	Ethanol	Lipozyme IM-60	66.5
Rapeseed oil	Hexanol	Candida rugosa	97
Avocado oil	C4–C18	Lipozyme IM-20	86.8-99.2
Sunflower seed oil	Ethanol	Lipozyme IM-20	93
Frying oil	Ethanol	Lipase PS-30	85.4
Fish oil	Ethanol	Candida antarctica	100
Lard		Lipozyme IM-60	95–99

Table 5.6. Some properties of biofuels

Fuel	Specific energy MJ/l	Ratio air : fuel	Specific energy MJ/kg of air	Heat of vaporization MJ/kg	Octane number	Kinematic viscosity at 20°C cSt
Petrol	32	14.6	2.9	0.36	91–99	0.4–0.8
Butanol	29.2	11.2	3.2	0.43	96	3.64
Ethanol	19.6	9.0	3	0.92	130	1.52
Methanol	16	6.5	3.1	1.20	136	0.64

The following resources are currently used for the production of biofuels: (a) Annual crops such as sunflower, soybean, canola, corn, (b) Wood: palm trees, olive trees, avocado trees, sugar cane, (c) Unicellular: phytoplankton, diatoms.

Vegetable oils and fats from nonfood sources such as Jaropha, Chinese tallow tree seeds, oil after frying and other waste fats, animal fat and grease have been investigated as potential oil-bearing material for biodiesel production. It should be noted that there are concerns that production of plants for fuels and chemicals will compete with the production of food and will soon replace food crops in cultivated fields.

Global production of biofuels (see Tab. 5.6) was 34 Mtoe in 2007, which is still only 1.5 % of total road transport fuel [393, 394]. The USA is now the world's leader in fuel ethanol production, with Brazil second. The total world production of biodiesel is around 25 % of ethanol production, and this is concentrated in the EU. Liquid biofuel production in the EU amounted to 12 Mtoe in 2009, which was only about 4 % of the road transport energy market. Biodiesel, mainly produced from rapeseed, predominated at about 9.6 Mtoe in 2009. Ethanol is mainly produced from wheat, and to a lesser extent sugar beet, with a total output of around 2.3 Mtoe in 2009 [394]. These two fuels are commonly referred to as first generation biofuels (see Fig. 5.36).

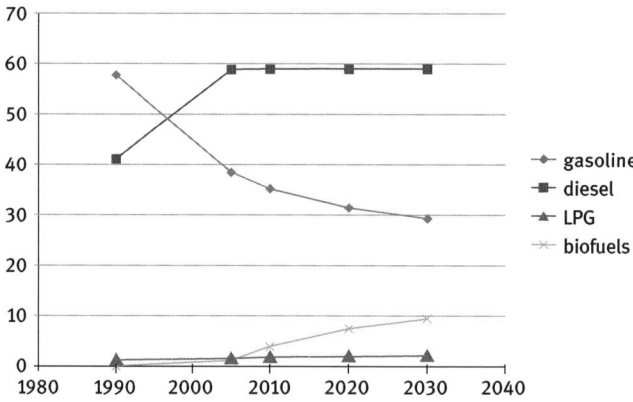

Fig. 5.36. Prospects for the use of different fuels in transport [Mt/y].

5.6.2 Energy fundamentals

The unit of 1 Joule (J) is a measure of the amount of energy in the International System of Units. The Joule is a derived unit of energy, work or amount of heat. It is named after the English physicist James Prescott Joule (1818–1889). One Joule is the work (equivalent to energy) required to produce one Watt of power for one second, or one "watt second" (Ws, compare kilowatt hour). This relationship can be used to define the watt:

$$J = W \times s. \tag{5.87}$$

One Joule is equal to the work necessary to apply a force of one newton for a distance of one meter:

$$J = N \times m. \tag{5.88}$$

One Joule is also the energy stored in 1 m^3 of gas compressed to 1 Pa:

$$J = Pa \times m^3. \tag{5.89}$$

The Joule is the work required to move an electric charge of one coulomb through an electrical potential difference of one volt, or one "'coulomb volt" (CV). The following relationship can be used to define the volt:

$$J = C \times V. \tag{5.90}$$

Or, to pass an electric current of one ampere through a resistance of one ohm for one second:

$$J = \Omega \times A^2 \times s. \tag{5.91}$$

we use multiples of the Joule to express the amount of energy on a global scale, for instance:

$$1\ EJ = 10^{18}\ J, \tag{5.92}$$

where EJ- means Exajoule. Kilowatt-hour is a common measure of energy consumption:

$$1\ kWh = 3.6 \times 10^6\ J. \tag{5.93}$$

The same units are used to describe many physical phenomena such as mechanical, thermal, electrical. If we wanted to warm up water on Earth by 1 °Celsius, a yottajoule (YJ) must be used, which corresponds to approximately septillion (10^{24} J).

Energy can be used for the benefit of man. Renewable energy could allow the survival and even the continuing development of our civilization for many generations. Otherwise, man is now capable of thoughtless destruction of its environment, either slowly, by gradual depleting of the fossil raw materials, or immediately, under the influence of irrational emotions. The largest thermonuclear bomb "Tsar" was built and detonated by the USSR on 30 October 1961 in the archipelago of Novaya Zemlya (73°51' N 54°30' E), located in the Arctic Ocean [390, 391]. The detonation of this bomb emitted energy of 2.1×10^{17} J in 39 nanoseconds, which is equal to 5.4×10^{24} W. This is 1 % of the power dissipated on the surface of the sun, and is equivalent to about 1,350–1,570 times the combined power of the bombs that destroyed Hiroshima and Nagasaki [390]. This bomb was of a magnitude stronger than the combined power of all the conventional explosives used in World War II, or one quarter of the estimated yield of the 1883 eruption of Krakatoa, and 10 % of the combined yield of all nuclear tests to date [391]. The earthquake (and tsunami) in Tahoka in 2011, which registered 9.0 on the Richter scale, together released 1.41 EJ of energy. The amount of energy used in the United States each year is roughly 94 EJ. The zettajoule (ZJ) is equal to one sextillion (10^{21}) Joules. Annual global energy consumption is approximately 0.5 ZJ.

To assess and compare values of different biofuels the new unit toe is used, which means ton oil equivalent. The energy content of 1 Toe = 41.87 GJ. Compared with gasoline (43.10 MJ/kg = 43.1 GJ/MT), ethanol has a lower energy content per unit weight (26.90 MJ/kg) than butanol (BtL = 33.50 MJ/kg). Pure vegetable oil (34.60 MJ/kg) and biodiesel (37.50 MJ/kg) have lower energy content than diesel (42.80 MJ/kg). The energy content of one metric ton of different fuels is given in Tab. 5.7.

Table 5.7. Comparison of energy content in different biofuels (in 1 ToE[*])

Fuel type	Volume [liters]	Energy [ToE]
Gasoline	1342	1.03
Ethanol	1267	0.64
Diesel	1195	1.02
Biodiesel	1136	0.90
Vegetable oil	1087	0.83
Butanol	1316	0.80

[*]Note that ethanol density is 0.789 MT/m^3, 1 Toe = 41.87 GJ is measure of the energy.

5.6.3 Biogas

A biogas, which is produced from different types of wastes, is a renewable fuel. Biogas is emitted naturally from places in which organic matter has been stored without oxygen. Biogas is released from the bottoms of natural water reservoirs and marshes, closing the carbon cycle in nature. Biogas is accompanied by the spontaneous decomposition of organic matter and can be produced very simply from landfills, agricultural waste and wastewater treatment plants. Biogas can help increase the efficiency of the expensive and energy-intensive drying process of sludge, in wastewater treatment plants, which can then be burned under control. Combustion energy can be further used for the purification of exhaust gases. Energy production from biogas has been an upward trend for several years (see Fig. 5.37).

5.6.4 Bioethanol

Bioethanol is an alcohol made by fermentation, mostly from carbohydrates produced by sugar or starch crops such as corn or sugarcane. Cellulosic biomass derived from nonfood sources, such as trees and grasses, is also being developed as a feedstock for ethanol production. Ethanol can be used as a fuel for vehicles in its pure form, but it is usually used as a gasoline additive to increase octane and improve vehicle emissions. Bioethanol is widely used in the USA and in Brazil. Current plant design does

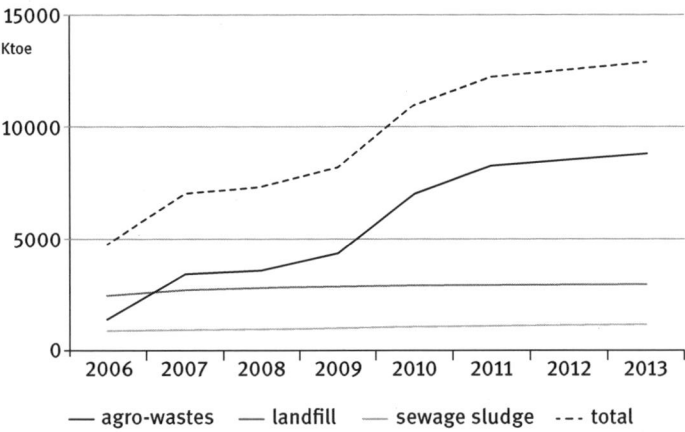

Fig. 5.37. Energy production (Ktoe = 41.871012 J) from biogas for heat and electricity in the EU.

not allow conversion of the lignin portion of plant raw materials to fuel components by fermentation.

While plants in the United States and Brazil are predominantly located in the feedstock production regions, and are focused on a single feedstock, plants in the EU are often located close to the end-market and designed as multi-feedstock plants (see Fig. 5.38). In the EU, bioethanol is mainly produced from wheat, corn, rye, barley and sugar beet derivatives.

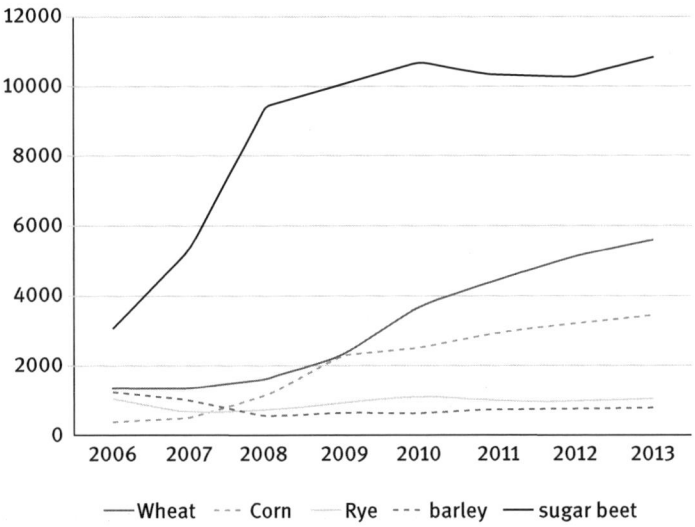

Fig. 5.38. Feedstock used for fuel ethanol production (per 1000 Mt ethanol).

Bio-ethanol and bio-methanol are clearly cleaner fuels than gasoline. In addition, they may be fuel to power fuel cells, the latest generation.

Bioethanol is also a precursor for the production of ethyl acetate as the esterification product of acetic acid. Note that acetic acid may also be produced from ethanol, and thus from the same materials as ethanol. The production of ethyl acetate can be used for transesterification in oils, so that biodiesel may be generated. It should be emphasized that such transesterification is very advantageous because it is not necessary (in this case) to separate the two phases (biodiesel and glycerol), as in the conventional method. Biodiesel from ethyl acetate is a solution of the two esters. The advantage of this transesterification is the simplicity of production and lack of waste. The raw materials are the same as for conventional biodiesel. The disadvantage of the technology is the production of ethyl acetate on a column of a heterogeneous catalyst in the reactive distillation.

An additional advantage of such production in biorefineries is the extremely high FLEXIBILITY of such production for three reasons:
1. plurality of possible raw materials for the production of ethanol;
2. plurality of different possible fatty acids for transesterification;
3. the wide variety of other applications for ethyl acetate.

Ethyl acetate is the ester of ethanol and acetic acid; it is manufactured on a large scale for use as a solvent. Global annual production of biofuels is growing rapidly. The combined annual production in 1985 of Japan, North America, and Europe was about 400,000 tons [286], but in 2004, an estimated 1.3M tons were produced worldwide [287] (see Fig. 5.39).

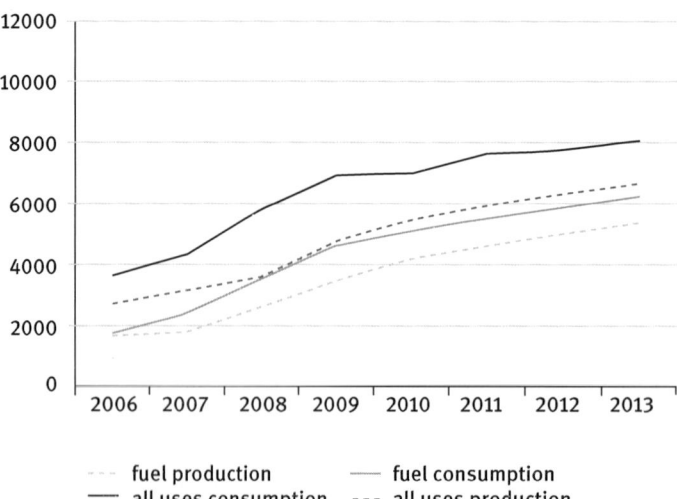

Fig. 5.39. Ethanol production and consumption in million liters (1 Mt Ethanol = 1,267 Liters = 0.64 Toe).

5.6.5 Biodiesel

The EU is the world's largest biodiesel producer on the basis of volume, representing about 70 % of the total biofuel market in the transport sector. Biodiesel can be used as a fuel for vehicles in its pure form, but it is usually used as a diesel additive to reduce levels of particulates, carbon monoxide and hydrocarbons from diesel-powered vehicles. Biodiesel is produced from oils or fats using transesterification (see Figs. 5.40 and 5.41). Biodiesel reduces emissions of carbon dioxide by 50 % and carbon monoxide by 78 %. Biodiesel contains less aromatic hydrocarbons as benzofluoranthene (about 56 %), and benzopyrene (about 71 %), the exhaust fumes contain less harmful substances.

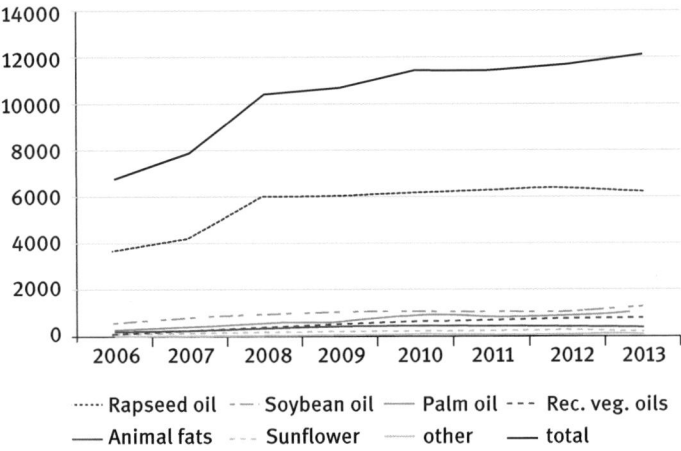

Fig. 5.40. Conventional feedstock use [1000 Mt].

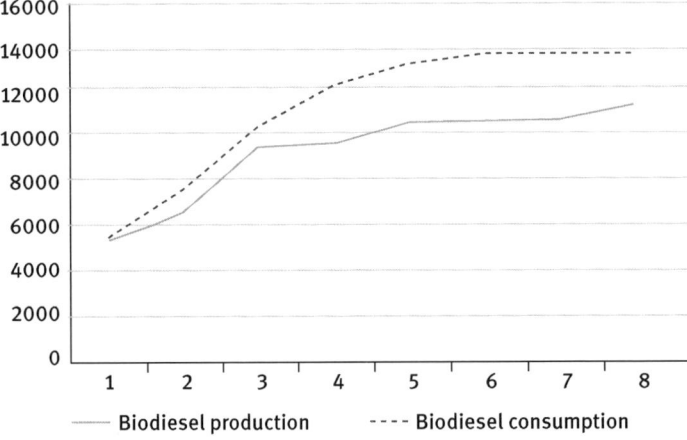

Fig. 5.41. Conventional and advanced biodiesel in the EU (million liters).

However, biodiesel emissions include more than 20 % of NOx and require additional catalysts. Biodiesel has a higher cetane number than petrodiesel, is biodegradable and nontoxic. The biodiesel flash point (> 150 °C) is much higher than that of crude oil (64 °C) and gas (-45 °C). The flash point of flammable liquid is the lowest temperature at which flammable mixtures occur with air. Pure B100 biodiesel can be used in any diesel engine, although it is mostly used mixed with oil in order to improve its properties. The addition of 2 % already improves the lubricity of the fuel.

Criteria taken into consideration when choosing feedstock do not only include the price of grains but also other costs such as logistics and the value of the byproducts obtained in the distillation process. Raw materials for conventional ethanol production are corn, wheat and barley. However, advanced biofuels are produced from lignocellulosic feedstock. The oils for diesel are presented in Tab. 5.8.

Table 5.8. Raw materials for the production of biodiesel.

Vegetable oil	Capacity l/ha
Soybean oil	375
Rapeseed oil	1000
Mustard oil	1300
Jaropha	3000
Palm oil	5800
Oil from algae	95000

Currently, in most industrial transesterification processes, triglycerides are catalyzed by basic homogeneous catalysts such as sodium hydroxide, potassium hydroxide and sodium methoxide. However, these types of catalysers have many drawbacks:
1. the re-use of homogeneous catalysts is restricted;
2. the use of continuous processing methods is limited;
3. side reactions are possible, such as hydrolysis and saponification;
4. such catalysts are extremely corrosive.

The aqueous phase after esterification contains glycerin and catalysts which must be separated from each other. The latter drawback in particular can significantly jeopardize the quality of the product because there is a risk that the very fine droplets of the emulsion in the aqueous phase may contain residues of the catalyst. This could be very dangerous when the catalyst is in contact with the aluminum engine block. Moreover, the purification process of glycerol with the catalyst increases production costs. Triglycerides derived from algae are frequently contaminated [297, 395] with free fatty acids (FFA), water, phospholipids and other impurities. Special catalysts are used for the transesterification of contaminated FFA [396–403]. The three most important properties of solid acid catalysts in the transesterification of triglycerides are their acidity, hydrophobicity and pore size. The number of reports in the literature on the catalysts which fulfill these criteria is limited [400–403].

The reaction of alcoholysis of triglycerides of fatty acids make a new ester and new alcohol and have the following sequence:

Triglyceride + alcohol = esters of fatty acids + glycerol

$$
\begin{array}{ll}
CH_2 - OOC - R_1 & R_1 - COO - R \quad CH_2 - OH \\
| & | \\
CH \ - OOC - R_2 + 3R - OH \longleftrightarrow R_2 - COO - R + CH \ - OH \qquad (5.94) \\
| & | \\
CH_2 - OOC - R_3 & R_3 - COO - R \quad CH_2 - OH
\end{array}
$$

Improvements to the transesterification reaction may be new materials and new methods of processing these materials. The processing may be via physical treatment or chemical operations, in addition to biological gene mutations or cytoplasmic in cells. The search for new efficient technologies can also deal with brand new materials and reactions.

A large improvement can be achieved in the transesterification process by forming a microemulsion of 1–150 nanometer, wherein the oil is stably dispersed in a solvent such as methanol, ethanol or butanol. Transesterification in the system of fine emulsion increases the "specific surface area", which significantly accelerates the process, reduces the energy required for mixing and pumping in addition to costs. The authors of this technique show that this reduces the viscosity and the cost of pumping [404].

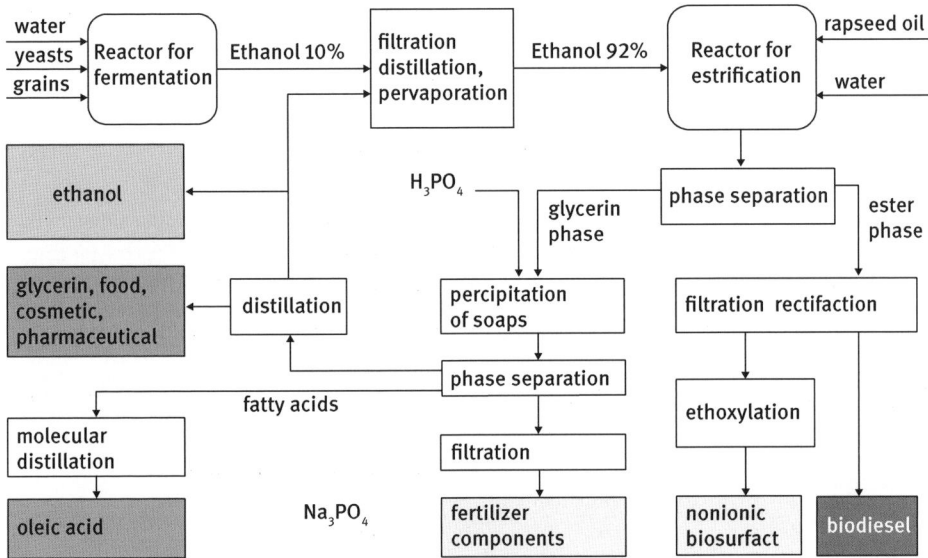

Fig. 5.42. Flow chart showing the production of clean biofuels.

The molar ratio of alcohol to triglyceride is one of the most important parameters of the process. Although an excess of alcohol in the range of from 6:1 to 30:1 favors rapid and efficient reaction there is also a greater amount of consumption of alkaline catalysts. Enzyme catalysis is a new challenge for researchers. Transesterification is carried out more easily in supercritical methanol under high temperature and pressure. However, there is no need for catalysts in the supercritical transesterification process, the reaction must be carried out at a temperature of 350 °C and a pressure of 45 MPa, and requires an excess of methanol [405–407]. Triglycerides of fatty acids may be subjected to thermal "cracking", resulting in a mixture consisting essentially of alkanes and alkenes with smaller amounts of other organic materials of a lower molecular weight, thus reducing viscosity. However, the disadvantages of this method of biofuel production are low throughput and the high cost of process equipment [408].

Considerable effort has been devoted to the investigation of lipases, carboxylesterases that catalyze hydrolysis and synthesis of long-chain acylglycerols, for the synthesis of biodiesel [409]. These remarkable enzymes currently constitute the single most widely-used biocatalyst class in biotechnology; they often possess high chemoselectivity, regioselectivity and stereoselectivity; they are natural extracellular enzymes and many are secreted in great quantity by fungi and bacteria, allowing relatively simple purification from culture media; they require no cofactors; they typically catalyze without undesirable side reactions. The commercial lipase market is currently worth approximately $1 billion per year, involving applications in detergents and in the production of food ingredients and enantiopure pharmaceuticals [410]; this status is fortunate for biodiesel applications because it provides preexisting incentives for improvements in lipase production and activity. A chart comparing the relative merits of the alkali-catalyzed and enzyme-catalyzed transesterification processes for biodiesel production is shown in Tab. 5.9.

Table 5.9. Comparison of alkali and enzymatically catalyzed transestrification for biodiesel production [411].

Reaction temperature	Alkali process	Lipase process
Free acids in raw material	60–70	30–40
Water in raw material	Formation of undesired saponified products	Formation of desired esters
Yield of methyl esters	Normal	Higher
Recovery of glycerols	Difficult	Easy
Purification of methyl esters	Cleaning required	No washing required
Cost of catalyst	Low	Relatively high

The direct use of whole cells, rather than the same enzymes, allows the difficulties associated with the sensitivity to a loss of enzyme activity to be overcome. Genetic modifications of cells are applied, causing the whole cells to catalyze the transesterification of fatty acids more efficiently, which further enhances the activity and stability of the enzymes (e.g., lipase enzyme).

Tallow and lard can also be used as raw material for biodiesel production. These are raw materials of small value and are derived from animal byproducts, and do not compete with food production [417, 418].

The innovative method of applying ethyl acetate to the transesterification reaction could solve many problems associated with traditional methods. Ethyl acetate is the ester of ethanol and acetic acid; it is manufactured on an increasingly large scale for use as a solvent. The combined annual production in 1985 of Japan, North America, and Europe was about 400,000 tons [419], but in 2004, an estimated 1.3M tons were produced worldwide [287]. The schematic diagram in Figure 5.43 presents a clean production of esters of fatty acids based on the transesterification of triglycerides with ethyl acetate. It is a concept for clean biodiesel technology for the future because it demonstrates complete zero-waste production. It should be noted that two different esters, which are appropriate fuels, are obtained at the same time during transesterification of ethyl acetate. They are soluble and there is no need for phase separation as in conventional technology. In addition, heterogeneous catalysts are unnecessary, which in cases of minimal residues (microemulsion) in biodiesel could damage the aluminum engine. Also of great benefit is a substantial technological simplification, because a water phase (with glycerol and catalyzers) is not required for separation from biodiesel. The only disadvantage is the need for ethyl acetate which, however, can be formed from ethanol. Flexibility and simplicity, these are further advantages of this technology which is very desirable in the design of biorefineries. The reaction of transesterification of triglycerides with alcohol results in new ester as follows:

Triglyceride + ethyl acetate = ethyl ester + glycerin acetate:

$$
\begin{array}{ll}
CH_2 - OOC - R_1 & \qquad R_1 - COO - C_2H_5 \quad CH_2 - OOC - CH_3 \\
\quad | & \qquad \qquad \qquad \qquad \qquad \qquad \quad | \\
CH - OOC - R_2 + 3CH_3COOC_2H_5 \longleftrightarrow R_2 - COO - C_2H_5 + CH - OOC - CH_3 \\
\quad | & \qquad \qquad \qquad \qquad \qquad \qquad \quad | \\
CH_2 - OOC - R_3 & \qquad R_3 - COO - C_2H_5 \quad CH_2 - OOC - CH_3
\end{array}
$$
$$\text{(5.95)}$$

The biodiesel formed during this clean technology is the solution of two esters and does not, as in the conventional two-phase production system, require phase separation and catalyst to be removed. This technology does not have byproducts because ethyl acetate is a complete product and has many applications discussed below. Acetate is formed entirely from ethanol in a catalytic reaction; moreover at least two options are possible. Such technology has the further advantage that it is more flexible

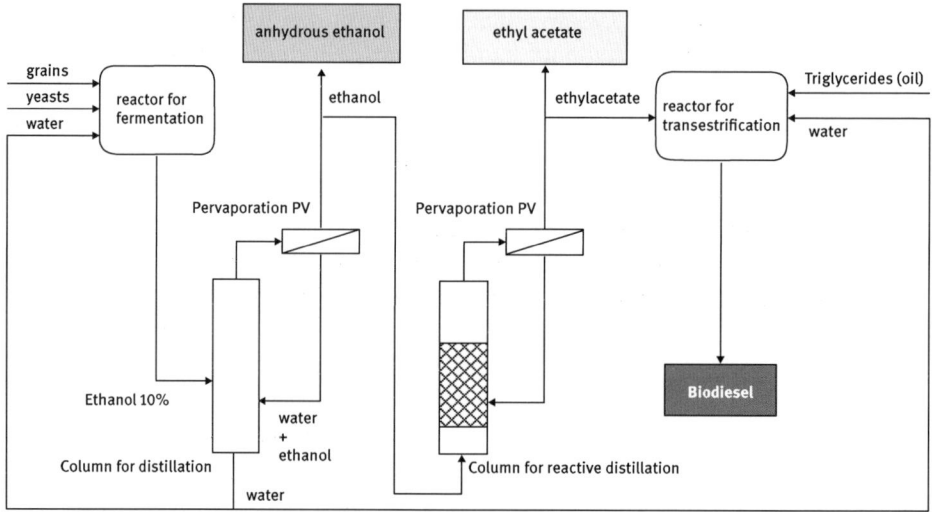

Fig. 5.43. Biodiesel clean technology based on ethyl acetate.

and dependant on the price of raw materials (cereals and oils), and the conditions of the products, such as biodiesel, ethyl acetate and ethanol. Production may be redirected immediately depending on actual needs.

The ethyl acetate is taking part in production of nitrocellulose, plastic, such as vinyl and ester resins. Ethyl acetate is used in glues, nail polish removers, decaffeinating tea and coffee, and cigarettes. It is used in many technologies as a low-toxicity solvent for paints, varnishes, enamels, isocyanates, adhesives, finishing substances (polyurethane), transparencies, coatings used for example in the production of artificial leather, ink, paint and varnish remover. It is also used in production of herbicides, oil, and grease. In organic synthesis it is used as a substrate or an intermediate, e.g. as solvent for extraction of N-nitrosamines. In food industry it is used as a gelling agent, the additive of flavoring to food, such as soft drinks grape. Ethyl acetate is also used in the perfume industry as a means fragrance ingredient and perfume essences. In the pharmaceutical industry it is used as an extractant in the manufacture of drugs and component of the essence. It may be also used as a soil stabilizer.

5.6.6 Production of biofuels from algae

Among algal fuels' attractive characteristics are the fact that they can be grown with minimal impact on fresh water resources [420], can be produced using sea and wastewater, and are biodegradable and relatively harmless to the environment. Algae cost more per unit mass (as of 2010, food-grade algae costs about $ 5000 due to high capital and operating costs), yet are claimed to yield between 10 and 100 times more fuel per unit area than other second-generation biofuel crops [421]. Microalgae is seen as a very attractive source because of the potential high yields of biomass (up to 250 t/ha/year), which is about 15–300 times more than in terrestrial plants. The companies RWE Energy and Glenturret (both with approx. 6 t/year dry algae mass on 600 m^2, which is equivalent to 100 t/ha/year, using tubular or flat bioreactors) and Enitech (three open ponds of 2.5 m^2, each and two tubular photo-reactors of the same surface) have small installations, far from the capacity of BFS with a full automatized semi-industrial plant with output near to 200 t/ha/year. BFS bioreactors' plants are located near large CO_2 emitting companies. The use of microalgae as a potential source of oil is attractive because of the faster growth rate, high oil content and the ability to harvest often for long periods of time. The composition of the fermentation broth in the bioreactor is typically: 25 % fatty acids, 30 % protein, 38 % carbohydrates and 7 % various fillers. Several solvents such as hexane or a mixture of chloroform and methanol are used for the extraction of lipids from the algal biomass. Further suitable solvents are ionic liquids and sub-critical solvents. In order to extract lipids from algae, he physical-chemical methods of cell disruption are used such as ultrasound, microwaves, salting etc.

Table 5.10. Oil production of selected vegetables

Oil source	Biomass (t/ha/yr)
Soybean	1–2.5
Rapeseed	3
Oil palm	19
Jaropha	7.5–10
Microalgae	140–250

Butanol can be made from algae or diatoms using only a solar-powered biorefinery. This fuel has 10 % less energy density than gasoline, and greater than that of either ethanol or methanol. Butanol can be used in place of gasoline in most gasoline engines without modifications. A number of tests have found that butanol consumption is similar to that of gasoline, and when blended with gasoline, provides better performance and corrosion resistance than that of ethanol or E85. The green waste left over from the extraction of oil from algae can be used to produce butanol. In addition, it has been shown that macroalgae (seaweeds) can be fermented by Clostridia genus bacteria to butanol and other solvents [422]. Biogasoline is gasoline produced from biomass such as algae. Like traditionally produced gasoline, it contains between 6 (hexane) and 12 (dodecane) carbon atoms per molecule and can be used in internal-combustion engines. Biogasoline is chemically different from biobutanol and bioethanol, as these are alcohols, not hydrocarbons. Methane, the main constituent of natural gas, can be produced from algae in various methods, namely gasification, pyrolysis and anaerobic digestion. Methane is extracted under high temperature and pressure in the gasification and pyrolysis methods. Anaerobic digestion is a straightforward method involving the decomposition of algae into simple components then transformation into fatty acids using microbes scuh as acidific bacteria, followed by removing any solid particles and finally adding methanogenic bacteria to release a gas mixture containing methane. A number of studies have successfully shown that biomass from microalgae can be converted into biogas via anaerobic digestion [423, 424]. In order to improve the overall energy balance of microalgae cultivation operations, recovering the energy contained in waste biomass via anaerobic digestion to methane for generating electricity has therefore been proposed.

5.7 Hydrogen production

5.7.1 Nonbiological hydrogen production

Hydrogen is known as the ideal fuel since it does not emit exhaust gases, in contrast to other well-known fuels. However, hydrogen is not a primary fuel at all, but only an energy carrier, because water is produced during combustion which can be turned back into hydrogen. This is why so much attention is being paid to hydrogen as the fuel of the future. The interest in hydrogen production has focused partners in an international forum in order to solve the following dilemma: on the one hand the realization of the absolute necessity of transition to hydrogen as soon as possible, and on the other it is very labor-intensive work, which as usual means economic difficulties. The organization (PATH) is engaged in partnerships to support the transition to hydrogen fuel. This is a nonprofit international coalition of hydrogen associations founded in 2002 which seeks to encourage international cooperation to help accelerate the transition to hydrogen as a potentially carbon-free energy carrier in order to solve environmental problems worldwide (www.path.org). Today, the organization PATH draws together 20 associations and partner organizations on 5 continents, representing 79 % of global GDP and over 40 % of the world's population.

However, it must be emphasized that it is not indifferent to how hydrogen is produced, i.e. whether the method of hydrogen production involves the use of fossil fuels or not, In addition, as the production of hydrogen must not become an additional cause of even faster depletion of fossil fuels. We have to remember that hydrogen is an energy carrier and should circulate globally. It is estimated that global production of hydrogen in 2013 amounted to 100 billion tons. In 2006, the U.S. had a hydrogen production capacity of 11 million tons of. Global revenues of hydrogen and fuel cells together could amount to $ 9.2 billion in 2015, and from 2020 could reach up to $ 38.4 billion. It is estimated that about 868 billion normal cubic meters of hydrogen will be consumed in 2018, more than 5 million tons for crude oil production. Hydrogen is used in the Haber process for ammonia and carbon monoxide production.

Approximately 0.4 million tons were a byproduct of the chlor-alkali process. China produces the most hydrogen, using steam reforming of natural gas and coal, the two main methods for producing large volumes of hydrogen. The hydrogen produced is then used mainly in the production of ammonia and methanol. China produces almost the same amount of hydrogen as in the US, Western and Eastern European production together. So far, nearly 96 % of all hydrogen comes from fossil fuels, including more than half from gas (49 %) followed by 29 % oil, 18 % carbon and the rest (about 4 %) from electrolysis and other byproducts. The following production methods are possible:

1. steam reforming;
2. photocatalyic water splitting;
3. thermolysis;
4. electrolysis;
5. biocatalysed water splitting;
6. enzymatic and microbial photofermentation;
7. algae photofermentation;
8. dark fermentation;
9. microbial fuel cells.

Currently, the most common technology is steam reforming of hydrocarbons, mainly in oil and gas. Hydrogen, however, does not occur in nature in sufficient amounts. Steam reforming of crude oil is a process that produces more than 5kg of carbon dioxide for every kilogram of hydrogen when methane is used, and even more if natural gas is used. Hydrogen produced in this costs about $ 4.50 per kilogram of hydrogen. One kilogram of hydrogen is an energy carrier of 143 MJ/kg or about 40 kWh/kg. The cost of hydrogen production using traditional methods is $ 4.00/kg and the energy of hydrogen costs $ 0.08/kWh.

Hydrogen production by electrolysis requires large amounts of energy. Creating the necessary heat and pressure is very energy-intensive and expensive. Hydrogen may result from high temperature electrolysis (HTE). High (950–1000 °C) temperature can be derived from nuclear reactors at a cost competitive with natural gas steam reforming. The defense and aeronautical contractor General Atomics predicts that hydrogen produced in high temperature gas cooled reactors (HTGR) would cost $ 1.53/kg. In 2003, the steam reforming of natural gas obtained hydrogen at a cost of $ 1.40/kg. In 2005, the cost of producing hydrogen from natural gas was $ 2.70/kg. At high temperature in the laboratory (not on an industrial scale) electrolysis absorbed 108 mega joules per kilogram of hydrogen produced. It is also lower quality hydrogen, which is not suitable for "commercial" use in fuel cells [425, 426].

A thermochemical hydrogen production process is one which requires only (i.e. mainly) water as a material input and mainly thermal energy, or heat, as an energy input. The process consists of many chemical reactions which result in the water being decomposed into hydrogen, oxygen and various residues, with. heat being developed in the process. Hydrogen production by this method is still in its infancy and installation has not reached a level beyond industrial research laboratories. There are more than 352 thermochemical cycles which can be used to separate of water, but only around a dozen of these cycles are currently being intensively investigated, such as the cycle as iron oxide, cerium (IV) oxide, cerium (III) of the cycle, zinc oxide, zinc oxide, sulfur-iodine cycle, cycle, copper-chlorine cycle and cycle hybrid sulfur. These processes could be more effective than high temperature electrolysis, typically in the range of 35 %–49 % LHV efficiency. Thermochemical hydrogen production using chemical energy from coal or natural gas is not generally considered as a direct chemical approach is more efficient.

Attempts were also made to use solar energy directly for hydrogen production in Spain in the pilot plant hydrosol 2 Solar de Almeria, with a capacity of 100 kilowatts. Heating water in the system is achieved through the direct use of solar energy, which is concentrated by the mirrors tracking the sun. Very high temperatures are required for the dissociation of water into hydrogen and oxygen (800 to 1200 °C). Pilot plant design was based on a modular concept which permitted easy increases in scale by multiplying.

Nonetheless, almost all hydrogen is produced nowadays from fossil fuels. In the future, however, biological methods need to be developed as a necessary condition for our survival. Methods of processing fossil fuels to hydrogen can continue to serve us for a long time. In particular method of underground combustion of coal using the underground water and with the absorption of carbon dioxide in underground rocks appears very promising. Underground deposits of lignite are very common all over the world, but they will also be used up and run out eventually. In addition, all nonbiological methods of hydrogen production require energy, which unfortunately comes from fossil fuels.

Photocatalycal water splitting is the simplest method, but it requires a catalyst, which is usually a semiconductor. The photon energy is converted to chemical energy accompanied by a largely positive change in Gibbs free energy through water splitting. This reaction is similar to the photosynthesis of green plants because they are both uphill reactions. Therefore, photocatalytic water splitting is regarded as artificial photosynthesis and is an attractive and challenging subject in chemistry. Photohydrogen is hydrogen produced with the help of artificial or natural light. This is how the leaf of a tree splits water molecules into protons (hydrogen ions), electrons (to make carbohydrates) and oxygen (released into the air as a waste product). Photohydrogen may also be produced by the photodissociation of water by ultraviolet light:

$$H_2O + h\nu \rightarrow H_2 + \frac{1}{2}O_2 + \Delta G = 238 \left[\frac{kJ}{mol} \right]. \tag{5.96}$$

The theoretical band gap separating water is at least 1.23 eV, which corresponds to light of about 1100 nm. However, the spectrum of solar UV radiation is small, so this simple method is rather impractical on Earth.

5.7.2 Biological production of hydrogen

It is obvious than only solar energy is free, which we have to learn. All of the biological methods are definitely associated with solar energy in some way. Solar energy, which is converted by photosynthesis into chemical energy, then circulates in many linked food chains on Earth, and they are therefore suitable for use as renewable energy sources.

Biohydrogen [427] is produced by algae, mainly as a result of the enzymatic reaction of a dehydrogenase [428, 429], bacteria and archaea. Dehydrogenase is the enzyme that converts protons to hydrogen, and is even more attractive than platinum to this end because it requires a relatively low overpotential and presents high catalytic activity [430]. [FeFe]-hydrogenase is considered to be the best for the production of H_2, as it has a high repetition frequency TOF (more than 9000 s-1) which results in good reaction kinetics [431]. However, compared to other biofuels, bio-hydrogen is still in its infancy. Biohydrogen technology is actually five distinct technologies: direct photolysis, indirect photolysis, photofermentation, dark fermentation and metabolic engineering.

Biohydrogen production can be based on four very different types of microorganisms. These are green algae, cyanobacteria, purple nonsulfur bacteria and anaerobic heterotrophic bacteria. The addition of heterologous pigments allows widening of the spectrum of photosynthesis, which is very desirable. The use of different sources of waste as substrate has been studied [432].

5.7.2.1 Biohydrogen from algae – photobiological water splitting

Photolysis [433, 434] plays an important role in photosynthesis, during which it produces energy by splitting water molecules into gaseous oxygen and hydrogen ions. Photolysis is one of the light-dependent reactions of photosynthesis. The general reaction of photosynthetic photolysis can be shown as

$$H_2A + 2\,hv \rightarrow 2e^- + A. \tag{5.97}$$

A chemical "A", can be oxygen or sulfur, depending on the type of body. In photosynthesis, oxygen, water (H_2O) is used as a substrate for the preparation of a result of the photolysis of diatomic oxygen (O_2) with carbon dioxide (CO_2). It is a process which draws oxygen into earth's atmosphere. Green algae generally also have a quite active uptake of H_2, however the enzymes responsible have not yet been identified. H_2 Fe-hydrogenases are known to produce H_2 and the uptake of hydrogenases of unknown composition.

All plants, algae and some bacteria are capable of photosynthesis: utilizing light as the source of metabolic energy. In oxygenic photosynthesis water (H_2O), serves as a substrate for photolysis resulting in the generation of diatomic oxygen (O_2) from carbon dioxide (CO_2). This is the process which returns oxygen to earth's atmosphere.

Green algae typically also have significant H_2 uptake activity, although the enzymes responsible have not yet been identified. In general terms, therefore, the primary enzymes of concern for H_2 production in green algae are Fe-hydrogenases, known to produce H_2, and uptake hydrogenases of unknown composition. Green algae, as well as cyanobacteria, must withstand periods of anaerobiosis, i.e. absence of oxygen.

The method of photobiological water splitting is performed in a closed photobioreactor with different algae mutants (C. reinhardtii) switch from oxygen production to hydrogen production [429]. The H_2 yield obtained in this manner is rather moderate, however, producing an estimated 0.07–0.08 mmol/l or 1.6–1.8 l of hydrogen per cubic meter of culture per hour. Enzymatic photolysis using the enzyme Fe-hydrogenase requires anaerobic conditions and is known as direct photolysis. The physiological role of hydrogenase-based H_2 production appears to be the discharge of excess reducing power, necessary when other suitable electron acceptors such as O_2 are absent [435–437]. Cyanobacteria can produce H_2 even by means of three different mechanisms simultaneously. This is indirect photolysis [438], when N_2 blue-green algae is used which can produce H_2. They use the enzyme nitrogenase, which uses photosynthesis to reduce nitrogen gas while generating H_2:

$$2H^+ + 2e^- \rightarrow H_2. \tag{5.98}$$

This feature is essential for the production of H_2 requiring anaerobic conditions for synthesis and dehydrogenase activity. Therefore, direct photolysis is a process in which there is a two-way flow of electrons activated by photosynthesis. Hydrogen production by algae and green plants, known as direct pholysis, is carried out using light and photosynthesis. In contrast, hydrogen can also be produced as a result of indirect photolysis involving cyanobacteria. The process is carried out using green algae such as Chlamydomonas reinhardtii [439, 440].

5.7.2.2 Hydrogen production from microbial fermentation

It has recently been shown that the cyanobacteria Synechocystis (strain PCC 6803 M55) also has the ability to direct photolysis [441–443]. Cyanobacteria are frequently mentioned as being capable of hydrogen production via oxygenic photosynthesis. However, only bacteria can produce H_2 by metabolic fermentation in large quantities. These bacteria include various strains of Bacillus, Clostridia and Enterobacter and the extremely thermophilic Caldicellulosiruptor, all heterotrophic and Thermotogae which feed on carbohydrates and are not usually capable of anaerobic respiration. Photofermentation is the conversion of organic substrates to bio-hydrogen with the participation of a diverse group of photosynthetic bacteria and enzyme reactions involving three steps similar to anaerobic metabolism.

Photofermentation [444] differs from dark fermentation [445] because it only occurs in the presence of light. Dark fermentation is the complex process of conversion of the fermentation of organic substrates to biohydrogen without light energy during the day and night. The process is performed by different groups of bacteria, including a number of biochemical reactions using three steps similar to anaerobic conversion. The dark fermentation of organic substances can be coupled with the treatment of industrial wastewater from various sectors of the food industry.

However, mixed organic waste substrates are complex and not always suitable for the production of hydrogen in the process of fermentation plants. Such trials have been performed with wastewater from dairy, sugar factories, lactic acid fermentation and from oil.The efficiency of conversion of the substrate into hydrogen can be described according to the equation

$$C_xH_yO_z + (2x - z) \cdot H_2O \rightarrow (y/2 + 2x - 2) \cdot H_2 + x \cdot CO_2. \tag{5.99}$$

Purple nonsulfur bacteria may use a variety of substrates for photoheterotrophic growth, only some of which are suitable for the production of H_2. Rhodobacter sphaeroides preferred butyrate, lactate and malate. The light output of light conversion is the ratio of the total energy or heat of combustion of hydrogen H_2, which is derived from the total light energy incident on the photobioreactor. Photofermentative production of H_2 varies between 1 and 5 % per cent on average:

$$CO + H_2O \rightarrow CO_2 + H_2. \tag{5.100}$$

This process is independent of light. The reaction is carried out by some bacteria such as Rhodospirillaceae photoheterotrophic in Rhodospirillum rubrum and Rubrivivax gelatinosus CBS using the enzyme known as carbon monoxide dehydrogenase (CODH) in combination with hydrogenase [446, 447].

Dark fermentation [448] refers to the light-independent production of H_2 by anaerobic heterotrophic bacteria and Gram-positive Clostridium bacteria. These microorganisms may be mesophilic (optimal metabolic conditions at 25–40 °C), thermophilic (40–65 °C), a thermophilic (65–80 °C) or hyperthermophilic (> 80 °C).

Most fermentation methods have a number of different culture conditions, which strongly affects the method of fermentation. Therefore not only hydrogen H_2 results, but also CO_2, CH_4 and volatile fatty acids, depending on the fermentation route used [449]. Organic wastes are an attractive option, but difficult to connect with the advanced approaches of metabolic engineering. The main limitation to the commercialization efforts according to many researchers, is the limited communication between scientists who study biohydrogen systems and engineers [253].

5.7.2.3 Microbial fuel cells, an ideal source of hydrogen and/or electricity

Microbial fuel cells (MFC) can be used to produce hydrogen, as well as other compounds such as methane or ethanol, and for electrochemically supported denitrification. But the biggest advantage of MFC is the ability to perform wastewater treatment [450] and electricity production simultaneously. When a corresponding electrical potential (voltage) is produced, hydrogen can also be obtained. Wastewater [451], sludge or other waste material may be used as a substrate for microbial growth. This is the technology of the future, because it allows the combination of waste degradation and the production of clean energy, i.e. electricity or hydrogen, thus meeting all requirements of clean technologies and sustainable development [452].

Microbial fuel cells (MFC) or biological fuel cells are semi-closed systems in which the bacteria produce protons in an anaerobic chamber during fermentation. These protons are then removed on flowing from the chamber through an ion-selective membrane. A typical MFC is composed of two compartments separated by the selective proton exchange membrane. In the anode chamber, there is organic material is decomposed by anaerobic bacteria and carbon dioxide, protons and electrons are released. The membrane allows the only the passage of protons, while the free electrons migrate to the cathode through an external electrical circuit.Oxygen reduction takes place in the cathode chamber with the participation of electrons, and water is formed from the oxygen and protons. Some MFCs have a chemical mediator which transfers electrons from the bacterial cells to the anode. MFCs are also employed without mediator, but with the use of special bacteria in the outer cell wall – specific redox proteins such as cytochrome, which can transfer electrons directly from the anode [453, 454]. According to new research, the conversion of energy to hydrogen using the new microbial fuel cells is 8 times higher than conventional hydrogen production technologies. A review of the known biohydrogen production methods is shown in Table 5.11.

Table 5.11. [445] Comparison of H_2 biosynthesis

Biohydrogen production	Synthesis rate mmol H_2/(liter hour)
Direct photolysis	0.07
Indirect photolysis	0.355
Photo-fermentation	0.16
CO oxidation by R. gelatinosus	96
Dark fermentation by mesophilic pure strains[A]	21
Dark fermentation by mesophilic undefined[B]	64.5
Dark fermentation by mesophilic undefined	121
Dark fermentation by mesophilic pure strains	8.2
Dark fermentation by mesophilic pure strains[C]	8.4

[A] Clostridium species #2
[B] A consortium of unknown microorganisms
[C] Caldicellulosiruptor saccharolyticus

6 Process engineering closer to man

The OECD report, which had to predict the future within the period 2005–2030 concluded that the world population will grow from 6.5–8.3 billion people, i.e. about 28 %. Also the global average income per capita will increase during this period by 57 %, i.e. from USD 5,900 to 8,600. The greater the number of wealthy people in the world, the greater the demand for health care services which improve quality and length of life and for healthy and natural products. Increased demand is expected for such products as organic foods and animal feeds, natural fibers for clothing and natural materials to build housing, manufacture clean water and clean energy, which generally increase the demand for clean technologies. the most important thing for every human being, however, is a long and healthy life. Therefore, the development of modern medicine is a prerequisite for human development.

6.1 Applications of process engineering to modern medicine

6.1.1 General view of medical cooperation with engineering

Cooperation between process engineering and medicine has recently become more successful. The need for an engineering approach to describe processes in the human body became obvious because of the advancement of medical technology, new methods of diagnosis and therapy. This multidisciplinary approach has been known and used in medicine for a long time because it has always sought the best solutions for the sake of the patient's health, regardless which discipline provided these solutions. Adolf Eugen Fick (born 1829 in Kassel), who formulated the law of diffusion mass transport was a German physiologist and professor at the University of Zurich and Würzburg. He is credited with the invention of contact lenses. Wilhelm Conrad Röntgen (1845–1923) German physicist and Nobel laureate built the foundations of medical diagnosis based on X-ray time distribution [455]. Eugene Cook Bingham (1878–1945) was a professor and head of the Department of Chemistry at Lafayette College. Bingham studied the rheological properties of blood and body fluids [456]. Researching residence time distributions which are routine in process engineering, Taylor began studying the stimulus-response method of the flow of blood in blood vessels. It remains a good and promising method for targeted drug delivery. In summary, the contribution of process engineering to medicine consists of treating the human body as a system in which there are flows, heat and mass transfer, i.e. the same physical laws as in other devices. The most important achievements of process engineering in medicine were the development of controlled drug delivery and artificial organs.

Dr. Willem Kolff is considered to be the father of dialysis. This young Dutch physician constructed the first dialyzer (artificial kidney) in 1943. Kolff gave relief and even saved many patients suffering from urease by using dialysis to remove urea from the blood. The artificial kidney was imperfect and inconvenient for patients. Artificial kidneys have been continuously upgraded in order to minimize patient comfort. Globally, dialysis membranes are still the biggest market for membranes. Membranes led to an introduction to medicine of other artificial organs such as artificial lung, artificial pancreas and artificial skin. Work has been performed on an artificial liver for many years.

Artificial organs are now bio-hybrid devices which are functional substitutes for human organs. All artificial substitutes must ensure both the proper microenvironment for activity in the metabolism of cells, endocrine islet and efficient mass transfer associated with the bio-active metabolites and their removal, thus allowing for homeostasis in the patient's whole body.

Recently, advanced physics has been involved in the development of such techniques as magnetic resonance imaging (MRI) to produce images of internal body structures. Therapeutic radiology, known as radiation oncology hereinafter, is a treatment for cancer involving the projection of high-energy radiation only onto the target tumor cells in order to reduce and eliminate tumors. Nuclear medicine is used for both imaging and therapy. In imaging, patients are given small amounts of radioactive materials called radiopharmaceuticals orally, intravenously or by inhalation; a special camera detects these radiopharmaceuticals.

The future of chemical engineering in medicine is in the development of bio-engineered hybrid devices using stem cells and the construction of a fully functional system with active transport and metabolic activity.

6.1.2 Drug delivery systems

Transport phenomena, in particular the diffusion of drug molecules in the organs are a perfect example of introducing chemical engineering to medicine and biomedical sciences. Formerly, drugs were delivered to patients orally or by injection. Based on the knowledge of the distribution of particles in the cardiovascular system and in organs or tissues, a variety of improved techniques of administration, such as nasal or oral inhalation or transdermal delivery have been developed. Methods also used are internal drug delivery, or targeted drug delivery techniques, which have been developed in collaboration between chemical engineering and medicine [457–459]. The application of advanced chemical engineering methods to drug delivery systems allows miniaturization of delivery devices from the macro-scale (> 1 mm) to the micro-scale ($100–0.1$ μm) or even nano scale (100^{-1} nm) [460, 461]. The desired concentration of the drug agent in the targeted organ or tissue, spatial position (i.e. diffusion characteristic) may be attained based on the drug delivery kinetics, i.e. rate of change of agent concentration with time together with the appropriate delivery devices, thus increased therapeutic effectiveness can be achieved.

With conventional methods of drug delivery, regardless of whether they are oral or intranasal, and irrespective of the method of delivery and dosage form a jump in concentration always occurs. The drug may even reach the level of toxicity before its concentration quickly falls below the therapeutic range. In comparison with conventional drug dosage, the main advantage of controlled drug delivery systems is maintenance of therapeutically optimum drug concentrations in the blood plasma through continuous zero-order kinetics without the need for frequent single dose administrations [462, 463]. Thus the drug is continuously supplied to the organism at a rate necessary to provide a constant, required drug concentration in the blood.

Diffusion-controlled dispensing of drugs has been studied since the early 1970s. The principle of controlled drug delivery of the active compound is combining a drug with a carrier, which may be, for example a hydrogel polymer in a concentrated lipid. Distributed in the carrier drug, it is then released from the carrier to the environment (e.g., tissue) at a specified rate. The dosage system is effective when the drug is released in a specified range of concentration between the minimum and maximum desired levels. Traditionally, the thickness of the capsule wall (e.g. a layer of starch) plays a key role in controlling the dose of the medication when the capsule is swallowed by the patient. Such a system is not always sufficient, however, specifically in cases of serious chronic disease such as. diabetes, depression, epilepsy, malignant lymphoma etc. The active diffusion controlled system, prevents irregular dosage of medicines by patients. A typical dose of most drugs does not exceed a few hundred micrograms of biologically active agent per adult patient with an average weight of 50–70 kg. Drugs administered in much larger quantities than those commonly used such as antibiotics (e.g. penicillin and amoxicillin), are typically dispensed in doses of less than 1g per day for adults. In addition, drugs for oral and nasal administration almost always contain pharmacologically unnecessary additives, such as carriers, flavoring agents and coatings. Regardless of the route of administration, its release and distribution in an organ or the whole body is related to the diffusion coefficient of drug molecules within the tissue or organ. In the case of clinical diffusion controlled release of the drug, several systems using continuous release of biochemically active compounds are actually recognized [464]. They are:

1. transdermal therapeutic systems (TTS);
2. oral (e.g. mechanical or nonmechanical mini-/micro-pumps);
3. parenteral (e.g. nano capsules);
4. subcutaneous (e.g. implanted drug-carriers);
5. intracavitary (e.g. intrauterine inserts); or
6. buccal (e.g. trans-mucosal release) drug delivery systems.

Some devices provide controlled release of the continuous diffusion of therapeutic agents such as derivatives of Norplant®, a Jadelle®, Implanon®, Nexplanon®, used for hormonal contraception. Gliadel® wafers are used against brain tumor (e.g., malignant glioma), and has proven a significant therapeutic and commercial success.

A number of other devices based on microfluidic dispensing systems and implants based on biomaterials are currently being developed.

6.1.3 Artificial kidney

Patients with renal insufficiency or chronic renal failure require dialysis, i.e. the treatment using an "artificial kidney" every third day. Blood flows through a special membrane dialyzer, which removes bothwaste metabolites and excess water, and is finally returned to the cardiovascular system. The whole process is maintained by a dialysis machine which is equipped with a hollow-fiber type dialyzer, a blood pump and monitoring systems to ensure safety. The key element of every dialyzer is the dialysis membrane. Small molecules dissolved in blood plasma (e.g. water, glucose, urea and creatinine) pass through pores of the membrane. The blood cells, i.e. red and white cells and thrombocytes, as well as most plasma proteins are retained by the dialysis membrane [459]. Cellulosic and synthetic membranes are used in dialyzers [465–467]. Cellulosic-based membranes are mostly prepared from cellulose acetate. Synthetic membranes are produced from hydrophilic blends or hydrophilic/hydrophilized copolymers.

The mass transfer though the dialysis membrane is made up of two components: diffusion and convection. Based on the mass transfer mechanism applied for solute removal, three main types of dialysis are commonly used [468]: (1) hemodialysis, (2) hemofiltration, and (3) hemodiafiltration. In hemodialysis, removal of dissolved small molecules from blood plasma is performed by diffusion and the driving force for mass transport is the concentration difference or osmotic gradient across the dialysis membrane, between the blood and the dialysate (purified and standardized electrolyte solution). In the hemofiltration process, the removal of waste metabolites is performed by convection and mass transport, which is driven by a hydrostatic pressure difference. In hemodiafiltration, the removal of the solute is done by a combination of both diffusion and convection, with concentration and pressure gradient as the driving forces for mass transport.

Regardless of the membrane performance, which is determined by the sieving coefficient, from a clinical point of view, the most important parameter characterizing dialysis is the clearance (K):

$$K = \frac{r_t}{C_B}, \tag{6.1}$$

which represents the relationship between the rate of toxin mass removal from the blood (r_t) and the concentration at incoming blood (C_B). The value of K not only depends on the membrane properties (i.e. dialyzer characteristics), but also on the process design (e.g. on volume flow rate of blood in hemodialysis and on the hydrostatic pressure gradient in hemofiltration). The hydrodynamics of the boundary layer may exert the predominant effect on the entire process, especially in hemofiltration or hemodiafiltration. From a functional point of view, the techniques of dialysis are inadequate for the removal of large molecules and protein-bound waste metabolites, which are therefore accumulated in the body (mainly in blood plasma) and lead to uremic toxicity. The selectivity and specificity of the mass transport through the dialysis membrane can be enhanced by the application of in vitro cultured renal cells within the dialyzer device [469–471]. The incorporation of renal cells into the dialyzer reconstructs the native biological metabolic function of the whole artificial kidney system. Such abio-hybrid idea, i.e. membrane combined with kidney cells, is known as "a bio-artificial kidney" and has been in development since the end of the 1980s.

6.1.4 Functional liver substitution

The liver is a vital organ performing a wide range of metabolic and detoxification functions toward endogenous substances and pharmacological agents which have been introduced into the organism. The functionality of the organ is maintained by hepatocytes and the loss of liver functionality directly leads to life-threatening complications. First attempts at artificial livers were made based on the detoxification of blood, such as hemodialysis, hemofiltration or hemoperfusion, and were developed based on membrane devices [459, 472]. Removal of small molecules of wastes by low-permeable membranes has been significantly enhanced in the hemodiadsorption system, combining hemodialysis with adsorption on charcoal or albumin as nonspecific or specific adsorbents for large molecules and protein-bound wastes [473]. Such systems are in widespread use for liver failure therapies and for brief periods because they do not support protein synthesis or biotransformations which are assured by the liver in vivo. In order to overcome these disadvantages, bio-artificial liver support systems (BALS) are currently in development. Such bio-hybrid systems of bio-artificial liver substitutes incorporate hepatocytes (human or animal) into the membrane system, therefore functionality of the liver can be fully recovered through native metabolic and biosynthetic functions of liver cells. The key element of commercial BALS systems is a bioreactor in which the hepatocytes are grown [471, 474, 475]. Several bioreactor designs have been proposed from the point of view of process engineering. They can be classified into three main categories: (1) membrane systems, (2) perfusion devices with immobilized cells on various supports, and (3) systems of entrapment [476]. Cells can be cultured as multicellular aggregate suspensions, or membranes with the cells attached (in hollow-fiber or flat sheet membranes) or immobilized (mainly encapsulated) in biomaterials. In the case of widespread use of membrane bioreactors with hollow-fibers, the blood flows either inside or outside the fiber lumen, whereas the hepatocytes are attached to the surface on the opposite side. The reliable use of BALS seems to be a key issue for efficient mass transfer in the case of toxins, metabolites and oxygen, rather than with respect to a tight immunological barrier. A BALS cartridge should provide the possibility for a 3D self-organization of hepatocytes and follow conditions. Iit must ensure efficient mass transfer of wastes from detoxified blood to the cells. Additionally, the validation of newly developed BALS systems should not only depend on clinical-trial results but much more significantly on fundamental studies needed to understand the mass transfer process between cells and the perfusion fluid, as well as to suggest means to optimize them. Of particular interest are the hydrodynamic conditions leading to better exchanges. To date all BALS systems have been proven fully successful in liver failure therapy. The spectacular scientific challenge is still the development of the bio-artificial liver which would allow patients to survive until liver transplantation is possible.

6.1.5 Artificial pancreas

The human pancreas has two main metabolic functions: 1. Endocrine gland producing several important hormones such as insulin, 2. Digestive organ producing digestive enzymes such as proteases, lipase and amylase. The functionality of the pancreas is focused on the controlled release of hormones directly into blood vessels and controlled release of enzymes into the duodenum. In the case of a malfunctioning pancreas, the patient takes insulin for proper regulation of glucose concentration in blood plasma and pancreatic enzyme supplements to aid digestion of proteins, lipids and carbohydrates. The most common causes of pancreatic failure are diabetes, a group of metabolic diseases in which a patient has elevated blood sugar caused by problems with insulin secretion. The insulin injection commonly applied as diabetes therapy can only imitate the pancreas function concerning insulin delivery, but continuous monitoring and active regulation of blood sugar and other pancreatic functions cannot be supported. Therefore, the development of a fully functional bio artificial pancreas (BAP) may be an important approach in the treatment of diabetic patients. Such a bio-hybrid device must coherently integrate two main functional aims:, i.e. medical and processing aspects [477, 478]. The BAP system must provide an adequate microenvironment for the active metabolism of endocrine islet cells (medical aspect) as well as for an efficient mass transfer in regard to the hormone (i.e. insulin) which is delivered intravenously, in a proper closed-loop system, for sustained blood sugar leveling (process engineering aspect).

The BAP devices are classified into two main groups: intra-vascular and extra-vascular systems. The intra-vascular systems are perfusion devices and are usually made of hollow-fiber membranes inserted as a blood vessel shunt in the body. The shell spaces between the membranes are inoculated with the islet cells and the blood perfuses the lumen of the inner membrane, allowing the transport of nutrients, oxygen, and bio-signaling molecules to the cells [479, 480]. Extra-vascular BAP devices are biomaterial-based (mainly hydrogel) micro-capsular or macro-capsular systems containing cells or cell clusters, respectively [481, 482, 489]. The cells are encapsulated by a polymer membrane which simultaneously allows diffusion of insulin and prevents contact with immune compounds from the host immune system. This enables the transplantation (intra-organism) of such micro-encapsulated islet cells derived from pig into the human blood system without continuous immunosuppressive therapy. Moreover, the crucial issues for the further success of capsular-based extravascular BAP systems are diffusional limitations, which are imposed by the encapsulating polymer and the capsule size. However, among several types of bio-hybrid BAP systems presented so far, DIABECELL® (Living Cell Technologies, New Zealand) microencapsulated beads containing porcine islets are already in use in late-stage clinical trials and may, perhaps, replace islet transplantation soon [483].

6.1.6 Liquid assisted ventilation (LAV) as an alternative therapy for pulmonary failure

Liquid assisted ventilation (LAV) with perfluorochemical (PFC) liquids has been investigated as an alternative respiratory modality for over 30 years [484]. This is a simple way of enhancing the mass transport of the gases O_2 and CO_2 in the lungs by filling them with a special kind of fluid instead the air. During breathing or ventilation, the fundamental function of the respiratory system is to provide a continuous gas exchange in the organism, by supplying oxygen from atmospheric air and removing carbon dioxide generated in the cells of the body. The gas exchange takes place by diffusion in the alveoli, which are densely covered by a network of capillaries. This requires an appropriate distribution of air in the lungs. The innovative strategies to support pulmonary exchange of respiratory gases (i.e. O_2 and carbon dioxide) in lung disease while preserving lung structure and functionality are particularly required in preterm infants. To address this issue, liquid assisted ventilation (LAV) with liquid carriers of respiratory gases has been investigated as an alternative pulmonary treatment [484–488]. In intrapulmonary LAV techniques the gas-liquid interfacial area is replaced with a liquid-liquid interface at the lung surface. Simplistically, in LAV therapy gaseous nitrogen is replaced by tracheal-applied liquid as the carrier of O_2 and CO_2 molecules in the inhalation and exhalation phases of breathing, respectively. In turn, high surface tension at the gas-liquid interface is eliminated. Because there is less resistance in mass transfer from the liquid-filled alveoli into the blood in the capillaries, blood flow through the intrapulmonary capillaries is more homogeneous in the fluid-filled lungs compared with the gas-filled organ. It can therefore be said that LAV undoubtedly assumes an integral role in clinical medicine concerning respiratory failure therapies.

The function of the respiratory system is closely related with that of the cardiovascular system. About 300–500 million alveoli co-create human lungs. A huge development of interfacial area (up to 70 m^2) is needed for efficient gas transfer between the gas phase and bloodflow in the capillaries. In the case of acute lung injury, surface tension at the air-liquid interface of the lung is increased because the pulmonary surfactant system is damaged, causing alveolar collapse which finally leads to hypoxemia. Widely used in lung diseases, conventional mechanical ventilation (CMV) techniques with oxygen-enriched atmospheric air commonly-used for long-term therapy of patients with respiratory failures have been shown to initiate lung injury, resulting in progressive structural damage of alveoli and release of inflammatory mediators within the lung.

Biologically and chemically inert, fully synthetic liquid perfluorochemicals (PFCs) possess high respiratory gas solubility (up to 0.5 and 2.1 dm^3 of gas/dm^3 for O_2 and CO_2, respectively) and a low surface tension (0.015–0.019 N/m) and are thus most often suited as uniquely pulmonary agents supporting an adequate alveolar reservoir for pulmonary gas exchange. The greater density of PFC compared to that of body fluids allows PFC to descend to injured regions of lungs and re-open alveoli. This recalls the studies of von Neergaard, who as early as the 1920s reported that the pressure necessary to expand a lung filled with atmospheric air is almost three times higher than that required for a lung filled with liquid. The kinematic viscosity profile of PFCs positively influences distribution and pressure requirements for LAV techniques. And finally, the low vapor pressure of PFCs increases the rate of evaporation from the lungs. Moreover, due to the high heat capacity of PFCs, the patient's body temperature can be regulated easily and thoroughly by PFC temperature during ventilation. Intrapulmonary LAV techniques differ with respect to methodology and expected effects of modality as well as the impact of the physicochemical profile of the PFC used, but from a process engineering point of view LAV can be realized as [489–493]:

- total tidal liquid ventilation or PFC lavage, wherein respiratory gases are transported solely in the dissolved form through tidal volume exchange of PFC to and from the lung;
- partial liquid ventilation, performed by filling the lungs with a functional residual capacity of PFC while continuous mechanical gas ventilation is performed;
- intrapulmonary delivery of PFC droplets via vaporization or aerosolization.

The future availability of improved biomedical-grade PFC-based LAV agents with specifically designed physicochemical characteristics will enable further tailoring of this alternative therapeutic approach to respiratory management for individual applications. In addition, with the aid of liquid-filled lungs, drugs can be administered with greater effectiveness in lung tissue diseases involving cancer. By using LAV as a carrier for pharmacologic agents, adverse side effects can be minimized because molecules of the drug are administered directly to the surface of the alveoli.

6.1.7 Red blood cell substitutes

Blood is a complex biological fluid which is regarded as a special form of connective tissue. Blood consists of cells and abiotic components, proteins and other ingredients which provide important functions for the whole body. The red blood cells (erythrocytes) are the natural oxygen carriers in blood that release oxygen into the flow, thereby replacing consumed oxygen and adding the necessary amounts of oxygen at the proper rate. With regard to rheological properties, the blood is the Bingham fluid. This means that viscosity decreases under the influence of shear forces, near the walls of blood vessels during blood flow. This results in the red blood cells flowing very close to the walls of blood vessels, which in turn decreases the diffusion path. If blood had the properties of a Newtonian fluid, these erythrocytes would flow axially so far away from the walls of blood vessels, mass transfer would be slower, and the entire circulatory system would be less efficient. The primary function of erythrocytes in the blood is to transport gas mass, namely the removal of oxygen from tissues and organs, and removal of the carbon dioxide released from the cells. The oxygen pressure in the air-filled vesicles is about 20.7 kPa, and decreases to almost 4 kPa within the capillaries, and about 0.13 kPa or less in human tissues [494]. Such a pO2 gradient produces a longitudinal negative gradient of oxygen concentration in plasma throughout the vascular system, which follows Fick's law and stimulates oxygen diffusion in vivo, from lungs to tissues. Molecules of oxygen are bound to hemoglobin in erythrocytes, and do not have any other physicochemical activity and so from the thermodynamic point of view they do not exist [494]. This consideration is important due to the fact that oxygen is a highly reactive compound which cannot be transported via the cardiovascular system in pure form, i.e. as O_2 molecules, may be dissolved in high concentrations in blood plasma. The range of oxygen concentration in blood plasma (i.e. aqueous fraction of blood without red cells) is only of 5–10 M. This low concentration is insufficient to provide the amount of oxygen necessary for the human mitochondrial metabolism, which requires a much more efficient transport of oxygen, approximately at the rate of 1 dm^3 per minute [495]. In native blood, the problem of oxygen level insufficiency is solved by the presence of hemoglobin, i.e. a protein with an extremely high affinity to oxygen. In synthetic solutions with the ability to bind, to transport, and to unload the oxygen in the tissues, oxygen carriers must be applied which release oxygen into the flow, thereby replacing metabolized oxygen and supplementing the necessary amounts of oxygen at the proper rate. Two kinds of oxygen carriers are currently available (Tab. 6.1): (1) cell-free bio-artificial modified human or animal hemoglobin-based oxygen carriers (HBOC) [496–498], and (2) fully synthetic red blood cell substitutes, mostly perfluorochemical-based oxygen carriers (PFCBOC) [499, 500].

Increasing amounts of blood are required for the ever more sophisticated surgical procedures or for patients with acute anemia or of hemostatic factor deficiencies. In all such cases a blood transfusion or an appropriate blood product is a lifesaving treatment. Unfortunately, increasingly stringent requirements to rule out potentially infective agents (e.g. AIDS, hepatitis agents, viruses, prions etc.) in blood have decreased the number of qualified donors and highly increased the complexity and cost of donor blood. A functional substitute for red blood cells (erythrocytes) which could overcome these limitations would be highly desirable. The ideal oxygen carrier has not

Table 6.1. Bio-artificial and totally synthetic O_2 carriers currently used as red blood cell substitutes

Bio-artificial oxygen carriers	Synthetic oxygen carriers
HBOCs: 　Stabilized Hb tetramers 　Polymerized Hb 　Conjugated Hbs 　Liposome encapsulated Hb	PFCBOCs: 　PFoctyl bromide (Oxygent®) 　PFdichlorooctan (OxyFluor®) 　PFdecalin/PFmethylcyclohexylpiperidin 　(Perftoran®)
Recombinant/transgenic Hb	Synthetic metal chelates
Albumin-heme hybrids	Lipid-heme vesicles
	Hb aquasomes

yet been developed, however its characteristics can be predicted. Effective red blood cells substitutes have to provide sufficient O_2 and CO_2 uptake at physiological concentrations of these gases. The application of PFCBOC necessitates delivery of high O_2 concentrations, since the absorption of gases into PFCs obeys Henry's law, thus the amount of gas dissolved in PFC is directly proportional to the partial pressure of gas delivered [501]. However, the level of oxygen delivered to the tissues cannot exceed physiological oxygen tension. The lack of any toxicity or adverse physiological effects as well as elimination of disease transmission (e.g. viruses, prions) is imperative for the ideal oxygen carrier and it must not interfere with capillary circulation. Universal compatibility, prolonged shelf-life, easy use, stability at room temperature and immediate availability are further desirable advantages. The oxygen carrier should have a molecular weight of 70–120 kDa, and a size of 10–20 μm. Because increased viscosity increases the likelihood of laminar flow, the viscosity of the ideal red blood cell substitute must be lower than the value for blood.

6.1.8 MEMS Micro-electro-mechanical systems for medical testing and diagnostics

The recent achievements in microfluidics and microreactor technology have been applied to modern medicine. An emerging application area for MEMS technologies is the researching of human cell-based substances and into toxicology and safety. Microreactors are used particularly in the research of newly developed drugs, in biosensing as well as in clinical pathology, especially in medical diagnostics. The obvious advantage of using microreactors in biomedical applications is the ability to handle small cell-based sample sizes and the resultant saving of valuable reagents used in assays. As medical costs grow continuously, inexpensive, reliable, and easy-to-use MEMS devices are very useful and may be considered as the next-generation tools anticipated by diagnostic users. Already MEMS instrumentation has been applied to the four most common laboratory techniques, i.e. for blood chemistries, immunoassays, nucleic-acid amplification tests and flow cytometer [502].

The microreactor technology essentially accomplishes the miniaturization of conventional process devices and enables the integration of reaction and unit operation elements directly with sensors and actuators. The integration of several micro-technologies onto a single, chip-based platform has given rise to micro-electro-mechanical systems (MEMS) or micro total analysis systems (µTAS). The development of modern microreactor engineering achieves small process volume, large heat and mass transport rates and high surface-to-volume ratios. The combination of these features results in extremely short response times, considerably simplifying process control, and eases the handling of large heat and mass fluxes if compared to classic large-scale chemical engineering processes [503, 504].

To date, only a small number of concepts combine a few different organotypic or tissue cultures in one microenvironment, thus allowing the effects of biochemically active molecules and pharmaceutical drugs on human tissue systems to be evaluated. Such devices have potential applications in the fields of toxicology, pharmacy and biomedicine by incorporating several steps of an assay into a single MEMS-based system [505–507]. A multipurpose microbioreactor integrates various unit operations including complex processes. This exhibits several distinct performance advantages, short diffusion distances, laminar flow regime, and high specific surface and high heat performance [508, 509]. The microfluidics MEMS-chip is appropriate for culturing cells isolated from organs of the patient, e.g. brain cortex, liver, bone marrow as well as endothelial cells, keratinocytes, fibroblasts or even stem cells. New concepts of integrated culture systems also cope with the original human counterparts can satisfy the requirements of efficient long-term toxicology testing.

The latest novelty in this area was the development of the hybrid (biotic/abiotic) micro-devices actuated actively by metabolizing cells which have been developed as "micro-bio-pumps" and "micro-bio-actuators" in which biological drive (i.e. cells) may be powered without external energy sources (i.e. without electric energy input). The spontaneous and autonomous contraction of muscle cells (e.g. cardiomyocytes) trigger the flow of fluid (e.g. culture medium or blood) through micro-channels in such micro-multi-organ-on-a-chip systems [510, 511]. The hybrid microactuators and micropumps are driven by cell contraction. They are micro-devices working only with chemical energy input. The only energy source is glucose input, without any electrical power supply or any other stimulus, unlike conventional microactuators and micropumps. Thus we can say that the collaboration of biology and medicine is beneficial for both parties but primarily for the future of humanity.

6.1.9 The future applications of process engineering to modern medicine

The design and production of improved enzymes is planned for a growing range of applications. Improved microorganisms which can produce an increasing number of products in one step, some of which build on genes identified through bio prospecting. Biosensors are being developed for real-time monitoring of environmental pollution, as well as biometric data for identification of individuals. Intensive research is planned on high-energy fuels, which are to be produced exclusively from plants. Also planned is a large market share of biodegradable biomaterials such as bioplastics. Forecasts are that the industry will soon makethe biggest jump in biotechnological applications.

Many new pharmaceuticals and vaccines based on biotechnological knowledge will receive marketing approval. Biotechnology will also be responsible for increasing the use of pharmacogenetics in clinical trials and in practice, with a decrease in the proportion of patients eligible for therapeutic treatment. The use of data and data binding will improve the safety and efficacy of therapeutic and long-term health outcomes. Extensive research is planned for the mutual links of several genetic risk factors for common diseases. Research into improved drug delivery systems is ongoing, taking the convergence of biotechnology and membranes into account. New nutraceuticals will be produced, some by microorganisms and plants or marine extracts. Research will also focus on the development of rapid tests for genetic risk factors for chronic diseases such as arthritis, type II diabetes, heart disease and some cancers. Regenerative medicine will be developed, providing better management of diabetes and replacement or repair of some types of damaged tissue.

Also foreseeable are the widespread use of marker assisted selection (MAS) in plant, livestock, fish and shellfish breeding. Genetically modified (GM) varieties of major crops and trees with improved starch, oil, and lignin content will be developed to improve industrial processing and conversion yields in the future. GM of plants and animals for the production of pharmaceuticals and other valuable compounds. Improved varieties of major food and feed crops with higher yield, pest resistance and stress tolerance developed through GM, MAS, intragenic or cisgenesis. New diagnostic methods for genetic traits and diseases of animals, fish and shellfish will also be developed, as well as high-grade cloning genetic material in animal husbandry.

6.2 Separation of enantiomers

6.2.1 Enantiomers and their role in pharmacotherapy

A number of industries such as the food, cosmetics and pharmaceutical industries often reach for natural resources that are a variety of precursors, preparations, extracts and essences. Sometimes we cannot rely only on processing raw materials, which are very expensive and sometimes even impossible. In such cases, complex chemical compounds such as medicines for humans and animals, substances used in agriculture, forestry, and generally those substances which are somewhat involved in animate nature and the environment, are produed synthetically. Modern chemistry is able to design and produce synthetic counterparts in such cases, and a number of different chemicals which are almost identical to their natural equivalents.

In all such cases, even if the synthesis of biologically active molecules is successfully completed, the new problems are racemic mixtures. If only the molecular structure allows this, i.e. the molecules are "sufficiently complex" to have mirror symmetry, then they have their mirror image counterparts, known as enantiomers. Enantiomers often represent different chemical reactions with other substances that are also enantiomers. In medicine, for example, often only one of the enantiomers of a drug are responsible for the desired physiological effects, while the other enantiomer is less active, inactive, or even harmful. These can be responsible for adverse reactions, aggravate the pharmacological efficacy and sometimes exert side effects. Although there is a great need, the separation of racemic mixtures is the greatest challenge of modern process engineering, because enantiomers are physically no different. The procedure for separation of an enantiomer into two types (left and right) from the given racemic mixture, must involve another enantiomer which reacts with only one type of enantiomer from the mixture. For the separation of the enantiomers A and B, which are homogeneously mixed in racemic mixture A + B, some kind of a third C-enantiomer is introduced into the mixture which reacts with only the one enantiomer from a mixture (e.g. A). The created molecule (A + C), can now be easily separated from B.

6.2.2 The characteristics and properties of enantiomers

A simple chemical formula does not contain information on the structure of a chemical compound. Many molecules are present in the form of isomers, i.e. compounds which have the same atomic composition, but may vary for example, in the sequence of bonding of atoms or the spatial arrangement of atoms within the molecules. Particular attention is paid to enantiomers due to their properties and biological activity.

Enantiomers [512, 513] are the nonsuperimposable mirror images of the given chemical compound (such as the left and right hand) and exist among chiral molecules. Chirality is a special feature of three-dimensional objects that do not have any plane, axis or any other center of symmetry. Most commonly, the asymmetrical center is chiral carbon, i.e. atom which has four nonidentical substituents. The tetrahedral geometry of the carbon atom gives rise to such molecules in two enantiomer forms differing in absolute configuration, i.e. the spatial arrangement of the carbon substituents.

To distinguish between two enantiomers, the prefixes R and S (according to Cahn-Ingold-Prelog) are placed before their names, depending on the configuration. An exception occurs in the case of sugars and amino acids in relation to which the older system, of the prefixes D or L, based on the comparison of the molecule to the absolute configuration of the standard glyceraldehyde is still applied.

Enantiomers in a chiral environment have the same chemical and physical properties, including boiling and melting points, as well as the value of free energy and spectral properties etc. The only difference between them is due to the optical activity of the direction in which the plane of polarized light is rotated, and therefore (+) or (−) is often placed before the name of the enantiomer. It should be noted that an equimolar mixture of enantiomers of the same substance, also known as racemic mixture or racemates, does not show any optical activity.

6.2.3 Biological activity of enantiomers

Chirality is a particularly important feature, not only of chemical compounds, but also one of the main characteristics of the living world [514]. The proteins, carbohydrates, nucleic acids, lipids all constitute the basic building blocks of living organisms that are composed of chiral monomers (amino acids, carbohydrates, nucleotides). Therefore, the fundamental metabolic and physiological processes occurring in cells are based on high stereoselectivity and are manifested as different interactions of exogenous and endogenous enantiomers with chiral structures including enzymes, receptors and ion channels. They are the result of stereoselective binding of chiral ligands to the receptor as a response of organisms to ketones and plant terpenes [515]. For example, the smell of (R) – (+)-carvone is the recognized smell of mint; while levorotary form has a distinctive aroma reminiscent of caraway. Similarly, (R)-terpineol, commonly used in cosmetics as a fragrance, has a pleasant smell similar to lilac, whereas the S enantiomer of this compound smells like cold metal.

The consequence of different interactions between enantiomers with different chiral building blocks of mammals is their different biological activity. This is particularly important in the case of substances with a therapeutic effect, because often only one enantiomer has the desired properties whilst the second one, metabolized by other means, may have a different influence on the organism, can be toxic or not exhibit any biological activity [516, 517]. The most well-known example of different biological activities of enantiomers is the case of thalidomide. The drug was administered unfortunately in the late 50s of the last century to pregnant women as a sedative. It was introduced to the market in the form of a racemic mixture, but after a few years it was found that the metabolite of (S)-thalidomide had potent teratogenic effect and caused deformation of limbs and appearance of abnormal body proportions. Several other examples of biologically active compounds, whose enantiomers (or their metabolites) have a noticeably different profile of activity in the human organism, are shown in Tab. 6.2.

Table 6.2. Examples of various different biological activities of enantiomers as drugs [518]

Trade name	Biological activity (+)-enantiomer	Biological activity (−)-enantiomer
Ketamine	anesthetic	toxic, hallucinogenic
Penicillamine	anti-rheumatic	highly toxic
Propranolol	β-blocker	contraceptive
Propoxyphene	analgesic	antitussive
Ethambutol	anti-tuberculitic	causes blindness
Naproxen	anti-inflammatory	toxic

Particularly noteworthy are nonsteroidal anti-inflammatory drugs, sold without prescription, the consumption of which nowadays is strongly increasing. Most are administrated in the racemic form. One exception is naproxen, used mainly in the treatment of arthritis pain. The beneficial medicinal effect is only provided by its S enantiomer, whilst the other causes some unwanted side effects, mainly strong liver poisoning.

In some cases, single enantiomers of drug substances exhibit greater biological activity than their racemic mixtures. Examples of such compounds are omeprazole and fluoxetine. Esomeprazole (S enantiomer of omeprazole) is more effective in the treatment of gastroesophageal reflux disease as a proton pump inhibitor than the racemic mixture of omeprazole. On the other hand, fluoxetine, having a strong antidepressant activity and still being administrated in the racemic form is metabolized in liver into another biologically active substance, norfluoxetine. However, it is well known that (S)-fluoxetine shows 1.5 times and (S)-norfluoxetine 20 times stronger action than their R counterparts [519].

The use of single enantiomers in pharmacotherapy does not always reduce or prevent the occurrence of side effects. In some cases, after the administration of an enantiomerically pure compound, it undergoes chiral inversion or racemization *in vivo*. As an example, this phenomenon is characteristic for the aforementioned thalidomide. The teratogenic effect of this drug was observed even when only one enantiomer, showing the desired activity, was used. *In vivo* inversion of configuration occurs in the case of ibuprofen [520]. Although (S)-ibuprofen is about 160 times more active than its mirror image, the R enantiomer from racemates of this drug is converted by isomerases into its active form in human body. For this reason, and due to the expensive process of obtaining a pure S enantiomer, most of the commercial drugs contain both enantiomers of ibuprofen.

The need to obtain medicinal substances in their enantiomerically pure form imposes a new direction on the development of pharmacology and pharmacotherapy, which is the replacement of racemic drugs with single enantiomers, commonly referred to as chiral switch [521]. Many pharmaceutical companies use it as a good method of getting a continuous extension of patent exclusivity. It is worthy of note that the replacement of racemic mixtures by optically pure compounds is not always dictated by the undesirable profile of one of the enantiomers but is often due to its low biological activity, which reduces the effectiveness of the racemic drug. An example, may be the case of anticancer drug produced by Bacillus laterosporus, 15-deoxyspergualin [522]. One enantiomer of this compound is biologically active, whilst the second is useless to the human body. Administration of the enantiomerically pure form of this drug has resulted in an increase of the therapeutic efficacy of the single dose. There are also drugs (e.g. paracetamol) containing a racemic mixture in which the biological activity of one enantiomer is sufficiently high and the drug is so cheap that the separation of the racemic mixture may be nonprofitable [523].

Although the optically active isomers may differ in therapeutic activity, some chiral synthetic drugs are still marketed in their racemic form. However, sales of drugs containing pure enantiomer are growing rapidly worldwide. In 2011 the global chiral technology market, comprising of synthesis, analysis and resolution segments, was nearly $ 5.3 billion and it is expected to approach $ 7.2 billion by the end of 2016, as reported in the latest report from BCC Research [524].

The continuous increase in the demand for individual enantiomers contributes to the growing interest in the development of new effective methods for their preparation. It has recently become one of the thriving fields of science, in which various available techniques, different chemicals and catalysts including enzymes are used [525–529].

6.2.4 Methods of preparation of enantiomerically pure compounds

The same properties of enantiomers cannot be distinguished in the absence of other chiral molecules. Conventional methods of organic synthesis, carried out without the involvement of chiral additives, catalysts etc. always lead to racemic mixtures. Moreover, they cannot be separated into pure enantiomers by conventional methods such as distillation, extraction etc. Pure enantiomers or enantiomerically enriched substances can be produced only in a chiral environment, i.e. in the presence of chiral factors. Many methods of preparing pure enantiomers have been developed so far. All require the use of chiral additives such as chiral auxiliaries, building blocks or catalysts. Nowadays, research into methods of preparing pure enantiomers is intended primarily to optimize the process conditions in order to upgrade the quality of the final product and to minimize the cost of the process. The main routes for obtaining pure enantiomers are shown schematically in Figure 6.1.

There are four main techniques applied in resolution of racemates: chiral chromatography, preferential crystallization, diastereomeric salt formation and kinetic resolution (chemical or enzymatic).

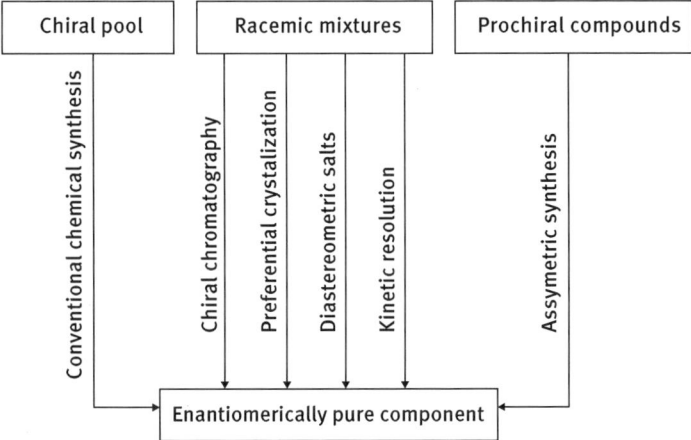

Fig. 6.1. The main methods of achieving enantiomerically pure compounds.

6.2.4.1 Chemical synthesis

The first approach is to use the enantiopure compounds provided by nature (e.g. carbohydrates, terpenes, alkaloids) as the starting materials (chiral source) for further conversion into desired new products. This strategy is one of the most commonly applied because of its simplicity and the availability of the starting compounds in large quantities. Unfortunately, in terms of structure and stereochemistry, the range of natural chiral synthons is very limited. For this reason, chiral pools do not entirely cover the demand of the chemical and pharmaceutical markets for enantiomerically pure compounds.

Apart from the application of natural chiral precursors, one of the most widely used sources of pure enantiomers is racemic mixtures, i.e. equimolar mixtures of two enantiomers. Most are obtained using traditional chemical synthesis carried out without any chiral agents, which makes them relatively cheap and available.

6.2.4.2 Chiral chromatography

Chiral chromatography [530] is one of the most versatile techniques, applied to access pure enantiomers from the racemic mixture. Separation occurs in this case, based on the different affinities of enantiomers to the chiral stationary phase (CSP) or to chiral mobile phase (CMP). A wide variety of CSPs allowing selective separation of racemic mixtures using high performance liquid chromatography or gas chromatography have been developed so far. In both cases transient diastereomeric complexes are formed between enantiomers and a chiral molecule of CSP. Due to the difference in their stability one of the isomers is more rapidly eluted. A large number of CSPs have been commercialized and are broadly used on a laboratory and a preparative scale. However, they are expensive and not universal and a lot of racemates still remain to be resolved.

Hence, the development of new CSPs is very important and desirable. Chromatography with CMP requires of expensive chiral solvents or additives and is therefore rarely used.

6.2.4.3 Preferential crystallization

Preferential crystallization [531, 532], also called crystallization by entrainment, is a process in which one enantiomer crystallizes preferentially whilst the other remains in solution. This technique is considered one of the simplest and most frequently used in the industry. However, it is particularly useful only in the case of enantiomer pairs which form conglomerates, i.e. mixture of crystals in which each consists of a single enantiomer (Fig. 6.2a). Unfortunately, about 90–95 % of chiral molecules crystalize in the form of the true racemate, i.e. a crystal lattice composed of both enantiomers (Fig. 6.2b).

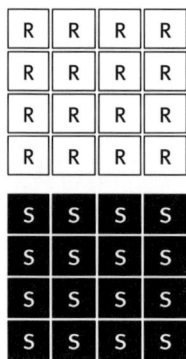 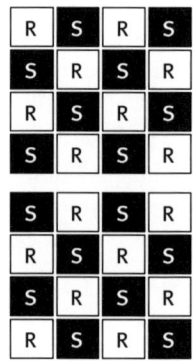

Fig. 6.2. Schematic representation of conglomerate (left) and true racemate (right) forming systems.

The resolution of racemic mixtures by preferential crystallization is strongly affected by the melting behavior of two enantiomers. Figure 6.3 shows the typical binary phase diagrams of the two types of racemic mixtures mentioned above. In the case of conglomerates, the region for individual enantiomers is much bigger than for true racemate and preferential crystallization in this case is quite easy. By adding seed crystals of only one enantiomer to the solution, only its crystals will preferentially grow at the beginning of the process.

6.2.4.4 Diastereomeric salts

Most of the methods used for the separation of enantiomers from their racemic mixtures are based on chemical reactions involving optically active auxiliary agents or catalysts. Chiral auxiliaries are applied mainly in the conversion of enantiomers (e.g. carboxylic acids or amines) into diastereomeric salts [533]. The general principle of

a)

b)

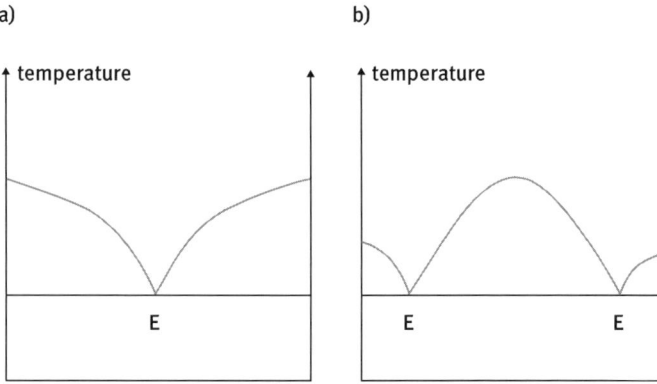

Fig. 6.3. Typical representation of binary phase diagram for conglomerate (a) and true racemate (b).

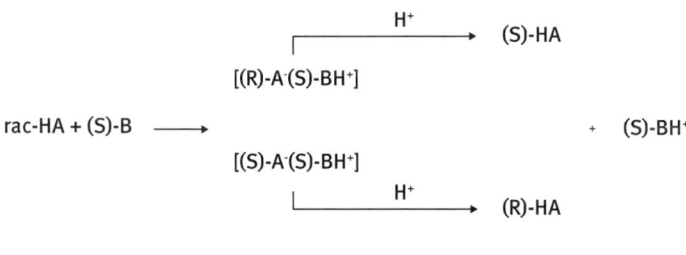

Acid separated enantiomers

Fig. 6.4. Schematic representation of the resolution of enantiomers by diastereomeric salt formation.

such a process is the acid-base reaction of the racemate with an optically active reagent (Fig. 6.4). In contrast to enantiomers, the diastereomeric salts formed have different properties, so they can easily be resolved by conventional physicochemical methods, among which the most common is crystallization. After resolution, chiral auxiliaries can be recovered by adding a strong mineral acid. The above-mentioned method is particularly useful in the case of direct formation of diastereomeric salts.

6.2.4.5 Kinetic resolution

A very useful and constantly developed method used for separation of racemic mixtures is kinetic resolution [534]. It is a process based on differences in the rate of transformation of enantiomers, shown schematically in Figure 6.5. The different rates of reactions are caused by the application of chiral additives such as optically active reagents, chiral organic or inorganic catalysts, enzymes or even whole cells of microorganisms capable of selectively oxidizing or reducing only one enantiomer from the racemic mixture.

(S)-substrate $\xrightarrow{\ \ k_S\ \ }$ (S)-product

$$k_S \neq k_R$$

(R)-substrate $\xrightarrow{\ \ k_R\ \ }$ (R)-product   **Fig. 6.5.** Scheme of kinetic resolution.

Among the many chiral catalysts, enzymes deserve special attention. This is due to their native stereo- and enantioselectivity. By appropriate selection of reaction conditions (e.g. temperature, solvents etc.) as well as enzyme, it is possible to perform resolution of enantiomers with high enantioselectivity. In some cases, enzymes catalyze the reaction of only one enantiomer of the racemic substrate, whilst the reaction rate constant of the second is incomparably smaller. The reaction mixture then contains equimolar amounts of the product formed from the reacting enantiomer and the unreacted enantiomer of the substrate. These compounds have different physicochemical properties and can therefore be separated by conventional methods. To determine enantiomeric purity of substrate and/or product of reaction, the enantiomeric excess (E_e) is defined as follows:

$$E_e = \frac{|C_R - C_S|}{C_R + C_S},\tag{6.2}$$

where C_R and C_S are concentrations of the R and S enantiomers, respectively.

The value of the enantiomeric excess is in the range of 0 (for the racemate) to 1 (for a single enantiomer), and is a function of the degree of conversion and strongly influenced by the enantioselectivity of kinetic resolution expressed by enantiomeric ratio (η_E) [535]:

$$\eta_E = \frac{r_S}{r_R},\tag{6.3}$$

where r_S and r_R are the rates of conversion of the R and S enantiomers, respectively.

Figure 6.6 shows the dependence of enantiomeric excess of product and substrate on reaction conversion for several different values of enantiomeric ratio.

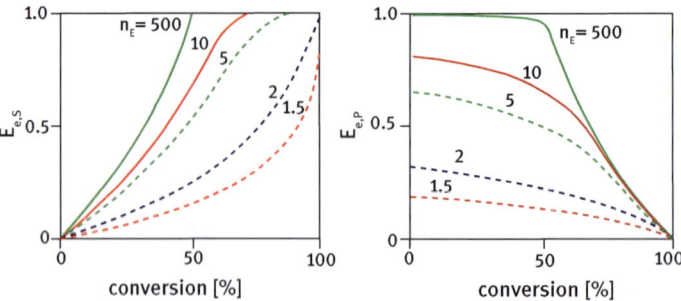

Fig. 6.6. Enantiomeric excess of substrate $(E_{e,S})$ and product $(E_{e,P})$ versus conversion for different values of enantiomeric ratio [537].

If the value of enantiomeric ratio is very high, both single enantiomers (substrate and product) can be obtained. On the other hand, in the case of low enantioselectivity (for example if $\eta_E = 1.5$), both substrate as well as product are composed of the mixture of R and S enantiomers, even at the beginning of the process.

Kinetic resolution is considered to be an attractive method of enantiomer separation, particularly if substrates are readily available and inexpensive, and if the catalyst is easily recovered and highly enantioselective, cheap and effective in small amounts. A major limitation of this method is that the maximum theoretical yield of one enantiomer is only 50 %.

The transformation of the racemic mixture into the desired enantiomer of the product in 100 % theoretical yield is possible by combining classical kinetic resolution with racemization *in situ* (Fig. 6.7). In such processes, called dynamic kinetic resolution [18], the most commonly used racemizing factors are chemical catalysts or racemates.

(S)-substrate $\xrightarrow{k_S}$ (S)-product $k_S \neq k_R$

$\updownarrow k_{rac}$

(S)-substrate $\xrightarrow{k_R}$ (R)-substrate $k_{rac} \gg k_s$ and k_R **Fig. 6.7.** Scheme of dynamic kinetic resolution.

Dynamic kinetic resolution is effective if the racemization rate constant is higher than the rate constant concerning the reaction of the substrate. Moreover, this procedure can be applied in cases of irreversible and nonspontaneous reaction (as in kinetic resolution) where the product of the reaction cannot undergo racemization. In contrast to the classical kinetic resolution, in this method the enantiomeric excess of the product does not depend on the conversion, but only on the enantioselectivity of the reaction.

6.2.4.6 Asymmetric synthesis

Another source of pure enantiomers are prochiral substrates [536], i.e. chemical substances that do not have any chiral center, but due to asymmetrical synthesis, are converted to the optically active chiral molecules. In the process of asymmetric synthesis (Fig. 6.8), the chiral agent (catalyst or reagent) must be involved in the reaction.

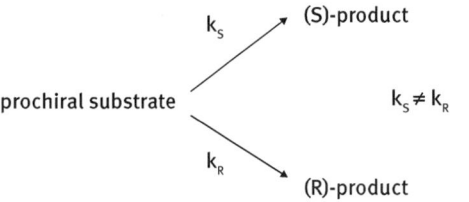

Fig. 6.8. General scheme of asymmetrical synthesis.

Methods used for the preparation of optically active compounds from prochiral molecules are now considered very elegant. However, the amount of pure enantiomers produced in this way is still approximately 4 times lower than in the case of resolution of enantiomers. The reason is economic considerations. Namely, the most commonly used catalysts for asymmetric synthesis are expensive and transition metal complexes are not sufficiently versatile. The same applies to chiral reagents and auxiliary agents, but additionally they must be used in stoichiometric amounts. Moreover, asymmetrical synthesis is generally a multistage process and often requires the application of drastic reaction conditions regarding pressure, pH and temperature.

Bibliography

[1]   Population Division of the Department of Economic and Social Affairs of the United Nations Secretariat, World Population Prospects: The 2008 Revision. (Accessed August 1, 2013 at http://www.un.org/esa/population/).

[2]   Malik K. Human Development Report 2013 The Rise of the South: Human Progress in a Diverse World, United Nations Development Programme (UNDP).

[3]   United Nations. World Summit on Sustainable Development. August 29, 2002, Department of Public Information – News and Media Services Division – New York, (Accessed June 20. 2013 at: http://www.un.org/events/wssd/summaries/envdevj10.htm).

[4]   Associated Press. Threat seen to half of earth's plant species. The Milwaukee Journal Sentinel (Milwaukee, WI). November 1, 2002.

[5]   EEA Technical report No 12/2013 EMEP/EEA Air pollutant emission inventory guidebook 2013 Technical guidance to prepare national emission inventories European Environment Agency Kongens Nytorv 6, 1050 Copenhagen K, Denmark, ISSN 1725–2237.

[6]   http://water.epa.gov/scitech/methods/cwa/pollutants.cfm, accessed June 1, 2013.

[7]   IPCC Fourth Assessment Report: Climate Change 2007, Climate Change 2007: Working Group I: The Physical Science Basis, (accessed August 1, 2013 at http://www.ipcc.ch/publications_and_data/ar4/wg1/en/tssts-2-5.html).

[8]   (http://www.epa.gov/climatechange/ghgemissions/sources.html, accessed August 1, 2013

[9]   http://www.epa.gov/climatechange/science/, accessed August 1, 2013

[10]   Likens GE, Wright RF; Galloway JN and Butler TJ. Acid rain. Scientific American 1979, 241 (4): 43–51.

[11]   Seinfeld J, Pandis H, Spyros N. Atmospheric Chemistry and Physics – From Air Pollution to Climate Change. John Wiley and Sons, Inc. 1998, ISBN 978–0-471–17816–3.

[12]   US EPA: Effects of Acid Rain – Surface Waters and Aquatic Animals (accessed October 5, 2013 at http://www.epa.gov/acidrain/effects/surface_water.html).

[13]   US EPA: Effects of Acid Rain – Forests (accessed October 5, 2013 at http://www.epa.gov/acidrain/effects/forests.html).

[14]   DeHayes DH, Schaberg PG and Strimbeck GR. Red Spruce Hardiness and Freezing Injury Susceptibility. In: Bigras F., ed. Conifer Cold Hardiness.. Dordrecht, The Netherlands: Kluwer Academic Publishers, 2001, ISBN 0–7923–6636–0.

[15]   Lazarus BE; Schaberg PG; Hawley GJ, DeHayes DH. Landscape-scale spatial patterns of winter injury to red spruce foliage in a year of heavy region-wide injury. Can. J. For. Res. 2006, 36: 142–152.

[16]   Mapping, sharing data, and growing awareness on eutrophication and hypoxia around the globe. Word Resources Institute (accessedNovember 23, 2013 at http://www.wri.org/our-work/project/eutrophication-and-hypoxia).

[17]   EPA, Primary National Ambient Air Quality Standards for Carbon Monoxide; Final Rule, Federal Register/Vol. 76, No. 169, 54293–54343/Wednesday, August 31, 2011/Rules and Regulations.

[18]   EPA, Primary National Ambient Air Quality Standards for Nitrogen Dioxide; Final Rule, Federal Register/Vol. 75, No. 26, 6474/Tuesday, February 9, 2010/Rules and Regulations.

[19]   Primary National Ambient Air Quality Standard for Sulfur Dioxide; Final Rule, Federal Register/Vol. 75, No. 11935520/Tuesday, June 22, 2010/Rules and Regulations.

[20]   National Ambient Air Quality Standards for Ozone; Final Rule, Federal Register/Vol. 73, No. 60, 16436/Thursday, March 27, 2008/Rules and Regulations.

[21]   EPA, Volatile Organic Compounds Emissions, (Accessed August 1, 2013 at
       http://cfpub.epa.gov/eroe/index.cfm?fuseaction=detail.viewlnd&lv=list.listByAlpha&r=
       219697&subtop=341).
[22]   Fetzer JC. The Chemistry and Analysis of the Large Polycyclic Aromatic Hydrocarbons. Poly-
       cyclic Aromatic Compounds 2011, 27 (2), 143.
[23]   Ortmann AC, Anders J, Shelton N, Gong L, Moss AG, Condon RH. Dispersed Oil Disrupts Mi-
       crobial Pathways in Pelagic Food Webs, PLoS ONE 2012, 7(7), e42548.
[24]   Schneyer J., U.S. oil spill waters contain carcinogens: Reuters report. 2010.
       http://blogs.reuters.com/joshua-schneyer/page/19/ (Accessed March 2013).
[25]   Oil from Deepwater Horizon disaster entered food chain in the Gulf of Mexico". Science-
       daily.com., http://www.sciencedaily.com/releases/2012/03/120320142100.htm (Accessed
       March 20, 2013).
[26]   Luch A. The Carcinogenic Effects of Polycyclic Aromatic Hydrocarbons. UK London: Imperial
       College Press. 2005, ISBN 1–86094–417–5.
[27]   Exposure to Lead: A Major Public Health Concern. World Health Organization. 2010 (accessed
       August 1, 2013 at http://www.who.int/entity/ipcs/features/lead.pdf).
[28]   Kant R. Textile dyeing industry and environmental hazard. Natural Science, 2012, 4(1), 22–26.
[29]   "Dye Manufacturing". Pollution Prevention and Abatement Handbook. World Bank Group,
       1998.
[30]   EPA "Pesticide Industry: A Profile – Draft Report." Research Triangle Institute. Prepared
       for the U.S. Environmental Protection Agency, December 1993 (accessed August 1, 2013 at
       http://www.epa.gov/ttnecas1/regdata/IPs/Agricultural %20Chemicals %20(pesticides)_
       IP.pdf).
[31]   Zhan W et al. Global pesticide consumption and pollution: with China as a focus. Proceedings
       of the International Academy of Ecology and Environmental Sciences, 2011, 1(2),125–144.
[32]   Environmental, Health, and Safety Guidelines for Pesticide Manufacturing, Formulation and
       Packaging. The World Bank Group. Washington, DC. April 2007 (accessed August 1, 2013 at
       http://www1.ifc.org/wps/wcm/connect/Topics_Ext_Content/IFC_External_Corporate_Site/
       IFC+Sustainability/).
[33]   Pullin AS. Conservation biology. UK: Cambridge University Press, 2002 ISBN 0–521–64482–
       8.
[34]   Zhu Y. Genetic diversity and disease control in rice. International Weekly Journal Of Science.
       2000, 406, 718–722.
[35]   www.emep.int, accessed November 11, 2013.
[36]   The Ozone Hole -The Montreal Protocol on Substances that Deplete the Ozone Layer.
       http://www.theozonehole.com/montreal.html (Accessed March 2013).
[37]   Speth JG. Red Sky at Morning: America and the Crisis of the Global Environment. New Haven.
       Yale University Press, 2004, p95.
[38]   South Sudan Joins Montreal Protocol and Commits to Phasing Out Ozone-Damaging Sub-
       stances. http://www.unep.org/newscentre/Default.aspx?DocumentID=2666&ArticleID
       =9010&l=en (Accessed March 2013).
[39]   Agenda 21 (accessed August 1, 2013 at http://sustainabledevelopment.un.org/content/
       documents/Agenda21.pdf).
[40]   Federal Water Pollution Control Act (accessed August 1, 2013 at http://www.epw.senate.gov/
       water.pdf).
[41]   Claudia C. et al. DC Congressional Research Service, Library of Congress, 1999.
[42]   Sustainable Development (accessed August 1, 2013 at http://sustainabledevelopment.un.
       org/content/documents/Agenda21.pdf)

[43]   Climate Justice Movement, (accessed November 16, 2013 at http://www.climate.org/climatelab/Climate_Justice_Movements).

[44]   EPA: Presidential Green Chemistry Challenge in Designing Safer Chemicals Award. Office of Pollution Prevention and Toxics (7406M), EPA 742-F-02–003, March 2002.

[45]   European Commission, Directorate-General Environment, LIFE FOCUS/Industrial pollution, European solutions: clean technologies, Directive on integrated pollution prevention and control (IPPC Directive), Luxembourg: Office for Official Publications of the European Communities, 2003, ISBN 92–894–6020–2.

[46]   Okamkia M et al. Water recycling using sequential membrane treatment in the electronics industry. Desalination, 2000, 131, 65–73.

[47]   Koltuniewicz AB. Process Engineering for Sustainability, Encyclopedia of Life Support Systems (UNESCO EOLSS 2011).

[48]   United Nations (1987 Brundland report), Report of the World Commission on Environment and Development: Our Common Future.

[49]   Frosch RA, Galloopoulos NE. Strategies for manufacturing. In: Planet Earth. Scientific American Special Issue September 1989.

[50]   Capra F. Ecologically conscious management. Environmental Law, 1992, 22, 529–37.

[51]   Van Weenen J. Towards sustainable product development, Journal of Cleaner Production, 1995, 3(1/2), 95–100.

[52]   Carslaw HS and Jager JC. Conduction of Heat in Solids, Bristol, UK: Oxford University Press, 1985.

[53]   Einstein HA and Li H. The viscous sublayer along a smooth Boundary. J. Eng. Mech. Div. Am. Soc. Civil Eng., April 1956, 82, EM2.

[54]   Hanratty TJ. Turbulent Exchange of Mass and Momentum with a Boundary. AIChE, 1956, 2,359–362.

[55]   Shaw DA, Hanratty TJ. Influence of Schmidt Number on the Fluctuations of Turbulent Mass transfer to a Wall, A.I.Ch.E.J., 1977, 23,(2), 160–169.

[56]   Molstad MC. Introduction – Process Kinetics. Ind. Eng. Chem., 1953, 45 (6), 1173–1173

[57]   Owen JS et al. Precursor conversion kinetics and the nucleation of cadmium selenite nanocrystals. J Am Chem Soc., Dec 292010,132(51), 18206–13.

[58]   Welty JR et all Fundamentals of Momentum, Heat and Mass transfer (5th edition). John Wiley and Sons, 2007. ISBN 978–0470128688.

[59]   Coulson C, Richardson F. Chemical Engineering Volume 1. Elsevier, 2000.

[60]   Chung S et al. Biological Membrane Ion Channels: Dynamics, Structure, And Applications, Springer-Verlag 2006.

[61]   Nelissen K et al. Single-file diffusion of interacting particles in a one-dimensional channel. EPL, Europhysics Letters, 2007, 80(5), 56004.

[62]   Bechinger C. Diffusion in reduced dimensions (Accessed August 1, 2013 at http://www.pi2.uni-stuttgart.de/contact/redaxo_research/more/more_www_sfd.pdf).

[63]   Trouton F. On Molecular Latent Heat. Philosophical Magazine Series 5, 1884, 18(110), 54–57 doi:10.1080/14786448408627563.

[64]   Maxwell JC. On the dynamical theory of gases. The Scientific Papers of JC Maxwell, 1965, 2, 26–78.

[65]   Mason EA, Londsdale AK. Statistical-mechanical theory of membrane transport. Journal of Membrane Science, 15 July 1990, 51, 1–2, 1–81.

[66]   IUPAC, Compendium of Chemical Terminology, 2nd ed. (the "Gold Book") (1997). Online corrected version: (2006).

[67]   Preparing for the Chemistry AP Exam. Upper Saddle River, New Jersey: Pearson Education, 2004. 131–134. ISBN 0–536–73157–8.

[68]   Weinhold F, Landis C. Valency and bonding. Cambridge. 2005, 96–100. ISBN 0–521–83128–8.

[69]   Campbell NA et al. Biology: Exploring Life. Boston, Massachusetts: Pearson Prentice Hall, 2006. ISBN 0–13–250882–6.

[70]   London F. Transactions of the Faraday Society, 1937, 33, 8–26.

[71]   Stahl WR. Dimensional Analysis in Mathematical Biology. Bulletin of Mathematical Biophysics, 1961, 23, 355.

[72]   Roche JJ. The Mathematics of Measurement: A Critical History, London: Springer, 1998, p. 203, ISBN 978–0-387–91581–4.

[73]   Buckingham E. On Physically Similar Systems; Illustrations of the Use of Dimensional Equations. The American Physical Society, Phys. Rev., 1914, 4, 345–376.

[74]   Kasprzak W et al. Dimensional Analysis in the Identification of Mathematical Models, World Scientific, 1990, ISBN 978–981–02–0304–7.

[75]   Schubert K et al. Microstructure Devices for applications in thermal and chemical process engineering. In: Microscale Thermophysical Engineering. Taylor & Francis, January 2001) 5 (1): 17–39. ISSN 1556–7265.

[76]   Watts P et al. Recent advances in synthetic micro reaction technology. Chem. Commun., 2007, 443–467.

[77]   Reschetilowski W. Microreactors in Preparative Chemistry, Practical Aspects in Bioprocessing, Nanotechnology, Catalysis and more. Wiley, 2013.

[78]   Brauner N, Maron DM. Identification of the range of 'small diameters' conduits, regarding two-phase flow pattern transitions, Int. Comm. Heat Mass Transfer, 1992, **19**, 29–39.

[79]   Triplet KA et al. Gas-liquid flow in microchannels Part I: two-phase flow patterns, Int. J. Multiphase Flow, 1999, **25**, 377–394.

[80]   Dziubiński M, Prywer J. Mechanika płynów dwufazowych, WNT, Warszawa, 2009, 475–486.

[81]   Shiells E. Chemistry World, (accessed August 1, 2013 at http://www.rsc.org/chemistryworld/2012/08/3d-printed-miniaturised-fluidic-devices.

[82]   Kockmann N. Transport phenomena in micro process engineering, Berlin, Germany: Springer, 2008.

[83]   (http://www.dsm.com/en_US/downloads/dpp/DSM_Micro_Reactor_Technology_Webinar_Presentation_24Sept09FINAL.pdf, accessed August 1, 2013.

[84]   Van den Berg A. and Lamerlink TSJ, Micro Total Analysis Systems: Microfluidic Aspects. Integration Concept and Applications.Top. Curr. Chem. 1998, 194, 21.

[85]   Harrison DJ. et al. Micromachining a miniaturized capillary electrophoresis-based chemical analysis system on a chip. Science,1993, 261, 895-7.

[86]   Wirth T. Microreactors in Organic Chemistry and Catalysis. Wiley VCH, 2013, ISBN-13: 978–3527332991.

[87]   Fletcher PDI et al. Microreactors: Principle and Applications in Organic Synthesis, Tetrahedron, 2002, 58, 4735–4757.

[88]   Doku GN. et al. Electric field-induced mobilisation of multiphase solution systems based on the nitration of benzene in a microreactor. Analyst 2001, 126, 14-20.

[89]   Burns JR, Ramslaw CG. AIChE Spring National Meeting March 5–9, Atlanta GA, USA, 2000 p 133.

[90]   Carruthers W. Some Modern Methods of Organic Synthesis. Cambridge, UK: Cambridge University Press, 1971, 81–90. ISBN 0–521–31117–9.

[91]   Skelton V. Lab-on-a-chip: Miniaturized Systems for (bio)chemical Analysis and Synthesis. Analyst, 2001, 126, 7–11.

[92]   Clayden J. Organic chemistry. Oxford, UK: Oxford University Press, 2001. ISBN 0–19–850346–6.

[93] Sands M. et al.The investigation of an equilibrium dependent reaction for the formation of enamines in a microchemical system. Lab on a Chip 2001, 1, 64.

[94] Wiles C. et al. The aldol reaction of silyl enol ethers within a microreactor. Lab on a Chip, 2001, 1, 100.

[95] Watts P. et al. The synthesis of peptides in microreactors. J. Chem. Soc. Chem. Commun. 2001, 990.

[96] Hisamoto H. et al. Fast and high conversion phase transfer synthesis exploiting the liquid-liquid interface formed in microchannel chip.J. Chem. Soc. Chem. Commun. 2001, 2662.

[97] Tucson University. Health & Safety in the Arts, A Searchable Database of Health & Safety Information for Artists. Tucson University Studies (Accessed May 2013 at http://archive.is/www.tucsonaz.gov/).

[98] Engel E et al. Azo Pigments and a Basal Cell Carcinoma at the Thumb. Dermatology, 2008, 216(1), 76–80.

[99] Dietz E. et al. Topical Proceedings, 4th Conference on microreaction Technology, AIChe Spring National Meeting, March 5–9, 2000, 89.

[100] Blumethal U. US PATENT US 3754036. Acyl-cyclododecenes,acyl-cyclododecanes and a process for producing them.

[101] Chambers RD, Spink RCH. Microreactors for elemental fluorine, J. Chem. Soc. Chem. Commun. 1993, 883.

[102] Chambers RD, Holling D, Spink RCH, Sandford G. Elemental Fluorine Part 13. Gas-liquid thin film microreactors for selective direct fluorination. Lab Chip 1, 2001, 132-137.

[103] deMas N, Jackman RJ, Schmidt MS, Jensen KF, Microchemical systems for direct fluorination of aromatics, in IMRET 5:. Procedings of the Fifth International Conference on Microreaction Technology, Ed. M Matlosz, W Ehrfeld, JP Baselt. Berlin, Germany: Springer, 2001.

[104] Srinvasan R. Micromachined reactors for catalytic partial oxidation. AIChE J. 1997, 43 (11), 3059–3069.

[105] Hessel V, Ehrfeld W, Golbig K, Hofman C, Jungwirth S, Löwe H, Richter T, Storz M, Wolf A. Microreactore: New Technology for Modern Chemistry, IMRET 3: Proceedings of the Third Conference on Microtechnology, 2000, p151.

[106] Sobieszuk P et al. Hydrodynamics and mass transfer ig gas-liquid flows in microreactors, Chem. Eng. Tech., 2012, 35 1346–1355.

[107] Hessel V. Gas–liquid and gas–liquid–solid microstructured reactors: contacting principles and applications, Ind. Eng. Chem. Res., 2005, **44**, 9750–9769.

[108] Ehrich H, Linke D, Morgenschweis K, Baerns M, Jahnish K, Application of microstructured reactor technology for the photochemical chlorination of alkylaromatics. Chimia, 2002, **56**, 647–653.

[109] Abolhasani M, Singh M, Kumacheva E, Günter A, Automated microfluidic platform for studies of carbon dioxide dissolution and solubility in physical solvents, Lab Chip, 2012, 12, 1611–1618.

[110] Lefortier SGR Hamersma PJ, Bardow A, Kreutzer MT. Rapid microfluidic screening of CO_2 solubility and diffusion in pure and mixed solvents. Lab Chip, 2012, **12**, 3387–3391.

[111] Jacobson et al. High Speed Separations on a Microchip. Anal. Chem. 1994, 66,1114–1118.

[112] Crabtree R, The Organometallic Chemistry of the Transition Metals. Wiley, 2005.

[113] Corriu RJP, Masse JP. Activation of Grignard reagents by transition-metal complexes. A new and simple synthesis of trans-stilbenes and polyphenyls. Journal of the Chemical Society, Chemical Communications 1972(3), 144a.

[114] Tamao K. et al. (1 June 1972). Selective carbon–carbon bond formation by cross-coupling of Grignard reagents with organic halides. Catalysis by nickel-phosphine complexes. Journal of the American Chemical Society 1972, 94(12), 4374–4376.

[115] Lu H, Schmidt MA, Jensen KF. Photochemical reactions and on-line UV detection in microfabricated reactors.Lab on a Chip 2001, 1, 22.

[116] European Commission, OPINION ON BENZOPHENONE-3 COLIPA No. S38, Opinion adopted by the SCCP during the 10th plenary of 19 December 2006.

[117] http://www.imm-mainz.de accessed October 22, 2013

[118] Series 4590 Micro Stirred Reactors. (Accessed August 1, 2013 at http://www.parrinst. com/products/stirred-reactors/series-4590-micro-stirred-reactors/?gclid= CK_B2vDOr7kCFQtY3godg0EAgA)

[119] Loeb S, Sourirayan S. Sea Water Demineralization by Means of an Osmotic Membrane, in: Saline Water Conversion – II, Chapter 9, pp 117–132, Vol. 38, 1963, ISBN13: 9780841200395.

[120] Hermia J. Constant pressure blocking filtration laws-application to power-law non-newtonian fluids, Trans. ICkemE, 1982, 60, 183–187.

[121] Koltuniewicz AB, Field RW. Process factors during removal of oil-in-water emulsions with cross-flow microfiltration. Desalination, 1996, 105, 79–89

[122] Danckwerts PV. Significance of Liquid-Film Coefficients in Gas Absorption, Ind. Eng. Chem., 1951, 43 (6), 1460–1467.

[123] Koltuniewicz AB. The yield of pressure-driven membrane processes in the light of the surface renewal theory (article in Polish). Oficyna Wydawnicza Politechniki Wroclawskiej, Wroclaw, 1996.

[124] Koltuniewicz AB. Predicting permeate flux in ultrafiltration on the basis of surface renewal concept. Journal of Membrane Science 1992,(68(1–2), 107–118.

[125] Koltuniewicz AB. et al, Method of yield evaluation for pressure-driven membrane processes. The Chemical Engineering Journal,1995, 58, 175–182.

[126] Koltuniewicz AB et al. Cross-flow and dead-end microfiltration of oily-water emulsion. Part I: Experimental study and analysis of flux decline. Journal of Membrane Science, 1995, 102, 193–207.

[127] Koltuniewicz et al. Dynamic Properties of Ultrafiltration Systems in Light of the Surface Renewal Theory. Ind. Eng. Chem. Res. 1994, 33, 1771–1779.

[128] Koltuniewicz AB, Bezak K. Engineering of membrane biosorption. Desalination, 2002, 144, 219–226.

[129] Koltuniewicz AB et al. Efficiency of membrane-sorption integrated processes. Journal of Membrane Science, 2004, 239, 129–141.

[130] Koltuniewicz AB et al. Designing of membrane contactors with cross-counter current flow. Chemical and Process Engineering, 2012, 33 (4), 573–583.

[131] Koltuniewicz AB, Drioli E. Membranes in Clean Technologies, Theory and Practice. Wiley-VCH, 2008.

[132] Koltuniewicz AB. Integrated Membrane Operations in various Industrial Sectors in: Comprehensive Membrane Science and Engineering, eds. E.Drioli and L. Giorno, Elsevier, 2010.

[133] Lipnizki F, Field RW. Pervaporation-based hybrid process. Journal of Membrane Science, 1999, 153, 183–210.

[134] Polish Patent PL 178665, granted on 31.05.2000.

[135] Drioli E. Membrane Contactors: Fundamentals, Applications and Potentialities, Elsevier, 2011, pp.516. ISBN 978-0-444-52203-0.

[136] Koltuniewicz AB, Submerged membrane. In: Encyclopedia of Membranes, Edition 1, Article ID: 310310, Chapter ID: 558. Eds: E. Drioli and L. Giorno, Springer, 2012 http://refworks.springer.com/membranes.

[137] Koltuniewicz AB. Submerged membrane bioreactors. In: Encyclopedia of Membranes, Edition 1, Article ID: 310311, Chapter ID: 559, Eds: E. Drioli and L. Giorno. Springer, 2012 http://refworks.springer.com/membranes.

[138] Koltuniewicz AB et al. Designing of membrane contactors with cross-counter current flow, Chemical and Process Engineering, 2012, 33(4), 573–583.

[139] Modelski S, Koltuniewicz AB. Kinetics of VOC absorption using capillary membrane contactor. Chemical Engineering Journal, 2011, 168, 1016–1023.

[140] Koltuniewicz AB. Membrane biosorption. In: Encyclopedia of Membranes, Edition 1, Article ID: 310110, Chapter ID: 358, Eds: E. Drioli and L. Giorno. Springer, 2012 http://refworks.springer.com/membranes.

[141] Yilmaz I et al. A submerged membrane–ion-exchange hybrid process for boron removal. Desalination, 2006, 198, 310–315.

[142] Koltuniewicz AB et al. p-Cresol removal using a membrane contactor enhanced by the micellar solubilization. Desalination, 2006, 200, 575–577.

[143] Koltuniewicz AB et al. Simultaneous removal of phenols and Cr3+ using micellar-enhanced ultrafiltration proces. Desalination, 2006, 191, 111–116.

[144] Koltuniewicz AB. Application of a membrane contactor for a simultaneous removal of p-cresol and Cr(III) ions from water solution. Desalination, 2009, 241, 91–96.

[145] Marcano JG. Catalytic Membranes and Membrane Reactors John Wiley & Sons, 2002, p251, ISBN: 3527302778.

[146] Directive 2000/69/CE, Diario Oficial de las Comunidades Europeas No. L313/12, 13 December 2000.

[147] Picasso G et al. Total combustion of methyl-ethyl ketone over Fe2O3 based catalytic membrane reactors. Applied Catalysis B: Environmental, 2003, 46, 133–143.

[148] Spivey JJ. Complete catalytic oxidation of volatile organics. Ind. Eng. Chem. Res., 1987, 26, 2165-2180.

[149] Moretti E, Mukhopadhyay N. Current practices and future trends. Chem. Eng. Prog., 1993, 89(7), 20.

[150] Casanave D et al. Control of transport properties with a microporous membrane reactor to enhance yields in dehydrogenation reactions. Catal. Today,1995, 25, 309.

[151] Ciavarella C. Isobutane dehydrogenation in a membrane reactor: influence of the operating conditions on the performance. Catal. Today, 2001, 67, 177.

[152] Illgen U et al. Membrane supported catalytic dehydrogenation of iso-butane using an MFI zeolite membrane reactor.,Catal. Commun., 2001, 2, 339.

[153] Frusteria F et al. Partial oxidation of light parafins on supported superacid catalytic membranes. Applied Catalysis A: General, 1999, 180, 325-333.

[154] Jghikafumi Y et al. Oxidative dehydrogenation of gaseous alkanes US Patent 6,548,447, 15 Apr 2003.

[155] Alfonso MJ et al. Oxidative dehydrogenation of butane on V/MgO catalytic membranes. Chemical Engineering Journal, 2002, 90, 131–138.

[156] Byeong H. Catalytic dehydrogenation of cyclohexane in an FAU-type zeolite membrane reactor. Journal of Membrane Science, 2003, 224, 151–158.

[157] Gokhale YV. Analysis of a membrane enclosed catalytic reactor for butane dehydrogenation. J. Membr. Sci., 1993, 77, 197.

[158] Yanglong G et al. Preparation and characterization of Pd/Ag/ceramic composite membrane and application to enhancement of catalytic dehydrogenation of isobutene. Separation and Purification Technology, 2003, 32, 271-279.

[159] Korchnak JD. Dunster M. Reduced Methanol Production Costs, World Methanol Production Conference, San Francisco, 1987.

[160] Mark MF, Maier WF. CO_2-reforming of methane on supported Rh and Ir Catalysts. J Catal 1996, 164, 122.

[161] Wang S Lu GQM. A comprehensive study on carbon dioxide reforming of methane over Ni = $-Al^2O^3$ catalysts. Ind Eng ChemRes, 1999, 38, 2615.

[162] Neomagus HWJP et al. The catalytic oxidation of H_2S in a stainless steel membrane reactor with separate feed of reactants. Journal of Membrane Science, 1998, 148, 147–160.

[163] Abashar MEE. Integrated catalytic membrane reactors for decomposition of Ammonia. Chemical Engineering and Processing, 2002, 41, 403–412.

[164] Choi JS et al. Design of H3PW12O40–PPO (polyphenylene oxide) composite catalytic membrane reactor and its performance in the vapor-phase MTBE (methyl tert-butyl ether) decomposition. Journal of Membrane Science, 2002, 198, 163–172.

[165] Yihang G. Mesoporous H3PW12O40-silica composite : Efficient and reusable solid acid catalyst for the synthesis of diphenolic acid from levulinic acid. Applied catalysis. B, Environmental, 2008, 81, 3–4, 182–191.

[166] http://www.aps.anl.gov/Science/Highlights/Content/APS_SCIENCE_20120814.php, accessed August 1, 2013.

[167] Kokhan DP. Elucidating the domain structure of the cobalt oxide water splitting catalyst by X-ray pair distribution function analysis. J Am Chem Soc. 2012,134(27), 11096–99.

[168] Kato H. Highly efficient water splitting into H_2 and O_2 over lanthanum-doped $NaTaO_3$ photocatalysts with high crystallinity and surface nanostructure. J Am Chem Soc., Mar 2003, 125(10), 3082–89.

[169] Sandru O. New Water-Splitting Catalyst 1000 Times Cheaper Than Platinum May Boost Hydrogen Economy (accessed August 1, 2013 at http://www.greenoptimistic.com/2012/06/05/nickel-molybdenum-catalyst-hydrogen/#.UjK_ztKpVsk).

[170] Chang WC. Decolorization of Reactive Red 2 by advanced oxidation processes: Comparative studies of homogeneous and heterogeneous systems. Journal of hazardous materials 2006, 128 (2–3), 265–72.

[171] Valentine JS. Active Oxygen in Biochemistry. New York: Blackie Academic and Professional, 1995.

[172] Fenton HJH. Oxidation of tartaric acid in presence of iron. J. Chem. Soc. Trans., 1984, 65 (65). 899–911.

[173] McNaught AD, Wilkinson A. IUPAC. Compendium of Chemical Terminology ("Gold Book"). Oxford: Blackwell Scientific Publications, 1997.

[174] http://geocleanse.com/fentonsreagent.asp, accessed August 1, 2013

[175] Haber F, Weiss J. (1932). "Über die Katalyse des Hydroperoxydes". Die Naturwissenschaften, 1932, 20(51), 948–950.

[176] Linsebigler AL et al. Photocatalysis on TiO_2 Surfaces: Principles, Mechanisms, and Selected Results. Chemical Reviews, 1995, 95(3), 735.

[177] Henschler D. Metabolism and mutagenicity of halogenated olefins–a comparison of structure and activity. Environ Health Perspect. December 1977, 21, 61–64.

[178] Rossberg M et al. Chlorinated Hydrocarbons in: Ullmann's Encyclopedia of Industrial Chemistry, Weinheim: Wiley-VCH, 2006.

[179] 1,1,2,2-Tetrachloroethane EPA (Accessed August 1, 2013 at http://www.epa.gov/ttn/atw/hlthef/tetrachl.html).

[180] Halmann MM. Photodegradation of Water Pollutants, CRC Press, 1995.

[181] Hoffmann MR et al. Environmental Applications of Semiconductor Photocatalysis. Chem. Rev. 1995, 95, 69–96.

[182] McCullagh C et al., The application of TiO2 photocatalysis for disinfection of water contaminated with pathogenic micro-organisms: a review. Research on Chemical Intermediates, 2007, 33(3–5); 359–375.

[183] Tan SS et al. Photocatalytic reduction of carbon dioxide into gaseous hydrocarbon using TiO2 pellets. Catalysis Today, 2006, 115, 269–273.

[184] Yao Y et al. Photoreactive TiO_2/Carbon Nanotube Composites: Synthesis and Reactivity. Environmental Science & Technology (American Chemical Society), 2008,42(13): 4952–4957.

[185] Linsebigler AL et al. Photocatalysis on TiO2 Surfaces: Principles, Mechanisms, and Selected Results. Chemical Review, 1995, 95(3), 735–758.

[186] Tributsch H. Dye sensitization solar cells: a critical assessment of the learning curve. Coordination Chemistry Reviews, 2004, 248(13–14), 1511.

[187] Ferrari L et al. Interaction of cement model systems with superplasticizers investigated by atomic force microscopy, zeta potential, and adsorption measurements. J Colloid Interface Sci., 2010, 347(1), 15–24.

[188] Volesky B, Biosorption of Heavy Metals. Ann Arbor: CRC Press, 1990.

[189] Koltuniewicz AB, Bezak K. Membrane biosorber: the new integrated system with biosorption and ultrafiltration. in: Euromembrane Proceedings, Jerusalem, Israel, 27–30 September 2000.

[190] Langmuir I. The constitution and fundamental properties of solids and liquids. Journal of the American Chemical Society. 1917, 38(11): 2221–2295.

[191] Masel RI. Principles of adsorption and reaction on solid surfaces.,New York: Wiley 1996. ISBN0471303925.

[192] Koltuniewicz AB, Bezak K. Removal of metal ions from contaminated waters by means of integrated system with biosorption and ultrafiltration. American Water Works Association, San Antonio, 3–7 March 2001.

[193] Koltuniewicz AB, Bezak K. Biosorption of cadmium, lead and copper ions – comparison between whole and disrupted cells of Saccharomyces cerevisiae. Chem. Agric., 2001, 2, 164.

[194] Pagnanelli F. Effect of equilibrium models in the simulation of heavy metals biosorption in single and two-stage UF/MF membrane reactor systems. Biochem. Eng. J., 2003, 15, 27–35.

[195] Altman J, Ripperger S. Particle deposition and layer formation at the cross-flow microfiltration, J. Membr. Sci., 1997, 124, 119.

[196] Blanpain B. A model for predict the cake layer rebuild-up after periodic backflushes, in: Proceedings of Euromembrane 1995, 1, 111–115.

[197] Howell JA et al. Yeast cell microfiltration: flux enhancement in baffled and pulsatile flow systems, J. Membr. Sci., 1993, 41, 59.

[198] Li H et al. An assessment of depolarisation models of cross-flow microfiltration by direct observation through the membrane, J. Membr. Sci., 2000, 172, 135.

[199] Howell JA. Sub-critical flux operation of microfiltration, J. Membr. Sci., 1995, 107, 165.

[200] Field RW. et al. Critical flux concept for microfiltration fouling, J. Membr. Sci. 1995. 100, 259.

[201] Madaeni SS,. Fane AG. Factors influencing critical flux in membrane filtration of biomass, in: Proceedings of IMSTEC '96, 1996, 142–143.

[202] Kwon DY et al. Experimental study on critical flux in cross-flow microfiltration, in: Proceedings of IMSTEC'96, 1996, 59.

[203] Wu D. Critical flux measurement for model colloids, J. Membr. Sci., 1999, 152, 89.

[204] Cheng TW. Enhancement of permeate flux by gas slugs for cross-flow ultrafiltration in tubular membrane module. Sep. Sci. Technol., 1998, 33, 2295.

[205] Lee CK et al. Air slugs entrapped cross-flow filtration of bacterial suspensions. Biotechnol. Bioeng., 1993, 41, 525.

[206] Cui ZF, Wright KIT. Gas–liquid two-phase cross-flow ultrafiltration of BSA and dextran solution. J. Membr. Sci., 1994, 90, 183.

[207] Mercier M et al. Delorme, Influence of the flow regime on the efficiency of a gas–liquid two-phase medium filtration., Biotechnol. Tech., 1995, 9, 853.

[208] Cui ZF,. Wright KIT. Flux enhancements with gas sparging in downwards cross-flow ultrafiltration: performance and mechanism. J. Membr. Sci., 1996, 117, 109.

[209] Cabassud C et al. How slug flow can improve ultrafiltration flux in organic hollow fibers. J. Membr. Sci., 1997, 128, 93.

[210] Mercier M. Yeast suspension filtration: Flux enhancement using an upward gas/liquid slug flow – application to continuous alcoholic fermentation with cell recycle. Biotechnology and Bioengineering, 1998, 58(1), 47–57.

[211] Metsämuuronnen M. Critical flux in ultrafiltration of myoglobin and baker's yeast, J. Membr. Sci., 2002, 196, 13.

[212] Parnham CS, Davies RH. Protein recovery from bacterial cell debris using cross-flow microfiltration with backpulsing, J. Membr. Sci., 1996, 118, 259–268.

[213] Kuberkar V et al. Flux enhancement for membrane filtration d bacterial suspensions using high frequency backpulsing. Biotechnol. Bioeng., 1998, 8(1), 45.

[214] Kołtuniewicz AB. The new concept of flux enhancement during cell separation with MF/UF processes. Water Sci. Technol., 2001, 5–6, 381.

[215] Moo-Young M, Chisti Y. Pure &Appl. Chern., 1994, 66(1), 117–136.

[216] ACS Chemistry for life, J. Am. Chem. Soc. (accessed July 19, 2013 at http://www.acs.org/content/acs/en/careers/whatchemistsdo/careers/biotechnology.html).

[217] Roels JA. Energetics and Kinetics in Biotechnology, Elsevier Biomedical Press, 1983, p330

[218] Schweitzer M. (Schmauder HP). Methods in Biotechnology. London UK: Taylor & Francis Ltd, 1997.

[219] Bonnet JABA, Roels JA. The growth of Sacharomyce Cerevisiae CBS 426 on mixtures of glucose and ethanol: Antonie von Leeuvenhoek, 1980, 46, 565–576.

[220] Giorno L, Drioli E. Biocatalytic membrane reactors: applications and perspectives, reviews, TIBTECH, August 2000, 18.

[221] Koshland DE Application of a Theory of Enzyme Specificity to Protein Synthesis. Proc. Natl. Acad. Sci. U.S.A. 1958,44(2), 98–104. doi:10.1073/pnas.44.2.98.

[222] Jencks WP. Catalysis in Chemistry and Enzymology New York: Dover, 1987.

[223] Warshel A. et al. Electrostatic Basis of Enzyme Catalysis. Chem. Rev., 2006, 106(8), 3210–3235.

[224] Warshel A, Levitt M. Theoretical Studies of Enzymatic Reactions: Dielectric Electrostatic and Steric Stabilization of the Carbonium Ion in the Reaction of Lysozyme. J. Mol. Biol. 1976, 103, 227.

[225] Stanton RVet al. Combined ab initio and Free Energy Calculations To Study Reactions in Enzymes and Solution: Amide Hydrolysis in Trypsin and Aqueous Solution. J. Am. Chem. Soc., 1998, 120, 3448–3457.

[226] Kuhn B, Kollman PA. QM-FE and Molecular Dynamics Calculations on Catechol O-Methyltransferase: Free Energy of Activation in the Enzyme and in Aqueous Solution and Regioselectivity of the Enzyme-Catalyzed Reaction. J. Am. Chem. Soc. 2000, 122, 2586–2596.

[227] Bruice TC, Lightstone FC. Ground State and Transition State Contributions to the Rates of Intramolecular and Enzymatic Reactions. Acc. Chem. Res., 1999, 32, 127–136.

[228] Warshel A et al. How do Serine Proteases Really Work? Biochemistry, 1989, 28, 3629.

[229] Marcus RA. On the Theory of Electron-Transfer Reactions. VI. Unified Treatment for Homogeneous and Electrode Reactions J. Chem. Phys., 1965, 43, 679–701.

[230] Warshel A. Energetics of Enzyme Catalysis. Proc. Natl. Acad. Sci. USA, 1978, 75, 5250.

[231] Toney MD. Reaction specificity in pyridoxal enzymes. Archives of biochemistry and biophysics 2005, 433, 279–287.

[232] Voet D, Voet J. Biochemistry. John Wiley & Sons Inc.. 2004, pp. 986–989. ISBN 0–471–25090–2.

[233] Garcia-Viloca M et al. How enzymes work: analysis by modern rate theory and computer simulations. Science, 2004, 303(5655), 186–95.

[234] Olsson MH et al. Simulations of the large kinetic isotope effect and the temperature dependence of the hydrogen atom transfer in lipoxygenase. Journal of the American Chemical Society, 2004, 126(9), 2820–8.

[235] Masgrau L et al. Atomic description of an enzyme reaction dominated by proton tunneling. Science, 2006, 312(5771), 237–41.

[236] Hwang JK, Warshel A. How important are quantum mechanical nuclear motions in enzyme catalysis. J. Am. Chem. Soc. 1996, 118, 11745–11751.

[237] Ball P. Enzymes: By chance, or by design? Nature, 2004 431, 396.

[238] Olsson MHM et al. Dynamical Contributions to Enzyme Catalysis: Critical Tests of A Popular Hypothesis. Chem. Rev., 2006, 105, 1737–1756.

[239] Volkenshtein MV et al. Theory of Enzyme Catalysis. Molekuliarnaya Biologia Moscow, 1972, 6, 431–439.

[240] Volkenshtein MV et. al. Electronic and Conformational Interactions in Enzyme Catalysis. In: E.L. Andronikashvili (Ed.) Konformatsionnie Izmenenia Biopolimerov v Rastvorakh. Moscow: Nauka Publishing House, 1973, 153–157.

[241] Chibata I et al. Immobilized aspartase-containing microbials cells, Appl. Microbiol., 1974, 27, 878–885.

[242] Oyama K et al. On the mechanism of the action of thermolysis kinetics study of the thermolysin – catalysed condensation reaction of N-benzyloxycarbonyl-1-aspartic acid with L-phenylanaline methyl ester. J. Chem. Soc., 1981, 11, 356–360.

[243] Takamatsu S et al. Production of L-alanine from ammonium fumarate using immobilized microorgansm- elimination of side reaction, Eur. J. Appl. Microbiol Biotechnol., 1982, 15, 147–149.

[244] Sato T, Tosa T. Optical resolution of racemic amino-acids by amino acilase in: Industrial applications of immobilized biocatalyst, ed. Tanaka et al. Marcel Dekker, 1993, pp. 3–14.

[245] Pastore M, Morisi I. Lactose reduction of milk by fiber entrapped β-galactosidase. Methods Enzymol. 1976, 44, 822–830.

[246] Carasik W, Carrol JO. Development of immobilizedenzymes for the production of high fructose corn sirup. Food Technol, 1983, 37, 85–91.

[247] Takata I, Tosa T. Production of L-malic acid in: Industrial applications of immobilized biocatalyst, ed. Tanaka et al. Marcel Dekker, 1993, pp.53–55.

[248] Bailey JE, Ollis DE. Biochemical engineering fundamentals. McGraw-Hill, 2000 pp157–227,

[249] Brown T et al. Chemistry: the central science. Upper Saddle River, NJ: Prentice Hall, 2003, p. 958. ISBN 0–13–048450–4.

[250] Bryant DA and Frigaard NU. Prokaryotic photosynthesis and phototrophy illuminated. Trends in microbiology. 2006, 14 (11), 488–96,

[251] Blankenship R. Molecular Mechanisms of Photosynthesis. Blackwell Science, Williston, VT: McGraw-Hill, 2000

[252] Whitmarsh J, Govindje E. (2004) The Photosynthetic Process, University of Illinois at Urbana-Champaign, http://www.life.uiuc.edu/govindjee/paper/gov.html.

[253] Hallenbeck PC, Benemann JR. Biological hydrogen production: fundamentals and limiting processes. Intl J Hydrogen Energy, 2002, 27, 1185–1193.

[254] Carroll BW, Ostlie DA. An Introduction to Modern Astrophysics. Addison-Wesley, 2007 p121. ISBN 0–321–44284–9

[255] Greenbaum E. Energetic efficiency of hydrogen photoevolution by algal water splitting, Biophys J, 1988, 54, 365–368.

[256] Boichenko VA et al. (1989). Efficiency of hydrogen photoproduction in algae and cyanobacteria, Fiziol Rast 36:239–247.

[257] Brand JJ et al. Hydrogen production by eukaryotic algae, Biotech Bioeng, 1989, 33, 1482–1488.

[258] Urbig TR et al. Inactivation and reactivation of the hydrogenases of the green algae Scenedesmus obliquus and Chlamydomonas reinhardtii, Z Naturforsch, 1993, 48c, 41–45.

[259] Boichenko VA et al. Hydrogen production by photosynthetic microorganisms, in: Photoconversion of Solar Energy: Molecular to Global Photosynthesis Eds: Archer MD, Barber J UK, London: Imperial College Press, 2000.

[260] Association of Energy Engineers (2004). Energy Dictionary: Fuel Conversion Efficiency (accessed August 1, 2013 at http://www.energyvortex.com/frameset.cfm?source=/energydictionary/energyvortex.html).

[261] Ghirardi ML et al. (1997). Oxygen sensitivity of algal H_2-production. Appl Biochem Biotechnol 1997, (63–65), 141–151.

[262] Brochure IEA Bioenergy Task 42 Biorefinery, 2009. (Accessed August 1, 2013 at http://www.biorefinery.nl/fileadmin/biorefinery/docs/Brochure_Totaal_definitief_HR_opt.pdf).

[263] BIOPOL, 2009. Final report. BIOPOL, EU-FP6-project, Deliverable D7.6, 68 pp. (Accessed August 1, 2013 at http://www.biorefinery.nl/fileadmin/biopol/user/documents/PublicDeliverables/BIO POL _D_7_6_-_Final_240609.pdf).

[264] DIRECTIVE 2001/77/EC OF THE EUROPEAN PARLIAMENT AND OF THE COUNCIL of 27 September 2001 on the promotion of electricity produced from renewable energy sources in the internal electricity market.

[265] Mantau U. et al. EU Wood Real potential for changes in growth and use of EU forests. Final report. http://ec.europa.eu/energy/renewables/studies/doc/bioenergy/euwood_final_report.pdf. 2010.

[266] EEA Report No 7, 2006. How much bioenergy can Europe produce without harming the environment? (Accessed June 2013 at http://www.eea.europa.eu/publications/eea_report_2006_7).

[267] Ericsson K, Nilsson LJ. Assessment of the potential biomass supply in Europe using a resource-focused approach. Biomass and Bioenergy, 2006, 30(1), 1–15.

[268] Doran PM. Bioprocess Engineering Principles. Elsevier, 1995, ISBN: 0122208552

[269] Shuler ML. Bioprocess engineering. In: Encyclopedia of Physical Science and Technology, vol 2, Ed., R.A. Meyers, Orlando Academic Press, 1987.

[270] OECD report: The Bioeconomy to 2030: designing a policy agenda, 2009, (Accessed August 1, 2013 at http://www.oecd.org/document/48/0,3343,en_2649_36831301_42864368_1_1_1_1,00.html.

[271] The Knowledge Based Bio-Economy in Europe: Achievements and Challenges. 2010 (background report for the KBBE Knowledge Based Bio-Economy towards 2020 Conference organised by the Belgian Presidency of the EU and the EC, accessed August 1, 2013 at http://sectie.ewi-vlaanderen.be/en/kbbe2010/about-kbbe/kbbe-study).

[272] US Biobased Products Market Potential and Projections Through 2025, February 2008, accessed August 1, 2013 at http://www.usda.gov/oce/reports/energy/BiobasedReport2008.pdf.

[273] Harrison RJ, Moyle M. 2-Furoic acid. Organic Syntheses. 1956, 36, 36.

[274] Ozer R. Vapor phase decarbonylation process. WIPO Patent Application WO/2011/026059 (2011).

[275] Kamm B et al. Biorefineries – Industrial Processes and Products. Status Quo and Future Directions. Vol. 1. Weinheim: Wiley-VCH, 2006 ISBN: 3-527-31027-4.

[276] Klingler FD, Ebertz W. Oxocarboxylic Acids. In: Ullmann's Encyclopedia of Industrial Chemistry Weinheim: Wiley-VCH, 2005.

[277] Huber GW et al. Synthesis of Transportation Fuels from Biomass: Chemistry, Catalysts, and Engineering Chemical Reviews, 2006, 106, 4044–4098.

[278] Nattrass L, Higson A. Renewable Chemicals Factsheet: Lactic Acid (accessed August 1, 2013 at http://www.nnfcc.co.uk/publications/nnfcc-renewable-chemicals-factsheet-lactic-acid)

[279] Zeikus JG et al. Biotechnology of succinic acid production and markets for derived industrial products. Applied Microbiology and Biotechnology, 1999, 51(5), 545.

[280] Nattrass L. et al. NNFCC Renewable Chemicals Factsheet: Succinic Acid (accessed August 1, 2013 at http://www.nnfcc.co.uk/publications/nnfcc-renewable-chemicals-factsheet-succinic-acid).

[281] US Food and Drug Administration, Inactive Ingredient Search for Approved Drug Products (accessed August 1, 2013 at http://www.accessdata.fda.gov/scripts/cder/iig/index.cfm).

[282] Lieberman HH, Lachman L, Schwartz LJB. (eds)Pharmaceutical Dosage Forms: Tablets. New York: Marcel Dekker, 1989 p. 18. ISBN 0–8247–8044–2.

[283] Suzuki TG et al. Silencing leaf sorbitol synthesis alters long-distance partitioning and apple fruit quality. Proceedings of the National Academy of Sciences of the United States of America 2006, 103(49), 18842–7.

[284] List of Registered Fuels and Fuel Additives (Accurate as of July 11, 2013; accessed August 1, 2013 at http://www.epa.gov/otaq/fuels/registrationfuels/registeredfuels.htm).

[285] White RD. Health effects of inhaled tertiary amyl methyl ether and ethyl tertiary butyl ether. In: Proceedings of the International Congress of Toxicology – VII, December 1995.

[286] Riemenschneider W, Bolt HM. Esters, Organic In: Ullmann's Encyclopedia of Industrial Chemistry. Weinheim: Wiley-VCH, 2005.

[287] Pankaj D. Ethyl Acetate: A Techno-Commercial Profile. Chemical Weekly, August 10, 2004, 184.

[288] IEA Bioenergy Task 42 Biorefinery, 2011 (accessed August 1, 2013 at http://www.iea-bioenergy.task42-biorefineries.com/).

[289] Biorefinery Euroview, 2009. D1.1: Report on the different concepts of existing European biorefineries. (Accessed September 2013 at http://www.biorefinery-euroview.eu/biorefinery/public/results.html).

[290] Frost & Sullivan Best Practices Awards. The Impact of green trends in the European Oleochemicals market, Jul-2007. (Accessed August 1, 2013 at http://www.frost.com/prod/servlet/market-insight-top.pag?docid=102601717)

[291] Sanders JPM, Meesters KPH. Method and installation for producing electricity and conversion products, such as ethanol, Octrooinummer: Patent EP2098596, ingediend: 2008–03–06.

[292] Kamm B, Kamm M. Principles of biorefineries. Appl Microbiol Biotechnol, 2004, 64, 137–145.

[293] Kamm B et al. The green biorefinery, concept of technology. (First international symposium on green biorefinery) Neuruppin, Society of Ecological Technology and System Analysis, Berlin, 1998.

[294] Kamm B et al. Green biorefinery Brandenburg, article to development of products and of technologies and assessment. Brandenburg Umweltber, 2000, 8, 260–269.

[295] Narodoslawsky M. Green biorefinery. (Second international symposium on green biorefinery) SUSTAIN, Feldbach, 1999.

[296] Wijffels RH, Barbosa MJ. An Outlook on Microalgal Biofuels. Science, 2010, 13, 796–799.

[297] Mata TM et al. Microalgae for biodiesel production and other applications: A review. Renewable and Sustainable Energy Reviews, 2010, 14, 217–232.

[298] Carlsson AS et al. Outputs from the EPOBIO project, 2007.

[299] Benemann J. NREL-AFOSR Workshop, Algae Oil for Jet Fuel Production, Arlington, VA, 2008.

[300] Kumari P et al. Tropical marinemacroalgae as potential sources of nutritionally important PUFAs. Food Chem, 2010, 120, 749–757.

[301] Spolaore P, Joannis-Cassan C, Duran E, Isambert A. Commercial applications of microalgae. J.Bioscience and Bioengineering, 2006, 101(2), 87–96.

[302] Posten C. Design principles of photo-bioreactors for cultivation of microalgae. Eng. Life Sci. 2009, 9(3), 165–177.

[303] Janssen M et al. Enclosed outdoor photobioreactors: Light regime, hotosynthetic efficiency, scale-up, and future prospects. Biotechnol. Bioeng., 2003. 81(2), 193–210.

[304] Hankamer B et al. Photosynthetic biomass and H_2 production by green algae: from bioengineering to bioreactor scale-up. Physiologia Plantarum, 2007. 131(1), 10–21.

[305] Pulz O. Performance Summary Report: Evaluation of GreenFuel's 3D Matrix Algae Growth Engineering Scale Unit. 2007.

[306] Waltz E. Biotech's green gold? Nature Biotechnology. 2009, 27(1), 15–18.

[307] Alias CB et al. Influence of power supply in the feasibility of Phaeodactylum tricornutum cultures. Biotechnol. Bioeng., 2004. 87(6), 723–733.

[308] Sierra E et al. Characterization of a flat plate photobioreactor for the production of microalgae. Chemical Engineering Journal, 2008. 138(1– 3), 136–147.

[309] Wang CH. Effect of liquid circulation velocity and cell density on the growth of Parietochloris incisa in flat plate photobioreactors. Biotechnol. Bioproc. Eng., 2005. 10(2), 103–108.

[310] Lopez MCGM et al. Comparative analysis of the outdoor culture of Haematococcus pluvialis in tubular and bubble column photobioreactors. J. Biotechnol., 2006. 123(3), 329–342.

[311] Gudin DC. Solar biotechnology study and development of tubular solar receptors for controlled production of photosynthetic cellular biomass. Eds. Palz W, Pirrwitz D. Proceedings of the Workshop E.C. Capri, Reidel, Dordrecht, 1983, 184–193.

[312] Torzillo G et al. Production of Spirulina biomass in closed photobioreactors. Biomass, 1986, 11, 61–74.

[313] Contreras-Flores C et al. Avances en el diseño conceptual de fotobioreactores para el cultivo de microalgas. Interciencia, 2003, 28, 450–456.

[314] Borowitzka MA. Commercial production of microalgae: ponds, tanks, tubes and fermenters. J. Biotechnol, 1999, 70, 313–321.

[315] Richmond A, Cheng-Wu Z. Optimization of a flat plate glass reactor for mass production of Nannochloropsis sp. outdoors. J. Biotechnol, 2001, 85, 259–269.

[316] Eriken NT. The technology of microalgal culturing. Biotechnol Lett, 2008, 30, 1525–1536.

[317] Chacón-Lee TL, González-Mariño GE. Microalgae for "healthy" Foods-Possibilities and Challenges. Comprehensive reviews in food science and food safety. 2010, 9, 655–675.

[318] Cuaresma M et al. Horizontal or vertical photobioreactors? How to improve microalgae photosynthetic efficiency. Bioresource Technology, 2011, 102, 5129–5137.

[319] Advancing the chemical Science Conferrece, RSC Royal Society of Chemistry,19 November 2013, RSC Chemistry Centre, London.

[320] Borowitzka MA, Micro-algae as sources of fine chemicals. Microbiol Sci. 1986, 3(12), 372–5.

[321] http://www.aquaflowgroup.com/capabilities-group/producing-fine-chemicals-from-wild-algae, accessed July 21, 2013.

[322] Plaza M et al. Innovative natural functional ingredients from microalgae. J. Agric Food Chem., 2009, 57, 7159–70.

[323] Griffiths TW. Cosmeceuticals: coming of age. Br J Dermatol, 2010, 162, 469–70.

[324] Reszko AE et al. Cosmeceuticals: practical applications. Obstet Gynecol Clin North Am., 2010, 37, 547–69.

[325] Kim SK et al. Prospective of the Cosmeceuticals Derived from Marine Organisms. Biotechnology and Bioprocess Engineering, 2008, 13, 511–523.

[326] Olaizola M. Commercial development of microalgal biotechnology: from the test tube to the marketplace. Biomol Eng., 2003, 20, 459–66.

[327] Lintner KF et al. Heat-stable enzymes from deep sea bacteria: a key tool for skin protection against UV-A induced free radicals. IFSCC, 2002, 5, 195–200.

[328] Lintner K et al. Cosmeticals and active ingredients. Clin Dermatol, 2009, 27(5), 461–468.

[329] Liu J et al. Protandim, a fundamentally new antioxidant approach in chemoprevention using mouse two-stage skin carcinogenesis as a model. PLoS ONE 4(4): e5284, 2009.

[330] Hartung J et al. Bromoperoxidase activity and vanadium level of the brown alga Ascophyllum nodosum. Phytochemistry, 2000, 69, 2826–2830.

[331] Lods LM et al. The future of enzymes in cosmetics. Int J Cosmet Sci., 2000, 22, 85–94.

[332] Regulation EC No 1334/2008 of the European Parliament and the Council of 16 December 2008 on flavourings and certain food ingredients with flavouring properties for use in and on foods and amending Council Regulation (EEC) No 1601/91. Regulations (EC) No 2232/96 and (EC) No 110/2008 and Directive 200/12/EC.

[333] Maksymiec W et al. The level of jasmonic acid in Arabidopsis thaliana and Phaseolus coccineus plants under heavy metal stress, J. Plant Physiology, 2005, 162, 1338–1346.

[334] Farmer EE et al. Jasmonates and related oxylipins in plant responses to pathogenesis and herbivory. Current Opinion in Plant Biology, 2003, 6, 372–378.

[335] Li HB et al. Evaluation of antioxidant capacity and total phenolic content of different fractions of selected microalgae. Food Chem, 2007, 102, 771–776.

[336] Yokoya NS et al. Endogenous Cytokinins, Auxins, and Abscisic Acid in Red Algae from Brazil. Journal of Phycology, 2010, 46, 1198–1205.

[337] Kowalczyk M, Sandberg G. Quantitative analysis of indole-3-acetic acid metabolites in Arabidopsis. Plant Physiol. 2001, 127, 1645–85.

[338] Jirásková D et al. High-Throughput Screening Technology for Monitoring Phytohormone Production in Microalgae. Journal of Phycology, 2009, 45, 108–118.

[339] John DM et al. Environmental fate of nonphenol ethoxylates: differential adsorption of homologs to components of river sediment. Environ Toxicol Chem, 2000, 19, 293–300.

[340] Ying GG. Fate, behaviour and effect of surfactants and their degradation products in the environment. Environ Int 2006, 32, 417–31.

[341] Verge C et al. Influence of water hardness on the bioavailability and toxicity of linear alkylbenzene sulphonate (LAS). Chemosphere, 2000, 44, 1749–57.

[342] Hashimoto K et al. Biodegradation of Nylon4 and Its Blend with Nylon6. J Appl Polym Sci, 2002, 86, 2307–2311.

[343] Dobrawa R. Method of producing water-soluble non-turbid copolymers of at least one hydrophobic co-monomer, US Patent 7,572,870.

[344] Huber GW, Cheng Y-T, Carlson T, VisputeT Jae J, Tompsett G. Catalytic pyrolysis of solid biomass and related biofuels, aromatic, and olefin compounds. University of Massachusetts September 2009: US 20090227823.

[345] James A. Stable, aqueous-phase, basic catalytsts and reactions catalyzed thereby. March 2008: US 20080058563,

[346] Haveren J. et al. Bulk chemicals from biomass, Biofuels, Biopord. Bioref., 2, 41–57, 2008.

[347] http://www.maya12–21–2012.com/2012olmec.html accessed August 1, 2013.

[348] Ryan D. One Word: Bioplastics. The Technology Gains Momentum, But Hurdles Remain, accessed August 1, 2013 at www.nerac.com.

[349] Petkov G, Garcia G. Which are fatty acids of the green alga Chlorella? Biochemical Systematics & Ecology, 2007, 35, 281–285.

[350] Yamasaki Y et al. Effects of alginate oligosaccharide mixtures on the growth and fatty acid composition of the green alga Chlamydomonas reinhardtii, J. Bioscience & Bioengineering,2012, 113(1), 112–116.

[351] Chua H. Accumulation of biopolymers in activated sludge biomass, Applied Biochemistry and Biotechnology, Spring 1999, 78(1–3), 389–399.

[352] (Accessed August 1, 2013 at http://www.daicelpolymer.com/en/news/2012/121129.html)

[353] Tserkia V. Biodegradable aliphatic polyesters. Part I. Properties and biodegradation of poly (butylene succinate-co-butylene adipate), Polymer Degradation and Stability, February 2006, 91(2), 367–376.

[354] Albertsson AC. Degradable Aliphatic Polyesters Series: Advances in Polymer Science 2002, 157 (XI), 179.

[355] Bioplastic innovations,accessed October 12, 2013 at: http://bioplastic-innovation.com/ 2010/10/16/roquette-nouvel-acteur-sur-le-marche-des-plastiques-lance-gaialene %C2 %AE-une-gamme-innovante-de-plastique-vegetal/)

[356] Wool RP, Sun XS. Bio-Based Polymers and Composites. Burlington, MA: Elsevier Academic Press, 2005.

[357] Skraly F. Bioplastics in: Encyclopedia of Environmental Microbiology. New York: John Wiley & Sons, 2002.

[358] Snell KD et al. Polyhydroxyalkanoate polymers and their production in transgenic plants. Metab Eng, 2002, 4, 29–40.

[359] Carothers WH et al. Studies of polymerization and ring formation. The reversible polymerization of six-membered cyclic esters, J of Am Chem Soc, 1932, 54, 761–772.

[360] Lowe CE. Preparation of High Molecular Weight Polyhydroxyacetic Ester. U.S. Patent 2,668,162, February 2, 1954.

[361] Gruber PR et al. Continuous Process for Manufacture of Lactide Polymers with Controlled Optical Purity. U.S. Patent 5,247,058, September 21, 1993.

[362] Gruber PR et al. Continuous Process for the Manufacture of a Purified Lactide from Esters of Lactic Acid. U.S. Patent 5,247,059, September 21, 1993.

[363] Gruber PR et al. Continuous Process for Manufacture of a Purified Lactide. U.S. Patent 5,274,073, December 28, 1993.

[364] Gruber et al. Continuous Process for Manufacture of Lactide Polymers with Purification by Distillation. U.S. Patent 5,357,035, October 19 1994.

[365] Vert M, Schwarch G, Coudane J. Present and Future of PLA Polymers, J.M.S.-Pure Appl Chem, 1995, A32(4), 787–796.

[366] Dorgan JR. Poly(lactic acid) Properties and Prospects of an Environmentally Benign Plastic, 3rd Annual Green Chemistry and Engineering Conference Proceedings, Washington DC, 1999, pp 145–149.

[367] Holmgren K, Henning D. Comparison between material and energy recovery of municipal waste from an energy perspective: a study of two Swedish municipalities, Res Cons Recyc, 2004, 43, 51–73.

[368] Kulkarni RK et al. Polylactic acid for surgical implants. Arch of Surg, 1966, 93, 839–843.

[369] Ikada YK et al. Stereocomplex formation between enantiomeric poly(lactides). Macromolecules, 1987, 20, 904–906.

[370] Vainionpää SP. Surgical applications of biodegradable polymers in human tissues. Progress in Polymer Sci, 1989, 14, 679–708.

[371] Eitenmüller J. Operative behandlung von sprunggelenksfrakturen mit biodegradablem schrauben und platten aus poly-L-lactide. Der Chirurg, 1996, 67, 413–418.

[372] Viljanen JT. Comparison of the tissue response to absorbable self-reinforced polylactide screws and metallic screws in the fixation of cancellous bone osteotomies: An experimental study on the rabbit distal femur. Journal of Orthopaedic Research, 1997, 15, 398–407.

[373] Richardson JB et al. (2000). The BiolokTM Screw A New Generation in Resorbables, Orthopaedic Product News: Online Edition, accessed August 1, 2013 at http://www.opnews.com/ole/archives/articles/mar.may00/items.html.

[374] Middleton JC, Tipton JA. Synthetic biodegradable polymers as medical devices. Med Plastics and Biomaterials, Mag 30, 1998.

[375] Pétas AM et al. The biodegradable self-reinforced poly-DL-lactic acid spiral stent compared with a suprapubic catheter in the treatment of post-operative urinary retension after visual Laser ablation of the prostate. British Journal of Urology, 1997, 80, 439–443.

[376] Vert M. Biomedical polymers from chiral lactides and functional lactones: properties and applications. Die Makromolekulare Chemie, Macromolecular Symposia, 1986, 6, 109–122.

[377] Griffith LG. Polymeric biomaterials, Acta Materialia, 2000, 48, 263–277.

[378] Ge HY et al. Preparation, characterization, and drug release behaviors of drug-loaded e-Caprolactone/ L-Lactide copolymer nanoparticles. Journal of Applied Polymer Science, 2000, 75(7), 874–882.

[379] Edlund U, Albertsson AC. Novel drug delivery microspheres from Poly (1,5-dioxepan-2-one-co-L-lactide). Journal of Polymer Science, Part A: Polymer Chemistry, 1999, 37, 1877–1884.

[380] Kricheldorf HR et al. Polylactides – synthesis, characterization and medical application, Macromolecular Symposium, 1996, 103, 85–102.

[381] http://www.gopolymers.com/plastic-types/cellulosic-plastics.html, accessed August 1, 2013

[382] Doi Y and Steinbuchel A. Biopolymers. Weinheim, Germany: Wiley-VCH, 2002

[383] Jacquel N et al. Solubility of polyhydroxyalkanoates by experiment and thermodynamic correlations. AlChE J., 2007, 53(10), 2704–2714.

[384] Kim BS. Production of poly(3-hydroxybutyrate) from inexpensive substrates. Enzyme Microb Technol., Dec 2000, 27(10), 774–777.

[385] Steinbüchel A. Biopolymers. Weinheim, Germany: Wiley-VCH, 2002. ISBN 3–527–30290–5.

[386] Jacquel N et al. Isolation and purification of bacterial poly(3-hydroxyalkanoates). Biochem. Eng. J., 2008, 39(1), 15–27.

[387] Poirier Y et al. Synthesis of highmolecular-weight poly([R]-3-hydroxybutyrate) in transgenic Arabidopsis thaliana plant cells. Int. J. Biol. Macromol., 1995, 17(1), 7–12.

[388] Hong J et al. Biopolymers from vegetable oils via catalyst- and solvent-free chemistry: effects of cross-linking density. Biomacromolecules, Jan 9, 2012;13(1), 261–6.

[389] Nakićenović N et al. Global Energy Perspectives UK: Cambridge University Press, 1998, pp. 299

[390] Sakharov A. Memoirs. New York: Alfred A. Knopf, 1990, pp. 215–225. ISBN 0–679–73595-X

[391] DeGroot GJ et al. A Life. Cambridge, Mass.: Harvard University Press, 2005, p. 254.

[392] Brundtland G. World Commission on Environment and Development. Our Common Future (The Brundtland Report) Oxford, UK: Oxford University Press, 1987

[393] IEA World Energy Outlook 2009, p. 87.

[394] Eurostat Bioenergy statistics, accessed August 1, 2013 at http://epp.eurostat.ec.europa.eu/tgm/table.do?tab=table&init=1&plugin=1&language=/en&pcode=t2020_31).

[395] Wilson K, Lee A. Rational design of heterogeneous catalysts for biodiesel synthesis. Catal. Sci. Technl., 2012, 2, 884–897.

[396] Vyas AP et al. A review on FAME production processes. Fuel, 2010, 89, 1–9.

[397] Krohn BJ et al. Nowlan Production of algae-based biodiesel using the continuous catalytic Mcgyan® process. Bioresource Technol., 2011, 102, 94–100.

[398] Borges ME, Díaz L. Recent developments on heterogeneous catalysts for biodiesel production by oil esterification and transesterification reactions: A review. Renewable and Sustainable Energy Reviews, 2012, 16, 2839–2849.

[399] Eugena Li E et al. MgCoAl–LDH derived heterogeneous catalysts for the ethanol transesterification of canola oil to biodiesel Original, Applied Catalysis B: Environmental, 2009, 88, 42–49.

[400] Chen SY et al. Sulfonic acidfunctionalized plated SBA-15 materials as efficient catalysts for biodiesel synthesis. Green Chem., 2011, 13, 2920–2930.

[401] Dacquin JP et al., Pore-expanded SBA-15 sulfonic acid silicas for biodiesel synthesis. Chem. Commun,, 2012, 48, 212–214.

[402] Harlin A et al. Method for producing olefinic monomers from biooil. World Intellectual Property Organization WO 2010/086507, 5.8.2010.

[403] Van Geem KM et al. Bio-Ethylene Production: Alternatives for Green Chemicals and Polymers. AIChE Spring Meeting Houston, TX, April 1–5, 2012.

[404] Dmirbas A. Biodiesel fuels from vegetable oils via catalytic and non-catalytic supercritical alcohol transesterifications and other methods: A survey. Energy Conv Mgmt, 2003, 44, 2093–2109.

[405] Kusdiana D, Saka S. Kinetics of transesterification in rapeseed oil to biodiesel fuel as treated in supercritical methanol. Fuel, 2001, 80, 693–698.

[406] Kusdiana D, Saka S. Methyl esterification of free fatty acids of rapeseed oil as treated in supercritical methanol. J Chem Eng Jpn, 2001, 34, 383–387.

[407] Saka S,Kusdiana D. Biodiesel fuel from rapeseed oil as prepared in supercritical methanol. Fuel, 2001, 80, 225–231.

[408] Ma F, Hanna MA. Biodiesel production: A review. Bioresource Technol, 1999, 70,1–15.

[409] King MW. (2003). Fatty Acid Oxidation, Indiana University School of Medicine, http://www.med.unibs.it/~marchesi/fatox.html, accessed September 2013.

[410] Jaeger KE, Eggert T. Lipases for biotechnology. Curr Opin Biotechnol, 2002, 13, 390–397.

[411] Fukuda HA et al. Biodiesel fuel production by transesterification of oils,. J Biosci Bioeng 2001, 92, 405–416.

[412] Matsumoto TS et al. Yeast whole-cell biocatalyst constructed by intracellular overproduction of Rhizopus oryzae lipase is applicable to biodiesel fuel production. Appl Microbiol Biotechnol, 2001, 57, 515–520.

[413] Hama SH et al. Effect of fatty acid membrane composition on whole-cell biocatalysts for biodiesel-fuel production. Biochem Eng J., 2004, 21, 155–160.

[414] Ban KM et al. Whole cell biocatalyst for biodiesel fuel production utilizing Rhizopus oryzae cells immobilized within biomass support particles. Biochem Eng J., 2001, 8, 39–43.

[415] Ban KS et al. Repeated use of whole-cell biocatalysts immobilized within biomass support particles for biodiesel fuel production, J Molec Catal B: Enz, 2002, 17, 157–165.

[416] Oda MM et al. Facilitatory effect of immobilized lipase-producing Rhizopus oryzae cells on acyl migration in biodiesel-fuel production. Biochem Eng J., 2005, 23, 45–51.

[417] Thamsiriroj T. The impact of the life cycle analysis methodology on whether biodiesel produced from residues can meet the EU sustainability criteria for biofuel facilities constructed after 2017. Renewable Energy, 2011, 36, 50–63.

[418] Ahmann D, Dorgan JR. Bioengineering for Pollution Prevention through Development of Biobased Energy and Materials – State of the Science Report, EPA/600/R-07/028, US Environmental Protection Agency, Office of Research and Development, National Center for Environmental Research, Washington DC, 20460.

[419] Riemenschneider W, Bolt HM. Esters, Organic in Ullmann's Encyclopedia of Industrial Chemistry. Weinheim: Wiley-VCH, 2005.

[420] Yang J et al. Life-cycle analysis on biodiesel production from microalgae: Water footprint and nutrients balance. Bioresources Technology, 2010, 10, 1016.

[421] Greenwell HC et al. Placing microalgae on the biofuels priority list: a review of the technological challenges. Journal of The Royal Society Interface, 23 December 2009, 7(46), 703–726.

[422] Potts T et al. The Production of Butanol from Jamaica Bay Macro Algae. Environmental Progress and Sustainable Energy, April 2012, 31(1), 29–36.

[423] Samson R, LeDuy A. Detailed study of anaerobic digestion of Spirulina maxima algal biomass. Biotechnology and Bioengineering, 1986, 28, 1014–1023.

[424] Yen, H-W, Brune DE. Anaerobic co-digestion of algal sludge and waste paper to produce methane. Bioresource Technology, 2007, 98, 130–134.

[425] Nuclear Hydrogen, R&D Plan Department Of Energy, Office of Nuclear Energy, Science and Technology, March 2004.

[426] DOE/Idaho National Laboratory (September 22, 2008). Steam Heat: Researchers Gear Up For Full-scale Hydrogen Plant. ScienceDaily, accessed June 1, 2013 at http://www.sciencedaily.com/releases/2008/09/080918170624.html

[427] Ahmann D, Dorgan JR. Bioengineering for Pollution Prevention Through Development of Biobased Materials and Energy, U.S. Environmental Protection Agency Office of Research and Development National Center for Environmental Research Washington, DC 20460, EPA/600/R-07/028.

[428] Akkerman I et al. Photobiological hydrogen production: photochemical efficiency and bioreactor design. Int J Hydrog Energy, 2002, 27, 1195–1208.

[429] Hemschemeier A et al. Analytical approaches to photobiological hydrogen production in unicellular green algae. Photosynth Res. December 2009, 102(2–3), 523–540.

[430] Hinnemann B et al. Biomimetic hydrogen evolution: MoS2 nanoparticles as catalyst for hydrogen evolution. J. Am. Chem. Soc. 2005, 127, 5308–5309.

[431] Madden C et al. Catalytic Turnover of [FeFe]-Hydrogenase Based on Single-Molecule Imaging. J. Am. Chem. Soc. 2012, 134, 1577–1582.

[432] Zaborsky O. BioHydrogen. New York: Plenum Press,1998.

[433] Akihiko K. et al. Strategies for the Development of Visible-light-driven Photocatalysts for Water Splitting. Chemistry Letters, 2004, 33(12), 1534.

[434] Happe TA et al. Hydrogenases in green algae: Do they save the algae's life and solve our energy problems? Trends Plant Sci, 2002, 7, 246–250.

[435] Happe T, Naber JD. Isolation, characterization, and N-terminal amino acid sequence of hydrogenase from the green alga Chlamydomonas reinhardtii. Eur J Biochem, 1993, 214, 475–481.

[436] Wu LF, Mandrand MA. Microbial hydrogenases: Primary structure, classification, signatures, and phylogeny. FEMS Microbiol Rev, 1993, 104, 243–270.

[437] Horner DS et al. Iron hydrogenases and the evolution of anaerobic eukaryotes. Mol Biol Evol, 2000, 17, 1695–1709.

[438] Pinto L et al. A brief look at three decades of research on cyanobacterial hydrogen evolution. Intl J Hydrogen Energy, 2002, 27, 1209–1215.

[439] Gfeller RP, Gibbs M. Fermentative metabolism of Chlamydomonas reinhardtii I: Analysis of fermentative products from starch in dark and light. Plant Physiol, 1984, 75, 212–218.

[440] Ghirardi ML et al. Microalgae: A green source of renewable H_2, Trends Biotech, 2002, 18, 506–511.

[441] Albracht SPJ. Nickel hydrogenases: in search of the active site. Biochim Biophys Acta, 1994, 1188, 167–204.

[442] Adams M. The structure and mechanism of iron hydrogenases. Biochim Biophys Acta, 1999, 1020, 115–145.

[443] Melis A. Green alga hydrogen production: Progress, challenges, and prospects, Intl J Hydrogen Energy, 2002, 27, 1217–1228.

[444] Koku H et al. Aspects of the metabolism of hydrogen production by Rhodobacter sphaeroides. Intl J Hydrogen Energy, 2002, 27,1315–1329.

[445] Levin DB et al. Biohydrogen production: Prospects and limitations to practical application. Intl J Hydrogen Energy, 2004, 29,173–185.

[446] Bonam D et al. Regulation of carbon monoxide dehydrogenase and hydrogenase in Rhodospirillum rubrum: Effects of CO and oxygen on synthesis and activity. J Bacteriol, 1989, 171, 3102–3107.

[447] Maness PC, Weaver P. Hydrogen production from a carbon-monoxide oxidation pathway in Rubrivivax gelatinosus. Intl J Hydrogen Energy, 2002. 27,1407–1411.

[448] Krupp M, Widmann R. Biohydrogen production by dark fermentation: Experiences of continuous operation in large lab scale. International Journal of Hydrogen Energy, May 2009, 34(10), 4509–4516.

[449] Clark DP. The fermentation pathways of Escherichia coli, FEMS Microbiol Rev 1989. 5, 223–234.

[450] Choi Y et al. Development of Microbial Fuel Cells Using Proteus Vulgaris. Bulletin of the Korean Chemical Society, 2000, 21(1), 44–48.

[451] Mohan V et al. Harnessing of bioelectricity in microbial fuel cell (MFC) employing aerated cathode through anaerobic treatment of chemical wastewater using selectively enriched hydrogen producing mixed consortia. Fuel, 2008, 87(12), 2667–2676.

[452] Sobieszuk P, Zamojska A, Kołtuniewicz AB. Harvesting energy and hydrogen from microbes Chemical and Process Engineering, 2012, 33 (4), 603–610

[453] Potter MC. Electrical effects accompanying the decomposition of organic compounds. Royal Society B, 1911, 84, 260–276

[454] Mohan V et al. Bioelectricity generation from chemical wastewater treatment in mediatorless (anode) microbial fuel cell (MFC) using selectively enriched hydrogen producing mixed culture under acidophilic microenvironment". Biochemical Engineering Journal, 2008, 39, 121–130.

[455] Ga-Young Suh et al. Quantification of Particle Residence Time in Abdominal Aortic Aneurysms Using Magnetic Resonance Imaging and Computational Fluid Dynamics. Ann Biomed Eng., February 2011, 39(2), 864–883.

[456] Bingham EC and Roepke R. The Rheology of Blood. Journal of General Physiology, 1944, 28, 79–93.

[457] LaVan D et al. Small-scale systems for in vivo drug delivery. Nature biotechnology, 2003, 21, 1184–1191

[458] Saltzman WM, Olbricht WL. Building drug delivery into tissue engineering. Nature Reviews Drug Discovery, 2002, 1, 177–186.

[459] Stamatialis DF et al. Medical applications of membranes: drug delivery, artificial organs and tissue engineering. Journal of Membrane Science, 2008, 308 (1–2) 1–34.

[460] Nisar A et al. MEMS-based micropumps in drug delivery and biomedical applications. Sensors and Actuators B: Chemical, 2008, 130, 917–942.

[461] Tsai NC. and Sue CY, Review of MEMS-based drug delivery and dosing systems, Sensors and Actuators A: Physical 134 (2007) 555–564.

[462] Vasilev AE et al. Transdermal therapeutic systems for controlled drug release (a review). Pharmaceutical Chemistry Journal, 2001, 35, 613–626.

[463] Guarang P et al. Controlled drug delivery system and polymers. International Journal of Pharmacy and Technology, 2013, 5, 2632–2644.

[464] Scholz OA et al. Drug delivery from the oral cavity: focus on a novel mechatronic delivery device. Drug Discovery Today, 2008, 13, 247–252.

[465] Buonomenna MG, Choi SH. Recent advances on polymeric membranes for membrane reactors, in Buonomenna MG and Golemme G (Eds.), Advanced Materials for Membrane Preparation, Bentham Science Publishers, 2012 (Chapter 14, 248–285).

[466] Gerard E et al. Surface modifications of polypropylene membranes used for blood filtration. Polymer, 2011, 52, 1223–1233.

[467] Chwojnowski A. Polysulphone and polyethersulphone hollow fiber membranes with developed inner surface as material for bio-medical applications. Biocybernetics and Biomedical Engineering, 2009, 29, 47–59.

[468] Drioli E,. and Giorno L (Eds.), Comprehensive Membrane Science and Engineering, Elsevier Science 2010.

[469] Humes HD et al. Replacement of renal function in uremic animals with a tissue-engineered kidney. Nature Biotechnology, 1999, 17, 451–455.

[470] Dankers PYW et al. The use of fibrous, supramolecular membranes and human tubular cells for renal epithelial tissue engineering: Towards a suitable membrane for a bioartificial kidney. Macromolecular Bioscience, 2010, 10, 1345–1354.

[471] De Bartolo L et al. Bio-hybrid organs and tissues for patient therapy: A future vision for 2030. Chemical Engineering and Processing, 2012, 51, 79–87.

[472] Kjaergard LL et al. Artificial and bioartificial support systems for acute and acute-on-chronic liver failure: a systematic review. JAMA – The Journal of the American Medical Association, 2003, 289, 217–222.

[473] Khuroo MS et al. Molecular adsorbent recirculating system for acute and acute-on-chronic liver failure: a meta-analysis. Liver Transplantation, 2004, 10, 1099–1106.

[474] De Bartolo L. Human hepatocyte functions in a crossed hollow fiber membrane bioreactor. Biomaterials, 2009, 30, 2531–2543.

[475] Gautier A et al. Hollow fiber bioartificial liver: Physical and biological characterization with C3A cells. Journal of Membrane Science, 2009, 341, 203–213.

[476] Legallais C et al. Bioartificial livers (BAL): current technological aspects and future developments. Journal of Membrane Science, 2001, 181, 81–95.

[477] Kobayashi N. Bioartificial pancreas for the treatment of diabetes. Cell Transplantation, 2008, 17, 11–17.

[478] Ricotti L et al. Wearable and implantable pancreas substitutes. Journal of Artificial Organs, 2013, 16, 9–22.

[479] Sumi S, Regenerative medicine for insulin deficiency: creation of pancreatic islets and bioartificial pancreas. Journal of Hepato-biliary-pancreatic Sciences, 2011, 18, 6–12.

[480] Ikeda H et al. A newly developed bioartificial pancreas successfully controls blood glucose in totally pancreatectomized diabetic pigs. Tissue Engineering, 2006, 12, 1799–1809.

[481] Cobelli C et al. Artificial pancreas: past, present, future. Diabetes, 2011, 60, 2672–2682.

[482] Opara EC et al. Design of a bioartificial pancreas. Journal of Investigative Medicine, 2010, 58, 831–837.

[483] Elliott RB et al. Live encapsulated porcine islets from a type 1 diabetic patient 9.5 yr after xenotransplantation. Xenotransplantation, 2007, 14, 157–161.

[484] Wolfson MR, Shaffer TH. Pulmonary applications of perfluorochemical liquids: Ventilation and beyond. Pediatric Respiratory Reviews, 2005, 6, 117–127

[485] Kaisers U et al. Liquid ventilation, British Journal of Anaesthesia, 2003, 91, 143–151.

[486] Wolfson MR, Shaffer TH. Liquid ventilation: an adjunct for respiratory management. Paediatric Anaesthesia, 2004, 14, 15–23.

[487] Haitsma JJ. Physiology of mechanical ventilation. Critical Care Clinics, 2007, 23, 117–134.

[488] Tawfic QA et al. Liquid ventilation. Oman Medical Journal, 2011, 26, 4–9.

[489] Shaffer TH et al. Liquid ventilation: current status. Pediatrics in Review, 1999, 20, e134-e142.

[490] Bull JL et al. Effects of respiratory rate and tidal volume on gas exchange in total liquid venti-lation, ASAIO (American Society for Artificial Internal Organs) Journal, 2009, 55, 373–381.

[491] Pasche A et al. Uitlité clinique du lavage bronchoalvéolaire [Clinical value of bronchoalveolar lavage]. Revue Médicale Suisse, 2012, 363, 2212–2218.

[492] Proquitté H et al. A comparison of conventional surfactant treatment and partial liquid venti-lation on the lung volume of injured ventilated small lungs. Physiological Measurement,2013, 34, 915–924.

[493] Kandler MA. Persistent improvement of gas exchange and lung mechanics by aerosolized perfluorocarbon. American Journal of Respiratory and Critical Care Medicine, 2001, 164, 31–35.

[494] Chang TMS et al. Red blood cell substitutes. Best Practice and Research Clinical Haematol-ogy, 2000, 13, 651–667.

[495] Kim HW. Engineering blood cells and proteins as blood substitutes: A short review. Biotech-nology and Bioprocess Engineering, 2007, 12, 43–47.

[496] Bucci E. Thermodynamic approach to oxygen delivery in vivo by natural and artificial oxygen carriers. Biophysical Chemistry, 2009, 142, 1–6.

[497] Tsai AG et al. Oxygen gradients in the microcirculation. Physiological Reviews, 2003, 83, 933–963.

[498] Kim HW, Greenburg AG. Artificial oxygen carriers as red blood cell substitutes: A selected review and current status. Artificial Organs, 2004, 28, 813–828.

[499] Henkel-Hanke T, Oleck M. Artificial oxygen carriers: A current review. American Association of Nurse Anesthetists Journal, 2007, 75, () 205–211.

[500] Habler O. et al. Artificial oxygen carriers as an alternative to red blood cell transfusion. Anaesthesist, 2005, 54, 741–754.

[501] Sobieszuk P, Pilarek M. Absorption of CO_2 into perfluorinated gas carrier in the Taylor gas-liquid flow in a microchannel system. Chemical and Process Engineering, 2012, 33, 595–602

[502] Yager P et al. Microfluidic diagnostic technologies for global public health. Nature, 2006, 442, 412–418.

[503] Rivet C et al. Microfluidics for medical diagnostics and biosensors, Chemical Engineering Science, 2011, 66, 1490–1507.

[504] Marx U. How drug development of the 21st century could benefit from human micro-organoid in vitro technologies, in: Marx U, Sandig V (Eds.), In Vitro Drug Testing – Breakthroughs and Trends in Cell Culture Technologie., Weinheim Wiley-VCH, 2007, pp. 271–282.

[505] Sonntag F et al. Design and prototyping of a chip-based multi-micro-organoid culture system for substance testing, predictive to human (substance) exposure. Journal of Biotechnology, 2010, 163 70–75.

[506] Pilarek M et al. Biological cardio-micro-pumps for microbioreactors and analytical micro-systems, Sensors and actuators B: Chemical, 2011, 156, 517–526.

[507] Tanaka Y. Biological cells on microchips: new technologies and applications, Biosensors and Bioelectronics, 2007, 23, 449–458.

[508] Whitesides GM. The origins and the future of microfluidics. Nature, 2006, 442, 368–373.

[509] Dittrich P et al. Micro Total Analysis Systems, Latest Advancements and Trends. Analytical Chemistry, 2006, 78, 3887–3907.

[510] Spatz JP. Bio-MEMS: building up micromuscles. Nature Materials 4, 2005, 115–116.

[511] Kim DH. Microengineered platforms for cell mechanobiology, Annual Review of Biomedical Engineering, 2009, 11, 203–233.

[512] Eliel EL et al. Basic organic stereochemistry. New York: Wiley-Interscience, 2001.

[513] Sarker SD, Nahar L. Chemistry for pharmacy students: general, organic and natural product chemistry. New York: Wiley–Interscience, 2007.

[514] Rouhi AM, Chirality at work. Chemical Engineering News, 2003, 81, 56–61.

[515] Brenna E et al. Enantioselective perception of chiral odorants. Tetrahedron: Asymmetry, 2003, 14, 1–42.

[516] Tucker GT. Chiral switches. The Lancet, 2000, 355, 1085–1087.

[517] Maier NM et al. Separation of enantiomers: needs, challenges, perspectives- review, Journal of Chromatography A, 2001, 906, 3–33.

[518] Stinson CS. Chiral pharmaceuticals, Chemical and Engineering News, 2001, 79, 79–97.

[519] Hiemke Ch. Härtter S. Pharmacokinetics of selective serotonin reuptake inhibitors, Pharmacology & Therapeutics, 2000, 85, 11–28.

[520] Reichel Ch et al. Molecular cloning and expression of a 2–arylpropionyl–coenzyme A epimerase: A key enzyme in the inversion metabolism of ibuprofen. Molecular Pharmacology, 1997, 51, 576–582.

[521] Hutt AJ, Valentova J, The chiral swich: the development of single enantiomer drugs from racemates. Acta Facultatis Pharmaceuticae Universitatis Comenianae, L, 2003, 7–23.

[522] Patel RN, Enzymatic preparation of chiral pharmaceutical intermediates by lipases. Journal of Liposome Research, 2001, 11, 355–393.

[523] Bosch EM, et al, Determination of paracetamol: Historical evolution. Journal of Pharmaceutical and Biomedical Analysis, 2006, 42, 291–321.

[524] BCC Research Market Forecasting, 2012
http://bccresearch.wordpress.com/tag/biotechnology-market/

[525] Subramanian G, Chiral separation techniques: a practical approach., Weinheim: Wiley–VCH Verlag GmbH & Co, 2001.

[526] Gladiali S, Guidelines and methodologies in asymmetric synthesis and catalysis. Comptes Rendus Chimie, 2007, 10, 220–231.

[527] Toda F, Enantiomer separation: fundamentals and practical methods. Dordrecht, The Netherlands: Kluwer Akademic Publishers, 2004.

[528] Strauss UT et al, Biocatalytic transformation of racemates into chiral building blocks in 100 % chemical yield and 100 % enantiomeric excess. Tetrahedron: Asymmetry, 1999, 10, 107–117.

[529] Ghanem A, Enein HY, Application of lipases in kinetic resolution of racemates – review. Chirality, 2005, 17, 1–15.

[530] Gübitz G, Schmid MG, Chiral separation principles in chromatographic and electromigration techniques: A review: Biopharmaceutics and Drug Disposition, 2001, 22, 291–336.

[531] Lorenz H. et al. Crystallization based separation of enantiomers (review). Journal of the University of Chemical Technology and Metallurgy, 2007, 42, 5–16.

[532] Lorenz H, et al. Crystallization of enantiomers. Chemical Engineering and Processing. 2006 45, 863–873.

[533] Kozma D et al. Mechanism of optical resolutions via diastereomeric salt formation, Journal of Thermal Analysis and Calorimetry. 2000, 60, 409–415.

[534] Keith JM, Larrow JF, Jacobsen EN., Practical considerations in kinetic resolution reactions, Advanced Synthesis & Catalysis. 2001, 343, 5–26.

[535] Chen CS et al. Quantitative analyses of biochemical resolutions of enantiomers. Journal of the American Chemical Society 1982, 104, 7294–7299.

[536] Lin GQ et al. Principles and applications of asymmetric synthesis. New York, Wiley–Interscience, 2001.

[537] Dabkowska K. Dissertation: The use of enzymatic catalysis for the separation of the enantiomers of mandelic acid (in Polish), 2010.

[538] http://en.wikipedia.org/wiki/Electrolysis_of_water Accessed January 7, 2014.

[539] http://en.wikipedia.org/wiki/Electrocoagulation Accessed January 7, 2014.

[540] http://en.wikipedia.org/wiki/Electrophoresis Accessed January 7, 2014.

[541] http://en.wikipedia.org/wiki/Electrofiltration Accessed January 7, 2014.

[542] http://en.wikipedia.org/wiki/Electro-osmosis Accessed January 7, 2014.

[543] http://www.merriam-webster.com/dictionary/electroosmosis Accessed January 7, 2014.

[544] http://www.thefreedictionary.com/electro-osmotic Accessed January 7, 2014.

[545] http://en.wikipedia.org/wiki/Electrodialysis Accessed January 7, 2014.

[546] http://ipp.nasa.gov/innovation/Innovation_81/tpatented.html Accessed January 19, 2014.

[547] Fukuda H, Kondo A, Noda H, Biodiesel fuel production by transesterification of oils, J Biosci Bioeng., 2001, 92, 405–416.

[548] Yue J, Chen B, Yuan Q, Luo L, Gonthier Y, Hydrodynamics and mass transfer characteristics in gas–liquid flowthrough a rectangular microchannel Chemical Engineering Science, 2007, 62, 2096–2108.

[549] Heyouni A, Roustan M, Do-Quang Z, Hydrodynamics and mass transfer in gas–liquid flow through static mixers. Chemical Engineering Science, 2002, 57, 3325–3333.

[550] Herskowit, D, Herskowits V, Stephan K, Tamir A, Characterization of a two-phase impinging jet absorber II. Absorption with chemical reaction of CO_2 in NaOH solutions. Chemical Engineering Science, 1990, 45, 1281–1287.

[551] Charpentier JC, Mass-transfer rates in gas–liquid absorbers and reactors. Advances in Chemical Engineering, 1981, 11, 1–133.

[552] Kies FK, Benadda B, Otterbein M, Experimental study on mass transfer of a co-current gas–liquid contactor performing under high gas velocities. Chemical Engineering and Processing, 2004, 43, 1389–1395.

[553] Dluska E, Wronski S, Ryszczuk T, Interfacial area in gas–liquid Couette–Taylor flow reactor. Experimental Thermal and Fluid Science, 2004, 28, 467–472.

[554] Ioannides T, Gavalas GR, J Membr. Sci., 1993, 77, 207.

[555] Casanave D, Giroir-Fendler A, Sanchez J, Loutaty R, Dalmon J-A, Catal. Today 25, 1995, 309.

[556] Zhu Y, Minet RG, Tsotsis TT, Catal. Lett., 1993, 18, 49.

[557] Matsuda Z, Koike I, Kubo N, Kikuchi E, Appl. Catal. A: Gen., 1993, 96, 3.

[558] Shu J, Grandjean BPA, Kaliaguine S, Ciavarella P, Giroir-Fendler A, Dalmon J-A, Can. J. Chem. Eng., 1997, 75, 712.

[559] Raybold TM, Huf MCf, Catal. Today, 2000, 56, 35.

[560] Sheintuch M, Dessau RM, Chem. Eng. Sci., 1996, 51 (4), 535.

[561] Electrolysis. http://en.wikipedia.org/wiki/Electrolysis_of_water, accessed 20.02.2014.

Index